Oceanic island archipelagos are profoundly interesting ecosystems in which to ask questions about evolutionary patterns and processes, and may rightly be considered as one of the best places on earth to seek an understanding of the origin and elaboration of biological diversity. This volume brings together contributions covering a range of important issues in contemporary oceanic island plant biology, focusing on patterns and processes in Pacific and other islands (with emphasis on the Bonin, Hawaiian and Juan Fernandez Islands) to provide a stimulating view of the current state of research and a possible agenda for future investigations. Topics addressed include chromosomal variation, macromolecular divergence, island biogeography theory, isolating mechanisms, modes of speciation and evolution of secondary plant products, resulting in a volume which reveals the special opportunities offered by oceanic archipelagos for investigating evolutionary phenomena in vascular plants.

T0215190

Evolution and Speciation of Island Plants

EVOLUTION AND SPECIATION OF ISLAND PLANTS

Edited by

Tod F. Stuessy
University of Vienna

and

Mikio Ono
Tokyo Metropolitan University

CAMBRIDGE
UNIVERSITY PRESS

CAMBRIDGE UNIVERSITY PRESS
Cambridge, New York, Melbourne, Madrid, Cape Town, Singapore, São Paulo

Cambridge University Press
The Edinburgh Building, Cambridge CB2 8RU, UK

Published in the United States of America by Cambridge University Press, New York

www.cambridge.org
Information on this title: www.cambridge.org/9780521496537

First published 1998
This digitally printed version 2008

A catalogue record for this publication is available from the British Library

Library of Congress Cataloguing in Publication data

Evolution and speciation of island plants / edited by Tod F. Stuessy and Mikio Ono.
    p. cm.
    "Papers resulted from a symposium convened by the co-editors at the XV
International Botanical Congress in Yokohama, Japan, August, 1993" –Pref.
    Includes indexes.
    ISBN 0 521 49653 5 (hb)
    1. Island plants–Islands of the Pacific–Congresses. 2. Island plants–Evolution–
Islands of the Pacific–Congresses. 3. Island plants–Speciation–Islands of the Pacific–
Congresses.
I. Stuessy, Tod F. II. Ono, Mikio, 1932– . III. International Botanical Congress
(15th : 1993 : Yokohama-shi, Japan)
QK471.E96   1998
581.3′8–dc21   97-36118 CIP

ISBN 978-0-521-49653-7 hardback
ISBN 978-0-521-04832-3 paperback

This book is dedicated to Sherwin Carlquist, outstanding investigator of island plants whose works have stimulated numerous scientists (including both editors) to pursue studies in oceanic archipelagos

Contents

List of contributors   *page* xi
Preface   xiii
Acknowledgements   xv
Part one: Hawaiian Islands   1
Introduction
  1 Chromosome evolution and speciation in Hawaiian
     flowering plants. *Gerald D. Carr*   5
  2 Evolution in the endemic Hawaiian Compositae.
     Bruce G. Baldwin   49
Part two: Juan Fernandez Islands   75
Introduction
  3 Isolating mechanisms and modes of speciation in endemic
     angiosperms of the Juan Fernandez Islands.
     Tod F. Stuessy, Daniel J. Crawford, Clodomiro
     Marticorena and Mario Silva O.   79
  4 *Dendroseris* (Asteraceae: Lactuceae) and *Robinsonia*
     (Asteraceae: Senecioneae) on the Juan Fernandez Islands:
     similarities and differences in biology and phylogeny.
     Daniel J. Crawford, Tao Sang, Tod F. Stuessy,
     Seung-Chul Kim and Mario Silva O.   97
  5 Island biogeography of angiosperms of the Juan Fernandez
     archipelago. *Tod F. Stuessy, Daniel J. Crawford,*
     Clodomiro Marticorena and Roberto Rodríguez   121
Part three: Southern and western Pacific Islands   139
Introduction
  6 Genetic diversity of the endemic plants of Bonin
     (Ogasawara) Islands. *Motomi Ito, Akiko Soejima*
     and Mikio Ono   141

7   Evolution of cryptic dioecy in *Callicarpa* (Verbenaceae)
    on the Bonin Islands.   *Nobumitsu Kawakubo*   155
8   Conservation of the endemic vascular plant species of
    the Bonin (Ogasawara) Islands.   *Mikio Ono*   169
9   Preliminary observations on the evolution of endemic
    angiosperms of Ullung Island, Korea.
    Byung-Yun Sun and Tod F. Stuessy   181
10 Evolution in *Crossostylis* (Rhizophoraceae) on the South
    Pacific Islands.   *Hiroaki Setoguchi, Hideaki Ohba and
    Hiroshi Tobe*   203
**Part four: General evolutionary patterns and processes on
                oceanic islands**   231
Introduction
11 Secondary compounds and evolutionary relationships of
    island plants.   *Bruce A. Bohm*   233
12 Chromosomal stasis during speciation in angiosperms of
    oceanic islands.   *Tod F. Stuessy and Daniel J. Crawford*   307
13 The current status of our knowledge and suggested
    research protocols in island archipelagos.
    Tod F. Stuessy and Mikio Ono   325
Author index   333
Taxon index   340
Subject index   353

Contributors

Bruce G. Baldwin
Jepson Herbarium and Department of Integrative Biology, University of California, Berkeley, CA 94720, USA

Bruce A. Bohm
Department of Botany, University of British Columbia,Vancouver, BC V6T 1Z4, Canada

Gerald D. Carr
Department of Botany, University of Hawaii, Honolulu, HI 96822, USA

Daniel J. Crawford
Department of Plant Biology, Ohio State University, Columbus, OH 43210, USA

Motomi Ito
Department of Biology, Faculty of Science, Chiba University, Chiba 263, Japan

Nobumitsu Kawakubo
Biological Diversity and Resources, Faculty of Agriculture, Gifu University, Gifu 501-11, Japan

Seung-Chul Kim
Department of Biology, Indiana University, Bloomington, IN 47405, USA

Clodomiro Marticorena
Departamento de Botánica, Universidad de Concepción, Concepción, Chile

Hideaki Ohba
Department of Botany, University Museum, University of Tokyo, Hongo 7-3-1, Bunkyo-ku, Tokyo 113, Japan

Mikio Ono
Makino Herbarium, Tokyo Metropolitan University, Minami-Osawa, Hachioji, Tokyo 192-03, Japan

Roberto Rodríguez
Departamento de Botánica, Universidad de Concepción, Concepción, Chile

Tao Sang
Department of Botany and Plant Pathology, Michigan State University, East Lansing, MI 48824, USA

Hiroaki Setoguchi
Makino Herbarium, Tokyo Metropolitan University, Minami-Osawa, Hachioji, Tokyo 192-03, Japan

Mario Silva O.
Departamento de Botánica, Universidad de Concepción, Concepción, Chile

Akiko Soejima
College of Integrated Arts and Sciences, University of Osaka Prefecture, Sakai, Osaka 593, Japan

Tod F. Stuessy
Institut für Botanik, Universität Wien, Rennweg 14, A-1030 Vienna, Austria

Byung-Yun Sun
Department of Biology, Chonbuk National University, Chonju 560-756, Republic of Korea

Hiroshi Tobe
Department of Environmental Sciences, Faculty of Integrated Human Studies, Kyoto University, Yoshida-Nihonmatsu-machi, Sakyo-ku, Kyoto 606-01, Japan

Preface

Oceanic island archipelagos are profoundly interesting ecosystems in which to ask questions about evolutionary patterns and processes. Their isolation from other land masses restricts numbers of plausible hypotheses that can be advanced to explain evolutionary events, often resulting in stronger scientific inferences. Island archipelagos may rightly be considered one of the best places on earth to understand origins and elaborations of biological diversity.

In recent years, interest in the evolution of plants on oceanic islands has increased with work appearing on the Hawaiian Islands, the Canary Islands, Bonin Islands (Japan), Juan Fernandez Islands (Chile) and numerous additional archipelagos. It seemed pertinent and timely, therefore, to bring persons together with experience in these different island groups for a new view of island plant biology.

The present papers resulted from a symposium convened by the co-editors at the XV International Botanical Congress in Yokohama, Japan, August 1993. The response from that symposium was very positive and suggested that publication of the presentations (in modified and up-dated form) would be highly desirable. Cambridge University Press agreed to take on the project. In addition to papers from the symposium itself, which focused on Pacific archipelagos, several additional invited manuscripts were solicited to provide an even stronger and more comprehensive review of the status of studies of plant evolution on oceanic islands.

It is clear from the chapters of this book that oceanic archipelagos do offer special opportunities for investigating evolutionary phenomena in vascular plants. It is also clear that certain processes occur preferentially on islands in contrast to continental areas. As the concluding chapter outlines (Chapter 13), more rigorous approaches to solving these evolutionary problems are now recommended to help maximize insights. The results and discussions in this book clearly move us in the right direction, but neither are they the final word on any topic. The contents together, however, do emphasize the beauty, fascination and importance of plants of oceanic islands, and the general role they play in helping reveal origins and diversity of plant life in this biosphere.

Tod F. Stuessy
Mikio Ono

Acknowledgements

It is a pleasure to acknowledge the many persons who have helped bring this book to successful completion. First are the authors who worked so diligently in preparing their manuscripts and who tolerated intrusions of the editors. Second are the editors and production staff of Cambridge University Press, who helped us shape a rough product into a much more highly polished finished volume. Especially encouraging was Maria Murphy, Science Editor, who enquired about the possibility of book publication for our symposium, and who helped guide us through book proposal and onto eventual contract.

Several financial sources made this book possible. Partial initial support from the organizing committee of the XV International Botanical Congress helped assure attendance by participants of the symposium. Additional funds were provided for individual contributors from local sources as acknowledged at the end of respective chapters. Financial aid for the editors to work together on editing manuscripts in Tokyo during June 1995 was provided by the Office of International Education, Ohio State University, and Tokyo Metropolitan University. Final editing of the book was made possible when the senior editor was Deputy Director for Research and Collections of the Natural History Museum, Los Angeles County, California, USA.

Part one

Hawaiian Islands

Introduction

Few islands of the world have received as much attention for evolutionary patterns and processes as the Hawaiian Islands. Reading through Sherwin Carlquist's stimulating *Hawaii: A Natural History* (1970) always elevates our interest. We have also read about fascinating evolutionary phenomena in picture-wing *Drosophila*, studied so successfully by Hampton Carson and colleagues (e.g., Carson & Kaneshiro, 1976; Kaneshiro, Gillespie & Carson, 1995; DeSalle, 1995). It is fitting, therefore, that the initial two chapters of this book deal with the Hawaiian Islands. Recent studies have greatly increased our understanding of patterns and processes in the endemic vascular plant flora of the Hawaiian Islands. A monumental achievement was the publication of the two-volume *Manual of the Flowering Plants of Hawaii* (Wagner, Herbst & Sohmer, 1990) that established for the first time a consistent species concept for the entire archipelago. In the past, some taxa had been split into numerous microspecies and others had been treated broadly, depending upon the perspective of the particular taxonomist. These disparate treatments of plant diversity in the archipelago made it very difficult to approach questions of speciation and biogeography. In fact, publication of the new *Manual* made it possible to conceive and execute a very meaningful project on biogeography in the archipelago, involving both plants and animals (Wagner & Funk, 1995). This would have been impossible without the consistent foundation of species concepts provided by the flora project.

Along with significant recent floristic efforts, detailed biosystematic studies have been carried out on the Hawaiian Islands on many taxa by different workers. Of particular mention are those investigations on Compositae, such as: the bizarre silverswords by Carr and associates (Carr & Kyhos, 1981, 1986; Carr, 1985); *Bidens* by Gillett & Lim

1

(1970), Ganders & Nagata (1984) and Helenurm & Ganders (1985); *Lipochaeta* by Gardner (Gardner, 1976, 1977) and Rabakonandrianina & Carr (1981); and *Tetramolopium* by Lowrey & Crawford (1985) and Lowrey (1986).

This first part of the book builds on previous studies and summarizes and extends our understanding of the endemic plants of the Hawaiian Islands. Carr (Chapter 1) focuses on chromosomal evolution in the endemic angiosperms and discusses possible reasons for cytological change or lack thereof (stasis) within many groups. Baldwin (Chapter 2) summarizes and extends his recent macromolecular studies with colleagues (e.g., Baldwin *et al.*, 1991) that finally resolved ancestry of the Hawaiian tarweeds from California progenitors. This general biogeographic connection has been known for some time (e.g., Carlquist, 1959), but the particular continental group from which all the Hawaiian diversity came was never determined satisfactorily. Taken together, these two chapters provide an excellent view of recent work on the angiosperm flora of the Hawaiian Islands and they provide a very good beginning to our volume.

Literature cited

Baldwin, B. G., Kyhos, D. W., Dvořák, J. & Carr, G. D. (1991). Chloroplast DNA evidence for a North American origin of the Hawaiian silversword alliance (Asteraceae). *Proceedings of the National Academy of Sciences USA*, **88**, 1840–3.

Carlquist, S. (1959). Studies on Madiinae: anatomy, cytology and evolutionary relationships. *Aliso*, **4**, 171–236.

Carlquist, S. (1970). *Hawaii: A Natural History. Geology, Climate, Native Flora and Fauna above the Shoreline.* New York: Natural History Press.

Carr, G. D. (1985). Monograph of the Hawaiian Madiinae (Asteraceae): *Argyroxiphium, Dubautia* and *Wilkesia. Allertonia*, **4**, 1–123.

Carr, G. D. & Kyhos, D. W. (1981). Adaptive radiation in the Hawaiian silversword alliance (Compositae–Madiinae). I. Cytogenetics of spontaneous hybrids. *Evolution*, **35**, 543–56.

Carr, G. D. & Kyhos, D. W. (1986). Adaptive radiation in the Hawaiian silversword alliance (Compositae–Madiinae). II. Cytogenetics of artificial and natural hybrids. *Evolution*, **40**, 959–76.

Carson, H. L. & Kaneshiro, K. Y. (1976). *Drosophila* of Hawaii: systematics and ecological genetics. *Annual Review of Ecology and Systematics*, **7**, 311–45.

DeSalle, R. (1995). Molecular approaches to biogeographic analysis of Hawaiian Drosophilidae. In *Hawaiian Biogeography: Evolution on a Hot Spot Archipelago*, ed. W. L. Wagner & V. A. Funk, pp. 72–89. Washington DC: Smithsonian Institution Press.

Ganders, F. R. & Nagata, K. M. (1984). The role of hybridization in the evolution of *Bidens* on the Hawaiian Islands. In *Plant Biosystematics*, ed. W. F. Grant, pp. 179–94. Toronto: Academic Press.

Gardner, R. C. (1976). Evolution and adaptive radiation in *Lipochaeta* (Compositae: Heliantheae) of the Hawaiian Islands. *Systematic Botany*, **1**, 383–91.

Gardner, R. C. (1977). Chromosome numbers and their systematic implications in *Lipochaeta* (Compositae: Heliantheae). *American Journal of Botany*, **64**, 810–13.

Gillett, G. W. & Lim, E. K. S. (1970). An experimental study of the genus *Bidens* (Asteraceae) of the Hawaiian Islands. *University of California Publications in Botany*, **56**, 1–63.

Helenurm, K. & Ganders, F. R. (1985). Adaptive radiation and genetic differentiation in Hawaiian *Bidens*. *Evolution*, **39**, 753–65.

Kaneshiro, K. Y., Gillespie, R. G. & Carson, H. L. (1995). Chromosomes and male genetalia of Hawaiian *Drosophila*: tools for interpreting phylogeny and geography. In *Hawaiian Biogeography: Evolution on a Hot Spot Archipelago*, ed. W. L. Wagner & V. A. Funk, pp. 57–71. Washington DC: Smithsonian Institution Press.

Lowrey, T. K. (1986). A biosystematic revision of Hawaiian *Tetramolopium* (Compositae: Astereae). *Allertonia*, **4**, 203–65.

Lowrey, T. K. & Crawford, D. J. (1985). Allozyme divergence and evolution in *Tetramolopium* (Compositae: Astereae) on the Hawaiian Islands. *Systematic Botany*, **10**, 64–72.

Rabakonandrianina, E. & Carr, G. D. (1981). Intergeneric hybridization, induced polyploidy and the origin of the Hawaiian endemic *Lipochaeta* from *Wedelia* (Compositae). *American Journal of Botany*, **68**, 206–15.

Wagner, W. L. & Funk, V. A. (ed.) (1995). *Hawaiian Biogeography: Evolution on a Hot Spot Archipelago*. Washington DC: Smithsonian Institution Press.

Wagner, W. L., Herbst, D. R. & Sohmer, S. H. (1990). *Manual of the Flowering Plants of Hawaii*. Honolulu: University of Hawaii Press; Bishop Museum Press.

1

Chromosome evolution and speciation in Hawaiian flowering plants

GERALD D. CARR

Abstract

Chromosome numbers available for about 38% of the 956 native species of Hawaiian plants indicate that more than 80% are polyploid. However, support for the occurrence of autochthonous polyploidy is very limited, with fairly clear examples in *Peperomia, Portulaca,* and *Wikstromia*; less certain instances in *Bobea, Lepidium, Plantago* and *Psychotria*; and dubious examples in *Labordia* and *Polygonum*. Likewise, evidence of chromosome evolution in the form of gross structural changes or dysploidy is sparse and clearly demonstrated only in the silversword alliance of *Argyroxiphium, Dubautia* and *Wilkesia. Luzula* and *Peperomia* may provide additional examples of dysploidy. In contrast, a large number of groups, most notably *Bidens, Cyrtandra, Hibiscadelphus, Lipochaeta, Pipturus, Scaevola, Tetramolopium, Vaccinium* and the lobelioid genera *Brighamia, Clermontia, Cyanea, Delissea, Lobelia, Rollandia* and *Trematolobelia*, are seemingly characterized by complete chromosome stasis, at least with respect to gross structural alterations, dysploidy and polyploidy. There appears to be little or no indication that chromosome evolution on the Hawaiian Islands has proceeded in a manner particularly different from continental areas. Rather, the examples of insular chromosome evolution appear to reflect the tendencies inherent in their continental ancestors. In light of the evidence accumulating from molecular studies, it is concluded that the overall patterns of chromosome structural evolution and chromosome stasis observed in plants are most readily explained on the basis of structural variants having different selective values that are determined by the relative positions of critical genes in the genome.

5

Knowledge related to chromosome evolution in Hawaiian plants is very incomplete. Available data are primarily the result of very few general surveys of chromosome numbers (Skottsberg, 1953; Carr 1978, 1985a) and several studies of individual species or plant groups. Most of what is known about chromosome numbers of Hawaiian flowering plants was summarized recently by Wagner, Herbst & Sohmer (1990). These data have been extracted (with a few minor adjustments) and combined with several more recent determinations in Table 1.1. Based on the analysis of the Hawaiian flora by Wagner, Herbst & Sohmer (1990), the chromosome determinations for the 359 species recorded in Table 1.1 represent one or more counts for just over one-third of the native Hawaiian plant species (359/956 = 37.6%).

No determinations based on Hawaiian material are available for 53 genera that are native to Hawaii. Most notable in this respect are *Myrsine* (20 species), *Pritchardia* (19 species) and *Mariscus* (10 species). Furthermore, only one determination is available for the third largest genus in Hawaii, *Melicope,* with 47 species. It may also be noted that the chromosome determinations reported for 38 of the non-endemic species are based only on extra-Hawaiian populations. Thus, in addition to filling in the gaps for species and genera with no counts, it would be desirable to have counts for Hawaiian populations of non-endemic taxa.

Based on the counts available for Hawaiian populations, more than 80% (288/359) of the native plant species are polyploid. The frequency of polyploidy is nearly 9% higher in monocots compared to dicots. This estimate relies on the criterion of $n > 13$ denoting polyploidy (Grant, 1963; Goldblatt, 1980) but specifically includes the species of *Dubautia* with $2n = 26$ because they exhibit duplicate gene expression indicative of polyploidy (Witter, 1990). As will be discussed more fully below, the high incidence of polyploidy reported here has not been derived autochthonously. Rather, it reflects mainly paleopolyploidy inherent in the ancestors of Hawaiian species. Nevertheless, the high frequency of polyploidy reported here for the Hawaiian flora considerably exceeds most other general estimates of polyploidy among flowering plants. For example, using the same criterion, Grant (1963) estimated that 47% of the 17 138 flowering plant species he sampled were polyploid (dicots 43%, monocots 58%) and Goldblatt (1980) estimated that 55% of the 10 580 species of monocots he sampled were polyploid. Although the sample is very small, the same criteria applied to the data provided by Sanders, Stuessy & Rodríguez R. (1983) for the flora of the Juan Fernandez Islands produces an estimate of 75% polyploidy. Perhaps the high

Table 1.1. *Chromosome numbers in Hawaiian flowering plants.*

An asterisk(*) identifies determinations based on populations outside the Hawaiian Islands, which are not referenced

Taxon	No. spp/genus counted : No. spp/genus native	Diploid chromosome number
DICOTS		
AIZOACEAE		
Sesuvium portulacastrum (L.) L.	1:1	(8,36,48)*,*c*.48[a]
AMARANTHACEAE		
Achyranthes splendens Mart. ex Moq.	1:3	76[b],78[b]
Amaranthus brownii Christoph. & Caum	1:6	34[c]
Charpentiera ovata Gaud.	2:5	52[d]
C. tomentosa Sohmer		*c*.52[e]
ANACARDIACEAE		
Nototrichium (A. Gray ex Hillebr.) Hillebr.	0:2	
Rhus L.	0:1	
APIACEAE		
Peucedanum sandwicense Hillebr.	1:1	66[f]
Sanicula sandwicensis A. Gray	1:4	16[c]
Spermolepis hawaiiensis Wolff	1:1	22[h]
APOCYNACEAE		
Alyxia oliviformis Gaud.	1:1	*c*.36[b],*c*.39[b]
Ochrosia Juss.	0:4	
Pteralyxia K. Schum.	0:2	
Rauvolfia sandwicensis A. DC.	1:1	44[a]
AQUIFOLIACEAE		
Ilex anomala Hook. & Arnott	1:1	80[a]

Table 1.1. (cont.)

Taxon	No. spp/genus counted : No. spp/genus native	Diploid chromosome number
ARALIACEAE		
Cheirodendron trigynum (Gaud.) A. Heller	1:5	24[b]
Munroidendron racemosum (C. Forbes) Sherff	1:1	48[a,i]
Reynoldsia A. Gray	0:1	
Tetraplasandra oahuensis (A. Gray) Harms	1:6	48[a]
ASTERACEAE		
Adenostemma lavenia (L.) Kuntze	1:1	20*
Argyroxiphium caliginis C. Forbes	4:5	28[f]
A. grayanum (Hillebr.) Degener		26[b],28[f,g]
A. kauense (Rock & M. Neal) Degener & I. Degener		28[a]
A. sandwicense DC.		28[a,f,g]
Artemisia australis Less.	2:3	18[a,f]
A. mauiensis (A. Gray) Skottsb.		18[a,f]
Bidens conjuncta Sherff	13:19	c.70[b]
B. cosmoides (A. Gray) Sherff		72[j]
B. forbesii Sherff		72[j]
B. hawaiiensis A. Gray		72[j]
B. hillebrandiana (Drake) Degener		72[j]
B. macrocarpa (A. Gray) Sherff		72[j]
B. mauiensis (A. Gray) Sherff		72[j]
B. menziesii (A. Gray) Sherff		72[j]
B. micrantha Gaud.		72[j]
B. molokaiensis (Hillebr.) Sherff		72[a,j]
B. sandvicensis Less.		72[b]

Species		
B. torta Sherff		72[j]
B. wiebkei Sherff		72[j]
Dubautia arborea (A. Gray) Keck	21:21	26[a,g]
D. ciliolata (DC.) D. Keck		26[a,g]
D. dolosa (Degener & Sherff) G. Carr		26[a,g]
D. herbstobatae G. Carr		26[g]
D. imbricata St John & G. Carr		28[g]
D. knudsenii Hillebr.		26[b], 28[a,g]
D. laevigata A. Gray		28[g]
D. latifolia (A. Gray) D. Keck		28[a]
D. laxa Hook. & Arnott		28[a,f,g]
D. linearis (Gaud.) D. Keck		26[a,g]
D. menziesii (A. Gray) D. Keck		26[a,g]
D. microcephala Skottsb.		28[a,g]
D. paleata A. Gray		28[a]
D. pauciflorula St. John & G. Carr		28[g]
D. plantaginea Gaud.		28[a,g]
D. platyphylla (A. Gray) D. Keck		26[g]
D. raillardioides Hillebr.		28[g]
D. reticulata (Sherff) D. Keck		26[a]
D. scabra (DC.) D. Keck		28[a,g]
D. sherffiana Fosb.		26[a]
D. waialealeae Rock		28[g]
Gnaphalium L.	0:1	
Hesperomannia arbuscula Hillebr.	1:3	20[a,k]
Lagenifera helenae C. Forbes & Lydgate	2:3	54[c]
L. maviensis H. Mann		54[a]
Lipochaeta connata (Gaud.) DC.	15:20	52[a,l]
L. heterophylla A. Gray		52[l]
L. integrifolia (Nutt.) A. Gray		30[a,l]

Table 1.1. (*cont.*)

Taxon	No. spp/genus counted : No. spp/genus native	Diploid chromosome number
L. kamolensis Degener & Sherff		30[l]
L. lavarum (Gaud.) DC.		30[a,l]
L. lobata (Gaud.) DC.		52[a,l]
L. micrantha (Nutt.) A. Gray		30[l]
L. remyi A. Gray		30[a,l]
L. rockii Sherff		52[a,l]
L. subcordata A. Gray		30[l]
L. succulenta (Hook. & Arnott) DC.		52[a,d,l]
L. tenuifolia A. Gray		30[l]
L. tenuis Degener & Sherff		30[m]
L. venosa Sherff		30[m]
L. waimeaensis St John		30[l]
Remya mauiensis Hillebr.	1:3	36[c]
Tetramolopium consanguineum (A. Gray) Hillebr.	7:11	18[n]
T. filiforme Sherff		18[a,n]
T. humile (A. Gray) Hillebr.		14[f],18[a,f,n]
T. lepidotum (Less.) Sherff		18[n]
T. remyi (A. Gray) Hillebr.		18[n]
T. rockii Sherff		18[n]
T. sylvae Lowrey		18[n]
Wilkesia gymnoxiphium A. Gray	2:2	c.24[f],28[a-g]
W. hobdyi St John		28[g]
BEGONIACEAE		
Hillebrandia sandwicensis Oliver	1:1	48[f]
BORAGINACEAE		

Heliotropium anomalum Hook. & Arnott	2:2	26*,28[a]
H. curassavicum L.		(24,26,28,52)*,26[c]
BRASSICACEAE		
Lepidium bidentatum Montin	2:3	48[e]
L. serra H. Mann		32[a]
CAMPANULACEAE		
Brighamia insignis A. Gray	1:2	28[o]
Clermontia calophylla F. Wimmer	8:22	28[o]
C. clermontioides (Gaud.) A. Heller		28[o]
C. drepanomorpha Rock		28[b,o]
C. grandiflora Gaud.		28[b,o]
C. kakeana Meyen		28[a]
C. montis-loa Rock		28[b,o]
C. oblongifolia Gaud.		22[f],28[a]
C. parviflora Gaud. ex A. Gray		28[o]
Cyanea angustifolia (Cham.) Hillebr.	4:52	24[f],28[a,o]
C. degeneriana F. Wimmer		28[o]
C. leptostegia A. Gray		28[o]
C. remyi Rock		28[p]
Delissea rhytidosperma H. Mann	1:9	28[o]
Lobelia grayana F. Wimmer	3:13	28[a,b]
L. hypoleuca Hillebr.		28[o]
L. yuccoides Hillebr.		28[a]
Rollandia lanceolata Gaud.	1:8	28[a]
Trematolobelia kauaiensis (Rock) Skottsb.	3:4	28[o]
T. macrostachys (Hook. & Arnott) A. Zahlbr.		28[a]
T. singularis St John		28[a]
CAPPARACEAE		
Capparis sandwichiana DC.	1:1	40[a]
Cleome spinosa Jacq.	1:1	(18,20,24,38,44)*

Table 1.1. (*cont.*)

Taxon	No. spp/genus counted : No. spp/genus native	Diploid chromosome number
CARYOPHYLLACEAE		
Alsinidendron trinerve H. Mann	1:4	60[b]
Schiedea verticillata F. Brown	1:22	60[c]
Silene hawaiiensis Sherff	2:7	24[a]
S. struthioloides A. Gray		c.22[q],24[c]
CELASTRACEAE		
Perrottetia Kunth	0:1	
CHENOPODIACEAE		
Chenopodium oahuense (Meyen) Aellen	1:1	36[a]
CONVOLVULACEAE		
Bonamia menziesii A. Gray	1:1	30[a]
Cressa truxillensis Kunth	1:1	28*
Ipomoea imperati (Vahl) Griseb.	3:5	30*
I. indica (J. Burm.) Merr.		30*
I. pes-caprae (L.) R. Br.		30*,60*
Jacquemontia ovalifolia (Choisy) H. Hallier	1:1	18*,20[a]
CUCURBITACEAE		
Sicyos pachycarpus Hook. & Arnott	2:14	24[c]
S. waimanaloensis St John		24[a]
CUSCUTACEAE		
Cuscuta sandwichiana Choisy	0:1	
DROSERACEAE		
Drosera anglica Huds.	1:1	40*
EBENACEAE		
Diospyros sandwichensis (A. DC.) Fosb.	1:2	c.48[b]
ELAEOCARPACEAE		

Species		
Elaeocarpus L.	0:1	
EPACRIDACEAE		
Styphelia tameiameiae (Cham. & Schlechtend.) F. v. Muell.	1:1	20[a]
ERICACEAE		
Vaccinium calycinum Sm.	2:3	24[b]
V. reticulatum Sm.		24[a,b]
EUPHORBIACEAE		
Antidesma L.	0:2	
Chamaesyce arnottiana (Endl.) Degener	4:15	c.38[c]
C. celastroides (Boiss.) Croizat & Degener		c.38[c]
C. clusiifolia (Hook. & Arnott) Arth.		c.38[c]
C. multiformis (Hook. & Arnott) Croizat & Degener		c.38[c]
Claoxylon sandwicense Müll. Arg.	1:1	44[a]
Euphorbia L.	0:1	
Flueggea Willd.	0:1	
Phyllanthus L.	0:1	
FABACEAE		
Acacia koa A. Gray	1:1	26[b],52[a,b]
Caesalpinia bonduc (L.) Roxb.	2:2	24*
C. kavaiensis H. Mann		22[a]
Canavalia Adans.	0:6	
Entada phaseoloides (L.) Merr.	1:1	28*
Erythrina sandwicensis Degener	1:1	42[a,f]
Mucuna Adans.	0:1	
Senna gaudichaudii (Hook. & Arnott) H. Irwin & Barneby	1:1	28[a]
Sesbania tomentosa Hook. & Arnott	1:1	24[a]
Sophora chrysophylla (Salisb.) Seem.	1:1	24[a]
Strongylodon Vogel	0:1	16[b]

Table 1.1. (*cont.*)

Taxon	No. spp/genus counted : No. spp/genus native	Diploid chromosome number
Vicia menziesii Spreng.	1:1	14^r
Vigna adenantha (G. Mey.) Maréchal, Mascherpa & Stainier	3:3	22^*
V. marina (J. Burm.) Merr.		22^*
V. o-wahuensis Vogel		22^f
FLACOURTIACEAE		
Xylosma hawaiiense Seem.	1:2	20^a
GENTIANACEAE		
Centaurium sebaeoides (Griseb.) Druce	1:1	$c.44^a$
GERANIACEAE		
Geranium arboreum A. Gray	1:6	$c.50^a$
GESNERIACEAE		
Cyrtandra calpidicarpa (Rock) St John & Storey	17:53	34^s
C. confertiflora (Wawra) C. B. Clarke		34^s
C. cordifolia Gaud.		$34^{s,t}$
C. garnotiana Gaud.		34^t
C. grandiflora Gaud.		$34^{s,t}$
C. grayana Hillebr.		34^s
C. hashimotoi Rock		34^s
C. hawaiensis C. B. Clarke		$34^{s,t}$
C. kauaiensis Wawra		34^s
C. kaulantha St John & Storey		34^s
C. laxiflora H. Mann		$34^{a,s,t}$
C. longifolia (Wawra) Hillebr. ex C. B. Clarke		34^s

Species		
C. lysiosepala (A. Gray) C. B. Clarke		34[s]
C. oenobarba H. Mann		34[s]
C. paludosa Gaud.		34[s,t,u]
C. propinqua C. Forbes		34[t]
C. sandwicensis (H. Lév.) St John & Storey		34[s,t]
GOODENIACEAE		
Scaevola chamissoniana Gaud.	9:9	16[a,b,v,w]
S. coriacea Nutt.		16[a,w]
S. gaudichaudiana Cham.		16[a,b,w]
S. gaudichaudii Hook. & Arnott		16[a,w]
S. glabra Hook. & Arnott		32[a,c,w]
S. kilaueae Degener		16[w]
S. mollis Hook. & Arnott		16[v,w]
S. procera Hillebr.		16[b,c,w]
S. sericea Vahl		16[a,b,w],32*
GUNNERACEAE		
Gunnera L.	0:2	
HYDRANGEACEAE		
Broussaisia arguta Gaud.	1:1	32[a]
HYDROPHYLLACEAE		
Nama sandwicensis A. Gray	1:1	14[a,f]
LAMIACEAE		
Haplostachys A. Gray	0:5	34[e]
Lepechinia hastata (A. Gray) Epling	1:1	66[a]
Phyllostegia glabra (Gaud.) Benth.	4:27	64[a]
P. grandiflora (Gaud.) Benth.		64[b]
P. mollis Benth.		c.66[a]
P. velutina (Sherff) St John	1:1	34*
Plectranthus parviflorus Willd.	4:20	64[b]
Stenogyne calaminthoides A. Gray		64[a]
S. kaalae Wawra		

Table 1.1. (cont.)

Taxon	No. spp/genus counted : No. spp/genus native	Diploid chromosome number
S. microphylla Benth.		64[a]
S. purpurea H. Mann		64[b]
LAURACEAE		
Cassytha filiformis L.	1:1	24*,48[a]
Cryptocarya R. Br.	0:1	
LOGANIACEAE		
Labordia degeneri Sherff	4:15	>40<60[s]
L. helleri Sherff		40–44[s]
L. hirtella H. Mann		*c.*80[a]
L. waiolani Wawra		*c.*80[a]
LYTHRACEAE		
Lythrum L.	0:1	
MALVACEAE		
Abutilon incanum (Link) Sweet	2:4	14[a],14*
A. menziesii Seem.		28[c]
Gossypium tomentosum Nutt. ex Seem.	1:1	52[a]
Hibiscadelphus distans L. Bishop & Herbst	3:6	40[a]
H. giffardianus Rock		40[a]
H. hualalaiensis Rock		40[a,f]
Hibiscus arnottianus A. Gray	7:7	80[f],84[f]
H. brackenridgei A. Gray		144[x],*c.*140[f]
H. clayi Degener & I. Degener		84[x]
H. furcellatus Desr.		72[f]
H. kokio Hillebr.		82[f],84[x]
H. tiliaceus L.		(40,80,92)*,(92,96)[b]

Taxon		
H. waimeae A. Heller	2:4	84[d,f]
Kokia cookei Degener		24[b,e]
K. drynarioides (Seem.) Lewton	1:1	24[b]
Sida fallax Walp.	1:1	28[a]
Thespesia populnea (L.) Sol. ex Corrêa		(24,26,28)*
MENISPERMACEAE		
Cocculus trilobus (Thunb.) DC.	1:1	50*,52*,78[a]
MORACEAE		
Streblus pendulinus (Endl.) F. v. Muell.	1:1	28[a]
MYOPORACEAE		
Myoporum Banks & Sol. ex G. Forster	0:1	
MYRSINACEAE		
Embelia N. L. Burm.	0:1	
Myrsine L.	0:20	
MYRTACEAE		
Eugenia L.	0:2	
Metrosideros polymorpha Gaud.	3:5	22[a,b]
M. tremuloides (A. Heller) P. Knuth		22[a]
M. waialealae (Rock) Rock		22[a]
Syzygium sandwicensis (A. Gray) Nied.	1:1	22[a]
NYCTAGINACEAE		
Boerhavia repens L.	1:3	52[a]
Pisonia brunoniana Endl.	2:5	136[a]
P. umbellifera (G. Forster) Seem.		c.112*
OLEACEAE		
Nestegis sandwicensis (A. Gray) Degener, I. Degener & L. Johnson	1:1	44[b]
PAPAVERACEAE		
Argemone glauca (Nutt. ex Prain) Pope	1:1	56[e]
PHYTOLACCACEAE		
Phytolacca L.	0:1	

Table 1.1. (*cont.*)

Taxon	No. spp/genus counted:No. spp/genus native	Diploid chromosome number
PIPERACEAE		
Peperomia alternifolia Yuncker	18:25	44[y]
P. cookiana C. DC.		88[y]
P. eekana C. DC.		c.36[b],44[y]
P. ellipticibacca C. DC.		44[y]
P. expallescens C. DC.		66[b]
P. globulanthera C. DC.		88[y]
P. hesperomannii Wawra		44[y],48[b],66[b]
P. hypoleuca Miq.		44[y],42-46[b]
P. kokeana Yuncker		88[y]
P. latifolia Miq.		44[y]
P. leptostachya Hook. & Arnott		44[y]
P. macraeana C. DC.		28[b],c.42[b],44[y],48[y]
P. membranacea Hook. & Arnott		88[y]
P. oahuensis C. DC.		44[y],96[y]
P. ovatilimba C. DC.		28[b]
P. remyi C. DC.		44[y]
P. sandwicensis Miq.		44[y]
P. tetraphylla (G. Forster) Hook. & Arnott		22*,40*
PITTOSPORACEAE		
Pittosporum glabrum Hook. & Arnott	1:10	24[a]
PLANTAGINACEAE		
Plantago pachyphylla A. Gray	2:3	24[b,u]
P. princeps Cham. & Schlechtend.		12[u]
P. princeps var. *anomala* Rock		24[s]

PLUMBAGINACEAE		
Plumbago zeylanica L.	1:1	16*,28*,28[a]
POLYGONACEAE		
Rumex albescens Hillebr.	3:3	36[b],40[u],60[c]
R. giganteus W. T. Aiton		c.54[b],c.56[b],60[u]
R. skottsbergii Degener & I. Degener		60[z]
PORTULACACEAE		
Portulaca lutea Sol. ex G. Forster	4:4	40[aa]
P. molokiniensis Hobdy		40[aa]
P. sclerocarpa A. Gray		18[aa]
P. villosa Cham.		18[aa]
PRIMULACEAE		
Lysimachia glutinosa Rock	3:10	72[a]
L. hillebrandii J. D. Hook. ex A. Gray		c.54[b],72[a]
L. mauritiana Lam.		20[a],20*
RANUNCULACEAE		
Ranunculus hawaiensis A. Gray	2:2	28[d]
R. mauiensis A. Gray		(c.34,c.36,c.39)[b]
RHAMNACEAE		
Alphitonia Reissek ex Endl.	0:1	
Colubrina oppositifolia Brongn. ex H. Mann	1:2	48[c]
Gouania hillebrandii Oliver	2:3	(46)–48[i]
G. meyenii Steud.		46[a]
ROSACEAE		
Acaena L.	0:1	
Fragaria chiloensis (L.) Duchesne	1:1	56*
Osteomeles anthyllidifolia (Sm.) Lindl.	1:1	34[a]
Rubus hawaiensis A. Gray	1:2	14[bb]
RUBIACEAE		
Bobea elatior Gaud.	2:4	66–70[i]
B. timonioides (J. D. Hook.) Hillebr.		88[i]

Table 1.1. (cont.)

Taxon	No. spp/genus counted: No. spp/genus native	Diploid chromosome number
Canthium Lam.	0:1	
Coprosma ernodeoides A. Gray	6:13	c.220[b,s] (185,192,223)[b]
C. foliosa A. Gray		44[s]
C. longifolia A. Gray		44[s]
C. montana Hillebr.		32[b],44[a,s]
C. ochracea W. Oliver		44[s]
C. rhynchocarpa A. Gray		44[s]
Gardenia brighamii H. Mann	2:3	22[i]
G. remyi H. Mann		22[i]
Hedyotis centranthoides (Hook. & Arnott) Steud.	3:20	c.100[b]
H. hillebrandii (Fosb.) W. L. Wagner & Herbst		c.72[b]
H. terminalis (Hook. & Arnott) W. L. Wagner & Herbst		(c.93,95,96,98,100,102–105)[b]
Morinda trimera Hillebr.	1:1	22[i]
Nertera granadensis (L. fil.) Druce	1:1	44[b],44[*]
Psychotria greenwelliae Fosb.	4:11	66–88[s]
P. hobdyi Sohmer		88[i]
P. kaduana (Cham. & Schlechten.) Fosb.		66 ± 2[i]
P. mariniana (Cham. & Schlechten.) Fosb.		66[s]
RUTACEAE		
Melicope elliptica A. Gray	1:47	36[a]
Platydesma cornuta Hillebr.	2:4	36[a]
P. rostrata Hillebr.		36[c]
Zanthoxylum hawaiiense Hillebr.	1:4	136–144[i]

SANTALACEAE		
Exocarpos gaudichaudii A. DC.	1:3	20[a]
Santalum ellipticum Gaud.	3:4	40[a]
S. freycinetianum Gaud.		40[a]
S. paniculatum Hook. & Arnott		40[a]
SAPINDACEAE		
Alectryon macrococcus Radlk.	1:1	32[cc]
Dodonaea viscosa Jacq.	1:1	28[a],(28,30,32)*
Sapindus L.	0:2	
SAPOTACEAE		
Nesoluma Baill.	0:1	
Pouteria sandwicensis (A. Gray) Baehni & Degener	1:1	48[b]
SCROPHULARIACEAE		
Bacopa monnieri (L.) Wettst.	1:1	64*,68[a]
SOLANACEAE		
Lycium sandwicense A. Gray	1:1	24[a]
Nothocestrum longifolium A. Gray	1:3	c.48[c]
Solanum americanum Mill.	3:4	12*,24*
S. nelsonii Dunal		24[c]
S. sandwicense Hook. & Arnott		24[q]
STERCULIACEAE		
Waltheria indica L.	1:1	(14,26,40)*
THEACEAE		
Eurya Thunb.	0:1	
THYMELAEACEAE		
Wikstroemia forbesii Skottsb.	8:12	18[dd]
W. furcata (Hillebr.) Rock		18[dd]
W. monticola Skottsb.		18[ee]
W. oahuensis (A. Gray) Rock		18[a,dd]
W. phillyreifolia A. Gray		18[dd]

Table 1.1. (*cont.*)

Taxon	No. spp/genus counted : No. spp/genus native	Diploid chromosome number
W. pulcherrima Skottsb.		36[b,dd]
W. sandwicensis Meisn.		18[dd]
W. uva-ursi A. Gray		18[b,dd],72[b]
URTICACEAE		
Boehmeria grandis (Hook. & Arnott) A. Heller	1:1	28[a]
Hesperocnide Torr.	0:1	
Neraudia ovata Gaud.	1:5	26[b]
Pilea Lindl.	0:1	
Pipturus albidus (Hook. & Arnott) A. Gray	2:4	28[b]
P. kauaiensis A. Heller		28[b]
Touchardia Gaud.	0:1	
Urera kaalae Wawra	1:2	26[a]
VERBENACEAE		
Vitex rotundifolia L. fil.	1:1	32*,34[a]
VIOLACEAE		
Isodendrion laurifolium A. Gray	2:4	16[c]
I. longifolium A. Gray		16[c]
Viola chamissoniana Ging.	3:7	*c.*76[b],80[a]
V. lanaiensis W. Becker		80[a]
V. maviensis H. Mann		(82,*c.*85,*c.*86)[b]
VISCACEAE		
Korthalsella Tiegh.	0:6	
ZYGOPHYLLACEAE		
Tribulus cistoides L.		1

Taxon		
MONOCOTS		
AGAVACEAE	1:1	12*,36[a]
Pleomele Salisb.	0:6	
ARECACEAE		
Pritchardia Seem. & H. A. Wendl.	0:19	
CYPERACEAE		
Bolboschoenus maritimus (L.) Palla	1:1	(40,76–77,80,86,90,96, 104,106,110,114)*
Carex echinata J. A. Murray	4:8	58*
C. macloviana Dum. d'Urv.		82*,86*
C. montis-eeka Hillebr.		48[b]
C. wahuensis C. A. Mey.		48[b]
Cyperus laevigatus L.	1:2	(72,86,88)*
Eleocharis calva Torr.	2:4	18*
E. obtusa (Willd.) Schult.		10*
Fimbristylis cymosa R. Br.	2:3	10*,28[b],40*,48[b]
F. dichotoma (L.) Vahl		(10,12,20,30)*
Gahnia beecheyi H. Mann	1:5	c.96[b]
Machaerina angustifolia (Gaud.) T. Koyama	1:2	c.60[b]
Mariscus Vahl	0:10	
Oreobolus furcatus H. Mann	1:1	42[b]
Pycreus polystachyos (Rottb.) P. Beauv.	1:1	96*,98*
Rhynchospora chinensis Nees & Meyen	1:3	48[b]
Schoenoplectus juncoides (Roxb.) Palla	2:2	c.70[b],76*
S. lacustris (L.) Palla		(38,40,42)*
Scleria Bergius	0:1	
Torulinium Desv.	0:1	
Uncinia uncinata (L. fil.) Kükenth.	1:2	88*
HYDROCHARITACEAE		
Halophila Thouars	0:1	
IRIDACEAE		
Sisyrinchium acre H. Mann	1:1	34[b]

Table 1.1. (*cont.*)

Taxon	No. spp/genus counted : No. spp/genus native	Diploid chromosome number
JOINVILLEACEAE		
Joinvillea Gaud. ex Brongn. & Gris	0:1	
JUNCACEAE		
Luzula hawaiiensis Buchenau	1:1	(14,16,24,38,42)[ff],28[b]
LILIACEAE		
Astelia argyrocoma A. Heller ex Skottsb.	2:3	c.68[b],c.70[b]
A. menziesiana Sm.		60[b]
Dianella sandwicensis Hook. & Arnott	1:1	c.32[a],c.40[b],c.70[b]
ORCHIDACEAE		
Anoectochilus Blume	0:1	
Liparis Rich.	0:1	
Platanthera Rich.	0:1	
PANDANACEAE		
Freycinetia Gaud.	0:1	
Pandanus tectorius S. Parkinson ex Z	1:1	(c.51,54,60)[b],60*
POACEAE		
Agrostis avenacea J. G. Gmelin	1:2	56*
Calamagrostis Adans.	0:2	
Cenchrus L.	0:1	
Chrysopogon aciculatus (Retz.) Trin.	1:1	20*
Deschampsia nubigena Hillebr.	1:1	26[b]
Dichanthelium cynodon (Reichardt) C. A. Clark & Gould	3:4	18[b]
D. hillebrandianum (Hitchc.) C. A. Clark & Gould		18[b]

D. isachnoides (Munro ex Hillebr.) C. A. Clark & Gould		18[b]
Digitaria setigera Roth	1:1	(18,36,54,72)*
Dissochondrus biflorus (Hillebr.) Kuntze ex Hack.	1:1	36[a]
Eragrostis grandis Hillebr.	2:10	44[b]
E. variabilis (Gaud.) Steud.		40[a]
Festuca L.	0:1	
Heteropogon contortus (L.) P. Beauv. ex Roem. & Schult.	1:1	(20,39–44,50,60,c.69–80,90)*
Isachne R. Br.	0:2	
Ischaemum L.	0:1	
Lepturus repens (G. Forster) R. Br.	1:1	42*,54*
Panicum torridum Gaud.	1:11	18[a]
Paspalum scrobiculatum L.	1:1	(20,40,42,50,54,60,80,c.90,120)*
Poa L.	0:2	
Sporobolus virginicus (L.) Kunth	1:1	(18,20,30,40)*
Trisetum Pers.	0:2	
POTAMOGETONACEAE		
Potamogeton foliosus Raf.	2:2	(14,26,28)*
P. nodosus Poir.		52*
RUPPIACEAE		
Ruppia maritima L.	1:1	(16,20,24,40)*
SMILACACEAE		
Smilax L.	0:1	

Source: [a]Carr, 1978; [b]Skottsberg, 1953; [c]Carr, 1985a; [d]Moore, 1974; [e]Carr, unpublished; ; [f]Fedorov, 1974; [g]Carr, 1985b [h]Constance & Affolter, 1990; [i]Kiehn & Lorence, 1996; [j]Gillett & Lim, 1970; [k]Moore, 1977; [l]Gardner, 1977; [m]Rabakonandrianina, 1980; [n]Lowrey, 1986; [o]Lammers, 1988; [p]Lammers & Lorence, 1993; [q]Source unknown; [r]Goldblatt, 1984; [s]Kiehn, unpublished; [t]Storey, 1966; [u]Moore, 1973; [v]Ornduff, 1968; [w]Gillett, 1969; [x]Goldblatt, 1988; [y]Sastrapradja, 1968; [z]Goldblatt, 1981; [aa]Kim & Carr, 1990; [bb]Thomson, 1995; [cc]Linney, 1988; [dd]Gupta & Gillett, 1969; [ee]Skottsberg, 1972; [ff]Coffey, 1990

incidence of polyploidy in these insular examples reflects a selective advantage of genetically amplified and potentially diversified polyploid genomes with respect to the effects of genetic bottlenecking that is associated with colonization after long-distance dispersal.

The data in Table 1.1 indicate that there are at least 38 genera which include interspecific or interpopulational variation in chromosome number based on Hawaiian populations. However, the balance of evidence indicates that apparent variation in several genera *(Argyroxiphium, Bidens, Clermontia, Cyanea, Tetramolopium* and *Wilkesia)* is almost certainly the result of faulty cytological determinations. This may also prove to be the case with *Achyranthes, Alyxia, Gouania, Pandanus, Silene, Viola* and *Zanthoxylum.* At least some of the variation in chromosome number in other genera *(Abutilon, Bobea, Coprosma, Dubautia, Heliotropium, Labordia, Lepidium, Lipochaeta, Luzula, Lysimachia, Scaevola, Peperomia, Plantago, Portulaca, Psychotria* and *Wikstroemia)* reported in Table 1.1 appears to reflect real chromosomal variation in Hawaiian plants. The remaining nine genera *(Acacia, Astelia, Dianella, Eragrostis, Hedyotis, Hibiscus, Phyllostegia, Ranunculus* and *Rumex)* probably constitute a mixture of examples of bona fide variation in chromosome number and apparent variation resulting from erroneous cytological determinations. It would be especially gratifying to have more data for *Acacia, Coprosma, Gouania, Hedyotis, Hibiscus, Lysimachia, Phyllostegia, Plantago, Psychotria, Ranunculus* and *Rumex.*

The interpretation of chromosome numbers reported for Hawaiian plants is enhanced by several studies that vary in scope and depth. In some instances these studies document considerable or moderate variation in chromosome number and/or structure within a genus or group (Table 1.2), while in others apparently complete chromosomal stasis is documented (Table 1.3). The most thoroughly documented of the examples of these alternative modes of chromosome evolution in Hawaiian plants are presented in some detail.

Examples of chromosome variation in Hawaiian plants

Chromosome evolution may involve structural changes such as deletions, duplications, inversions and translocations as well as changes in chromosome number that result from polyploidy or aneuploidy. Permanent aneuploid change in chromosome number in diploid or diploidized organisms is often termed dysploidy (cf. Garber, 1972). The studies

Table 1.2. *Hawaiian plant groups providing evidence for autochthonous chromosomal evolution*

Group or genus	Scope of relevant information	References
Argyroxiphium, Dubautia, Wilkesia (Asteraceae)	Chromosome numbers of all 27 extant species; detailed data on chromosome pairing and fertility of infra- and intergeneric hybrids; molecular and biosystematic data bearing on origin of group	Carr, 1985b; Carr & Kyhos, 1981, 1986; Baldwin et al., 1991; Baldwin & Robichaux, 1995; Carr, Baldwin & Kyhos, 1996
Bobea (Rubiaceae)	Chromosome number or approximation for 2 of 4 species	Kiehn & Lorence, 1996
Lepidium (Brassicaceae)	Chromosome numbers for 2 of 3 species; presumption that the native species are monophyletic	Carr, 1978; Carr, unpublished; Wagner, Herbst & Sohmer, 1990
Peperomia (Piperaceae)	Chromosome numbers for 18 of 25 species; detailed chromosome study	Sastrapradja, 1968
Plantago (Plantaginaceae)	Chromosome numbers for 2 of 3 species	Skottsberg, 1953; Moore, 1973; Kiehn, unpublished
Portulaca (Portulacaceae)	Chromosome numbers of all 4 native species; fertility and chromosome pairing data from several hybrids	Kim & Carr, 1990
Psychotria (Rubiaceae)	Chromosome numbers or approximations for 4 of 11 species	Kiehn & Lorence, 1996; Kiehn, unpublished
Wikstroemia (Thymelaeaceae)	Chromosome numbers of 8 of 12 species; fertility data from many artificially produced interspecific hybrids	Skottsberg, 1953, 1972; Gupta & Gillett, 1969; Mayer, 1991

reported herein are limited to the detection of gross chromosomal changes including translocations, dysploidy and polyploidy.

Structural and dysploid evolution

The only well-documented example of chromosome evolution in the Hawaiian flora that involves translocations and dysploidy is the silversword alliance, a group of three closely related genera with 27 extant species of exceedingly diverse morphological and ecological characteristics (Carr & Kyhos, 1981, 1986; Carr, 1985*b*). This assemblage of species includes mat- and cushion-forming subshrubs, and monocarpic and polycarpic shrubs, trees and lianas. Despite their impressive adaptive radiation into myriad habitats, these species retain the ability to produce interspecific and intergeneric hybrids under natural conditions, and in the laboratory. Cytogenetic analysis of hybrids has made it possible to assess in detail the nature of chromosome alterations that distinguish various species within the alliance. The chromosome studies have identified two dysploid clusters of species, 18 with $n = 14$ and 9 with $n = 13$ (Carr, 1985*b*). Moreover, the occurrence of at least seven chromosome races differentiated by reciprocal translocations has been documented among the 14-paired species (Carr & Kyhos, 1986). Interracial hybrids exhibit multivalents at meiosis and have pollen stainabilities that reflect reduction in fertility caused by faulty segregation of chromosomes. The chromosome configuration at meiosis in these hybrids includes 12 pairs and a chain or ring of four chromosomes, 11 pairs and a chain of six chromosomes or ten pairs and two chains of four chromosomes. The cytogenetic evidence indicates that certain chromosomes have been involved repeatedly in translocations to produce the known cytotypes in this alliance of species.

In contrast, the 13-paired genome is apparently structurally homogeneous such that hybrids among species with this genome are fully fertile and show no meiotic abnormalities. The simplest pairing configuration between the 13-paired genome and a 14-paired species is 12 pairs and a linear chain of three chromosomes. This configuration is produced in hybrids involving either *Dubautia scabra* ($n = 14$) or *D. laevigata* ($n = 14$) and any 13-paired species. All hybrids between other 14-paired species and the 13-paired genome likewise show the linear chain of three but also exhibit one or more additional multivalents. Many factors, including species distributions, geological ages of islands, patterns of genetic variation, ecological characteristics and molecular evidence,

Table 1.3. *Hawaiian plant groups providing evidence for chromosomal stasis*

Group or genus	Scope of relevant information	References
Bidens (Asteraceae)	Chromosome numbers of 13 of 19 species; comprehensive fertility data from artificial hybridization programs; data on chromosome analysis of hybrids	Gillette & Lim, 1970; Ganders & Nagata, 1984
Cyrtandra (Gesneriaceae)	Chromosome numbers of 8 of 53 species; fertility data for many putative natural hybrids; limited information on chromosome pairing in putative hybrids	Wagner, Herbst & Sohmer, 1990; Luegmayr, 1993; Kiehn, personal communication
Hibiscadelphus (Malvaceae)	Chromosome numbers of 3 of 6 species; meiotic analysis of spontaneous F_1 and F_2 hybrids	Carr & Baker, 1977
Lipochaeta (Asteraceae)	Chromosome numbers of 15 of 20 species; extensive fertility and chromosome pairing data from many interspecific and intersectional hybrids; data from induced polyploids	Gardner, 1976, 1977; Rabakonandrianina, 1980; Rabakonandrianina & Carr, 1981
Lobelioids, 7 genera (Campanulaceae)	Chromosome numbers of 21 of 110 species; limited data on meiosis in putative natural hybrids; molecular data bearing on evolutionary relationships	Lammers, 1988, 1990, 1995; Lammers & Lorence, 1993; Givnish *et al.*, 1995
Pipturus (Urticaceae)	Chromosome numbers of 2 of 4 species based on several population samples	Nicharat & Gillett, 1970
Scaevola (Goodeniaceae)	Chromosome numbers of all 9 species; chromosome pairing data in a natural interspecific hybrid; flavonoid data; cladistic analysis of morphological data	Gillett, 1966, 1969; Patterson, 1984*a*, 1984*b*, 1995

Table 1.3. (*cont.*)

Group or genus	Scope of relevant information	References
Tetramolopium (Asteraceae)	Chromosome numbers of 7 of 11 species; comprehensive fertility and chromosome pairing data from artificial interspecific hybrids	Lowrey, 1986
Vaccinium (Ericaceae)	Chromosome numbers of 2 of 3 species; fertility data from artificially produced interspecific hybrids	Vander Kloet, 1993

indicate the ancestral nature of the 14-paired genome and suggest an Hawaiian origin of the 13-paired genome from one like that possessed by *Dubautia scabra* (Carr, 1985*b*; Carr *et al.*, 1989; cf. Baldwin & Robichaux, 1994). The origin of the entire assemblage of species constituting the Hawaiian silversword alliance from a single introduction of an ancestral form of tarweed from the Pacific coast of North America is supported by long-term anatomical and biosystematic studies, and especially by recent molecular studies and experimental hybridization between *Dubautia* from Hawaii and *Raillardiopsis* and *Madia* from North America (Kyhos, Carr & Baldwin, 1990; Baldwin, Kyhos & Dvořák, 1990; Baldwin *et al.*, 1991; Baldwin, 1992; Baldwin & Robichaux, 1995; Carr, Baldwin & Kyhos, 1996).

A feature within the genus *Dubautia* that is correlated with the dysploid reduction in chromosome number from $n = 14$ to $n = 13$ is the ability of species with the 13-paired genome to survive in increasingly arid habitats. The morphological and anatomical modifications of the leaves of 13-paired species that promote this ability are rather obvious and include reduced leaf area, thickened cuticle and compact internal tissues (Carr *et al.*, 1989). Not so obvious is a physiological shift that has occurred, perhaps concomitant with dysploidy, that increases the ability of 13-paired species to maintain turgor pressure under conditions of decreasing tissue water content compared to 14-paired species of *Dubautia*. The mechanism of this adaptation to arid environments involves a modification of the tissue elastic properties and is apparently universal among the 13-paired species of *Dubautia* (Carr *et al.*, 1989).

A second example of dysploidy occurring among Hawaiian populations of an endemic species may be found in *Luzula hawaiiensis* (Table 1.1). However, these reports, which represent chromosome numbers rarely found or not otherwise known in the genus, need further documentation and substantiation. *Peperomia* represents still another possible instance of dysploidy occurring among Hawaiian populations of a single species or between closely related species but the occurrence of polyploidy and the uncertainty of relationships and identification of species in this genus make it a somewhat dubious example. The genus is discussed further under the section dealing with polyploidy. Other possible examples of dysploidy based on information in Table 1.1 are more likely the result of uncertain or erroneous counts or misidentifications.

Polyploid evolution

Well-documented occurrences of autochthonous polyploidy within the evolutionary milieu of Hawaii are found in *Peperomia, Portulaca* and *Wikstroemia* (Table 1.2). Additional examples of autochthonous polyploidy in the Hawaiian flora may exist in *Bobea, Labordia, Lepidium, Plantago* and *Psychotria*. More documentation is needed to characterize the situation in these genera.

Bobea is an endemic, presumably monophyletic, Hawaiian genus comprising four species (Darwin & Chaw, 1990). The chromosome counts $2n = 66\text{--}70$ and $2n = 88$ for *B. elatior* and *B. timonioides*, respectively (Kiehn, unpublished), suggests that $2n = 44$ also occurs in this genus (or did at some time in the past). Hybridization between the apparent octaploids and tetraploids would be the simplest explanation for presence of the apparent hexaploid in Hawaii if the presumption of a single common ancestor for the genus is correct.

Chromosome numbers available for *Labordia* (Table 1.1) make it appear that autochthonous polyploidy has occurred in this endemic Hawaiian genus. The report of $2n = 40\text{--}44$ for *L. helleri* is in close agreement with the observations of $2n = 40$ (Conn, 1980; Carr, unpublished) for *Geniostoma rupestre* J. R. & G. Forst., a possible ancestor of *Labordia*. However, only $2n = c.\,80$ has been observed in *L. helleri* and several other species of *Labordia* by Motley & Carr (unpublished). Thus, additional clarification and confirmation of the lower numbers determined by Kiehn (unpublished) (Table 1.1) would strengthen the case for autochthonous polyploidy in *Labordia*.

With two of the three native species of *Lepidium* determined, the chromosome numbers of $2n = 32$ and $2n = 48$ appear to represent autochthonous polyploidy in this genus, at least if the Hawaiian species are monophyletic, as suggested by Wagner, Herbst & Sohmer (1990). As it seems likely that the indigenous *L. bidentatum* gave rise to the two endemic species, one of which has $2n = 32$, the existence of a ploidy level lower than $2n = 48$ in the ancestor is to be expected. The detection of a lower ploidy level in *L. bidentatum* would certainly make a stronger case for autochthonous polyploidy in this genus. Interestingly, among the 125 species of *Lepidium* worldwide, only two (*L. graminifolium* and *L. latifolium*) reportedly share the chromosome number of $2n = 48$ with *L. bidentatum* (Moore, 1977; Goldblatt, 1984).

The concerted effort by Sastrapradja (1968) supports the occurrence of $2n = 44, 48, 88$ and 96 among Hawaiian species of *Peperomia* (Table 1.1) and casts doubt on some of the earlier reports by Skottsberg (1953). As Hawaiian species of *Peperomia* probably resulted from as many as three or four separate ancestral introductions (cf. Wagner, Herbst & Sohmer, 1990), some of this polyploid diversification probably occurred prior to dispersal of the ancestors to the Hawaiian archipelago. However, some polyploid (and possibly aneuploid) diversity occurs between populations of a single species or between closely related species. Thus, it seems highly probable that some of this polyploidization occurred within the Hawaiian archipelago.

The three species of *Plantago* native to Hawaii are believed to be derived from two separate introductions (Wagner, Herbst & Sohmer, 1990). The presence of $2n = 12$ and $2n = 24$ in one of these species, *P. princeps,* indicates either autochthonous origin of the tetraploid element or taxonomic misalignment of the subspecific taxon it represents. More information is needed to clarify this situation.

Two of the native Hawaiian species of *Portulaca* are diploid with $2n = 18$, and two others are of an unknown polyploid origin with $2n = 40$ (Table 1.1). According to Wagner, Herbst & Sohmer (1990), these two groups of species are derived from separate introductions to Hawaii. Therefore, it seems likely that the polyploidy distinguishing these groups occurred before dispersal of the ancestors to the Hawaiian archipelago. However, polyploidy of *Portulaca* has occurred within Hawaii, at least in the form of a naturally occurring fertile polyploid derivative ($2n = 94$) of a hybrid between the introduced *P. oleracea* ($2n = 54$) and the native *P. lutea* ($2n = 40$) from Molokai (Kim & Carr, 1990).

The 10 species of Hawaiian *Psychotria* are divided among two sections, each of which is believed to be independently derived from a separate ancestral colonizer in Hawaii (Wagner, Herbst & Sohmer, 1990). A chromosome determination of $2n = 88$ for *P. hobdyi* represents one of these sections, while determinations of $2n = 66$, 66 ± 2 and 66–88 correspond to *P. mariniana, P. kaduana* and *P. greenwelliae*, respectively, from the other section of the genus (Table 1.1). If $2n = 88$ (presumed octoploid) and $2n = 66$ (presumed hexaploid) are both ultimately documented in the same section, then tetraploids ($2n = 44$) may also be expected to occur in this section because, as mentioned for *Bobea*, autochthonous origin of the hexaploids is most easily explained on the basis of hybridization between tetraploids and octoploids. Clearly, more documentation is needed to characterize this situation.

Based on present knowledge, *Wikstroemia* is taken herein as the clearest example of autochthonous polyploid evolution of flowering plants in the Hawaiian archipelago. Chromosome numbers of $2n = 18$ and $2n = 72$ have been reported for *Wikstroemia uva-ursi*, while $2n = 36$ has been reported for *W. pulcherrima* (Skottsberg, 1953; Gupta & Gillett, 1969). The remaining six species that have been examined have $2n = 18$. All of these species are believed to have been derived autochthonously from a single colonizing ancestor (Wagner, Herbst & Sohmer, 1990). In addition to the polyploidy indicated above, naturally occurring triploid individuals have been detected on Oahu (Carr, unpublished).

More than one ploidy level is found in other Hawaiian genera, e.g., *Abutilon, Coprosma, Heliotropium, Hibiscus, Lipochaeta, Lysimachia* and *Scaevola,* but in these cases there is reason to believe that the different ploidy levels found in Hawaii represent separate colonizations and that in each instance polyploidization occurred before the ancestor reached Hawaii. The situation in each of these genera is discussed briefly elsewhere (Fosberg, 1948; Wagner, Herbst & Sohmer, 1990). The reports of $2n = 54$ and 72 for *Lysimachia hillebrandii* could represent autochthonous polyploidy but confirmation of the lower number reported by Skottsberg (1953) is needed. *Lipochaeta* and *Scaevola* are discussed in further detail under the section on chromosome stasis.

Examples of chromosome stasis in Hawaiian plants

The genus *Bidens* has been more thoroughly studied biosystematically than any other genus of Hawaiian plants. Most of the work on the Hawaiian species stems from the activities of George W. Gillett and Fred R. Ganders and various co-workers. In Hawaii, there are 19 species that range in habit from small, more or less herbaceous plants, to large, somewhat woody shrubs, or small trees up to four metres (m) high. The genus occupies very diverse Hawaiian habitats ranging from very dry to very wet, and from sea level to over 2200 m elevation. Despite these marked morphological and ecological differences, the Hawaiian species are very closely related and no doubt arose from a single ancestor (Ganders & Nagata, 1984). Gillett & Lim (1970) and Ganders & Nagata (1984) have collectively completed an exhaustive programme of hybridization involving virtually all of the available forms of the genus in Hawaii. Ganders & Nagata (1984) reported that all of the hybrids they produced were fully seed- and pollen-fertile. Gillett & Lim (1970) reported similar results but also made observations on chromosome pairing in hybrids, noting irregularities in only two instances. The only irregularities reported were the occurrences of up to four univalents found in some microsporocytes of two hybrids which nevertheless had high fertility. The comprehensive research on *Bidens* makes it perhaps the best example available to demonstrate apparently complete stasis of chromosome structure during otherwise striking evolutionary change and adaptation to varied Hawaiian habitats.

Cyrtandra is another example of a genus that has speciated profusely in Hawaii (53 species) but so far as available data indicate, this differentiation has not involved change in chromosome number or structure. This conclusion is based on preliminary investigation of some of the 67 putative interspecific hybrids identified (Wagner, 1986; Wagner, Herbst & Sohmer, 1990; Smith, Burke & Wagner, 1995). All 17 of the species counted have $2n = 34$ (Table 1.1), and limited observations of chromosomes of hybrids have failed to reveal any abnormalities that might be attributed to structural differences in the genomes of the parental species (M. Kiehn, personal communication). In addition, pollen stainability in ten different interspecific hybrids was 70–99%, suggesting complete or nearly complete chromosome homology among all the species involved (Luegmayr, 1993).

Hibiscadelphus is an endemic Hawaiian genus of six tree species of uncertain, but undoubtedly autochthonous origin from a single common ancestor (cf. Fosberg, 1948). The chromosome number of all known species is $2n = 40$. A cytogenetic study of spontaneous hybrids revealed no structural differences in the chromosomes of *H. giffardianus* and *H. hualalaiensis* (Carr & Baker, 1977).

Fosberg (1948) considered *Lipochaeta* an endemic genus of 55 species and varieties. The cytology of the genus was almost unknown until two workers reported chromosome numbers for most of the species in two papers (Gardner, 1977; Carr, 1978). The present treatment of *Lipochaeta* (Wagner, Herbst & Sohmer, 1990) recognizes 20 species, chromosome numbers for 15 of which are available (Table 1.1). Ten of these are apparent diploids with $2n = 30$, while the other five are apparent tetraploids with $2n = 52$. Gardner (1976) favoured the hypothesis that the tetraploids evolved autochthonously by autopolyploidy from the diploids in the Hawaiian archipelago and he recognized one diploid and one tetraploid section in the genus. This scenario of evolution necessitates aneuploid reduction from the autopolyploid number of $2n = 60$ to reach the chromosome number of $2n = 52$ that characterizes the 'tetraploid' species of *Lipochaeta*.

Studies on *Lipochaeta* by Rabakonandrianina (1980) and Rabakonandrianina & Carr (1981) emphasized cytogenetic analyses of intersectional and intergeneric hybrids to determine the relationships of the two sections with each other and with extra-Hawaiian taxa. These studies support the view that *Lipochaeta* is biphyletic, i.e., each section probably arose from a separate extra-Hawaiian ancestry; the diploid section from *Wollastonia biflora* (L.) DC. or a similar species, and the tetraploid section from some other wedelioid ancestor. The cytogenetic data further suggest that the genome of the tetraploid progenitor was allotetraploid with one genome similar or identical to that of *W. biflora* and the other genome from a wedelioid plant with $2n = 22$ (Rabakonandrianina & Carr, 1981). If this wedelioid ancestry is accurate, then the two sections of *Lipochaeta* probably align with different elements of the diverse assemblage of wedelioid genera and it may eventually be concluded that neither element is endemic at the generic level (cf. Strother, 1991).

Taken independently, the genomes of each of the two sections of *Lipochaeta* are completely homogeneous based on cytogenetic analysis of numerous interspecific hybrids (Rabakonandrianina, 1980). This extensive study found no evidence of abnormality other than occasional univalents during meiosis in hybrids between species within each section. Such hybrids had high pollen stainability of 80–100% and those examined also were seed-fertile. Thus, each section apparently exhibits complete chromosome stasis with respect to gross chromosome structural alterations and chromosome number. The only cytologically unusual feature within either section of *Lipochaeta* is the occurrence of supernumerary (B) chromosomes that is relatively common in some populations of the tetraploid species but is much more infrequent among the diploid species (cf. Gardner, 1977; Rabakonandrianina, 1980). Indeed, *Lipochaeta* may represent the only documented occurrence of supernumerary chromosomes in native Hawaiian plants.

The Hawaiian lobelioids represent a truly remarkable example of adaptive radiation of one or a few ancestral types into a great variety of forms occupying ecologically diverse habitats. The group comprises 110 species distributed among seven genera. While there is disagreement as to exactly how many ancestral types are involved, virtually all agree the number is very small and many feel that at least the 96 species comprising the endemic genera *Clermontia*, *Cyanea*, *Delissea* and *Rollandia* have evolved autochthonously from a single ancestral type that reached the Hawaiian archipelago (cf. Lammers, 1990, 1995; Givnish *et al.*, 1995).

Chromosome numbers have been reported for 21 species of Hawaiian lobelioids (Campanulaceae) (Table 1.1), including one or more determinations for each of the seven genera. For a few species, more than one population has been sampled. Except for two isolated reports of $2n = 22$ and $2n = 24$, the determinations have all been $2n = 28$. Each of the two exceptions involves a species that has otherwise been determined to have $2n = 28$. At this point, considering the diversity of species sampled, there is a high probability that all Hawaiian lobelioids have $2n = 28$ chromosomes and that other determinations reported earlier in the literature are in error (Lammers, 1988). Furthermore, an examination of meiosis in two different putative interspecific hybrids did not reveal any evidence of chromosome structural differentiation of the parental genomes (Lammers, 1988). Thus, although more cytogenetic studies clearly are needed, there is presently no evidence of chromosome evolution within this large group of species despite the extreme range of morphological and ecological diversification that has occurred.

Nicharat & Gillett (1970) cytologically investigated 31 plants from 23 populations representing two of the four species of *Pipturus* presently recognized (Wagner, Herbst & Sohmer, 1990). The nature of their study did not allow assessment of variation in chromosome structure but it does document the constancy of the chromosome number of $2n = 28$ among all 31 individuals of *Pipturus* examined.

Chromosome determinations for Hawaiian populations of *Scaevola* have all been $2n = 16$ except for those of *S. glabra* which all have been $2n = 32$ (Table 1.1). However, the polyploidy represented by the latter count was likely to have occurred before the ancestor of this taxon was dispersed to Hawaii. This interpretation is supported by observations of Gillett (1966) and more recently by chemical evidence and cladistic analysis (Patterson, 1984*a*, 1984*b*, 1995). These studies also support the hypothesis that seven of the nine species currently recognized have an autochthonous Hawaiian origin from a single common ancestor. Gillett (1966) studied meiosis in hybrids between two of these species and saw no evidence that the parental genomes were chromosomally distinct in any way. Thus, available evidence supports *Scaevola* as another example of chromosome stasis in a group of otherwise substantially differentiated species.

Chromosome numbers are known for all seven of the extant species of *Tetramolopium*. Four species are presumed extinct. All reports have been $2n = 18$ except for an apparently erroneous determination of $2n = 14$ for *T. humile* (Table 1.1). Due to the biosystematic efforts of Lowrey (1986), *Tetramolopium* provides one of the best documented examples of chromosomal stasis in the Hawaiian flora. Virtually all possible combinations

of interspecific hybrids involving all seven extant species and 23 populations of *Tetramolopium* were produced. The pollen stainabilities in F_1
hybrids were generally as high or higher than those of the parents.
Meiosis in all hybrid combinations was normal and all F_1 plants produced fertile achenes that led to the generation of large progenies of
vigorous F_2 plants. It is improbable that any major variations in chromosome structure would have gone undetected in this study.

Chromosome numbers apparently have been reported from only two
of the three presently recognized species of *Vaccinium* (Table 1.1).
However, based on the results of hybridization among all three species
(Vander Kloet, 1993), it is highly probable that all have $2n = 24$.
Furthermore, the high fertility of F_1 plants that he observed in all combinations of the Hawaiian species suggests that the parental genomes
harbour no major chromosome structural differences.

On one level of chromosome evolution, *Wikstroemia* provides what
may be the clearest example of autochthonous polyploidy in the
Hawaiian flora (see above). On another level, the biosystematic efforts
of Mayer (1991) established apparent uniformity of chromosome structure among the diploid Hawaiian species. In this study, 19 populations
were utilized to produce 11 different interspecific hybrid combinations
involving seven different species in the genus. As pollen stainability of
these hybrids ranged from 70–99%, compared to 77–99% for parental
plants, the level of apparent pollen viability in the hybrids appears to be
completely normal. Thus, although no meiotic analyses were performed,
Mayer's study indicates that it is unlikely that there has been any major
repatterning of chromosome structure among the diploid species of
Wikstroemia.

The Hawaiian milieu

All plant groups that have evolved in the Hawaiian archipelago share a
similar set of circumstances related to their early history on the islands.
These shared features mainly relate to the isolation of the archipelago
from the source of propagules and to the volcanic nature of the islands
(cf. Macdonald, Abbott & Peterson, 1983). The isolation of the Hawaiian
Islands by over 2000 miles of open ocean has prevented the successful
establishment of many families and genera of flowering plants that are
common in continental areas. Dispersibility of viable propagules and
stochastic events have determined the success of a given group in pene-
trating the barrier of great isolation of the islands. As a consequence of

this filtering effect, the composition of the flora encountered by any given successful immigrant was likely to be very different than that of its source area.

Other factors shared by successful colonizers of the islands include virtually complete isolation of an extremely limited gene pool, frequently consisting of the genetic diversity of one or very few individuals dispersed to a given site. In such instances the influence of stochastic events are maximized. This effect is repeated with each colonization of a new island after the initial founding event in the archipelago. Even after becoming well-established, many species have repeatedly been partitioned into effectively isolated colonies by lava flows resulting from frequent volcanic activity on the geologically younger islands. Likewise, erosional processes on the older islands have had the similar effect of dissecting populations into smaller and smaller isolated enclaves of few individuals. In these situations the effects of stochastic events should also be maximized.

Indeed, stochastic factors appear to permeate the evolutionary fabric in such insular ecosystems and they, along with other effects of population bottlenecks and the Hawaiian milieu, presumably help to account for the otherwise inexplicable morphological and ecological diversity of groups like the lobelioids and the silversword alliance. If, as many claim, chromosome evolution is generally largely or wholly the result of stochastic events, then it seems that the Hawaiian flora should also represent a showcase of chromosome evolution. This is hardly the case. In fact, almost all of the published studies that are sufficient in scope and depth to address the question with any meaning show quite the opposite. How can this paradox of stasis of chromosome structure in Hawaiian flowering plants be explained?

Natural selection for chromosome evolution

There are undoubtedly instances where new chromosome arrangements have become fixed in populations largely or wholly by chance. Moreover, many argue that this is the normal mode of origin of new races distinguished by chromosomal differentiation (cf. Lewis, 1973; Lande, 1979, 1984; Hedrick, 1981; Hall, 1983; Hedrick & Levin, 1984; Larson, Prager & Wilson, 1984). Others have asserted that natural selection plays a significant role in the fixation of new chromosome arrangements (cf. John & Lewis, 1959; Dobzhansky, 1970; Bengtsson & Bodmer, 1976; Bush, 1981; Baker *et al.*, 1983; Chesser & Baker, 1986).

Kyhos & Carr (1994) argue that there is a surplus of examples of chromosome evolution in groups that have biological characteristics counter-intuitive to models based on chance, e.g., groups comprising self-incompatible annual species. Conversely, they note several examples of selfing groups that have maintained structurally constant genomes among many species over long periods of time even though autogamy should maximally promote the fixation of new chromosome arrangements by chance. Kyhos & Carr (1994) appear to be the first to cite instances of chromosomal stasis as indicative of the possible action of natural stabilizing selection in maintaining a particular chromosome arrangement over a long period of time during which stochastic factors should favour chromosome evolution.

The flora of the Hawaiian Islands appears to be a very good example of this sort of chromosome stasis. As indicated above, stochastic factors have played a large role in the establishment and evolution of Hawaiian plant groups. Furthermore, only one example of self-incompatibility in the Hawaiian flora is known (Carr, Powell & Kyhos, 1986). The self-compatible nature of possibly all other plant groups in the Hawaiian flora should be another factor promoting the stochastic evolution of new chromosome races in the Hawaiian milieu. Ironically, however, the only well-documented example of autochthonous Hawaiian chromosome evolution other than polyploidy is found in what may be the only self-incompatible Hawaiian group, the silversword alliance (Carr, Powell & Kyhos, 1986). This and other apparent paradoxes of chromosome evolution are not explained by models based predominately on stochastic factors.

The basis of natural selection for chromosome evolution

Considerable evidence is accruing from molecular studies that provides a basis for the action of natural selection in the establishment of new chromosome races in populations of plants (cf. Kyhos & Carr, 1994). This evidence is basically that genes have a positional effect in the chromosome complement. When segments of chromosomes are repositioned within the genome, the loci near the ends of segments are brought into a new gene neighbourhood and react in unpredictably new and different ways. For example, in humans any chromosome translocation that brings the *c-myc* oncogene into the vicinity of immunoglobulin genes and their enhancers results in a great increase in the transcription level of the *c-myc* oncogene and Burkitt lymphoma is induced (cf. Dalla-Favera *et al.*, 1982; Erikson *et al.*, 1982; Taub *et al.*, 1982; Klein, 1983). Results such as these in humans and other animals have been detected because of their drastic and obviously lethal or sublethal effects. However, there appears ample reason to believe similar effects are more widespread. Most molecular studies of this sort have focused on animals, but similar evidence has been found in studies dealing with plants (Jones, Dunsmuir & Bedbrook, 1985; Nagy *et al.*, 1985; Napoli, Lemieux & Jorgensen, 1990; van der Krol *et al.*, 1990). In review, Manuelidis (1990) concluded: 'Genes are strategically positioned and hierarchically segregated in the interphase nucleus. Sequential compartmentalization of genes within domains of different size, ... and the nucleus as a whole may additively influence phenotypic expression'.

The mounting molecular evidence suggests that the repositioning of genes within the genome that would be the inevitable result of chromosomal restructuring provides ample basis for the action of natural selection. It seems apparent that most, if not all, such changes in an organism that has had ample time to evolve the 'perfect genome' would be suboptimal and be eliminated by what amounts to stabilizing natural selection. Conversely, a species that is still undergoing evolutionary refinement of its genome may occasionally experience enhanced genetic functionality due to repositioning of loci through chromosome restructuring. With the acceptance of a selective basis for chromosome evolution, the apparent paradoxes observed by Kyhos & Carr (1994), including that of chromosome stasis of the Hawaiian flora, disappear. After all, most elements of the Hawaiian flora are woody, perennial groups that have been evolving for long periods of time and presumably have finely adjusted genomes that may tolerate few changes of the sort that might be caused by chromosome restructuring. The Hawaiian silversword alliance, on the other hand, is from a youthful family (Asteraceae) and an immediate ancestry (Heliantheae, Madiinae) that is notoriously fraught with chromosome change (Kyhos, Carr & Baldwin, 1990). The chromosomal variation observed in the Hawaiian elements of this group may be viewed as a continuing genomic adjustment towards an optimum that theoretically will be eventually reached and maintained by stabilizing selection.

In conclusion, it seems there may be little that is special about chromosome evolution in the Hawaiian environment. Rather, it appears that the chromosome stasis or evolution exhibited by Hawaiian groups reflects these same tendencies in the continental ancestors of these groups. These 'tendencies' can be explained on the basis of natural selection operating to maintain the characteristics of a well-adapted genome or to adjust those that can be improved. Thus, an examination of chromosomally active groups in Hawaii reveals that similar patterns are generally seen in their mainland counterparts. The relatively modest degree of chromosome differentiation that may have accrued in any given chromosomally evolving Hawaiian group is readily traced to the geologic youth of the Hawaiian Islands in comparison with the continents.

Acknowledgements

I am grateful to Derral Herbst, Ken Marr, Tim Motley, Ann Sakai, Warren L. Wagner and Steve Weller for assistance in bringing materials together or for help in defining the scope of this paper. I thank Don Kyhos for many stimulating discussions on chromosome evolution and Robert Robichaux for comments on an early draft of this paper. I am also indebted to Michael Kiehn for sharing his unpublished chromosome determinations and thoughts on chromosome evolution in island environments.

Literature cited

Baker, R. J., Chesser, R. K., Koop, B. F. & Hoyt, R. A. (1983). Adaptive nature of chromosomal rearrangements: differential fitness in pocket gophers. *Genetica*, **61**, 161–4.

Baldwin, B. G. (1992). Phylogenetic utility of the internal transcribed spacers of nuclear ribosomal DNA in plants: an example from the Compositae. *Molecular Phylogenetics and Evolution*, **1**, 3–16.

Baldwin, B. G., Kyhos, D. W. & Dvořák, J. (1990). Chloroplast DNA evolution and adaptive radiation in the Hawaiian silversword alliance (Asteraceae–Madiinae). *Annals of the Missouri Botanical Garden*, **77**, 96–109.

Baldwin, B. G., Kyhos, D. W., Dvořák, J. & Carr, G. D. (1991). Chloroplast DNA evidence for a North American origin of the Hawaiian silversword alliance (Asteraceae). *Proceedings of the National Academy of Sciences USA*, **88**, 1840–3.

Baldwin, B. G. & Robichaux, R. H. (1995). Historical biogeography and ecology of the Hawaiian silversword alliance (Asteraceae): new molecular phylogenetic perspectives. In *Hawaiian Biogeography: Evolution on a Hot Spot Archipelago*, ed. W. L. Wagner & V. A. Funk, pp. 259–87. Washington DC: Smithsonian Institution Press.

Bengtsson, B. O. & Bodmer, W. F. (1976). On the increase of chromosomal mutations under random mating. *Theoretical Population Biology*, **9**, 260–81.

Bush, G. (1981). Stasipatric speciation and rapid evolution in animals. In *Evolution and Speciation*, ed. W. R. Atchley & D. S. Woodruff, pp. 201–18. New York: Cambridge University Press.

Carr, G. D. (1978). Chromosome numbers of Hawaiian flowering plants and the significance of cytology in selected taxa. *American Journal of Botany*, **65**, 236–42.

Carr, G. D. (1985*a*). Additional chromosome numbers of Hawaiian flowering plants. *Pacific Science*, **39**, 302–6.

Carr, G. D. (1985*b*). Monograph of the Hawaiian Madiinae (Asteraceae): *Argyroxiphium, Dubautia* and *Wilkesia*. *Allertonia*, **4**, 1–123.

Carr, G. D. & Baker, J. K. (1977). Cytogenetics of *Hibiscadelphus* (Malvaceae): a meiotic analysis of hybrids in Hawaii Volcanoes National Park. *Pacific Science*, **31**, 191–4.

Carr, G. D., Baldwin, B. G. & Kyhos, D. W. (1996). Cytogenetic implications of artificial hybrids between the Hawaiian silversword alliance and North American tarweeds (Asteraceae: Heliantheae–Madiinae). *American Journal of Botany*, **83**, 653–60.

Carr, G. D. & Kyhos, D. W. (1981). Adaptive radiation in the Hawaiian silversword alliance (Compositae–Madiinae). I. Cytogenetics of spontaneous hybrids. *Evolution*, **35**, 543–56.

Carr, G. D. & Kyhos, D. W. (1986). Adaptive radiation in the Hawaiian silversword alliance (Compositae–Madiinae). II. Cytogenetics of artificial and natural hybrids. *Evolution*, **40**, 959–76.

Carr, G. D., Powell, E. A. & Kyhos, D. W. (1986). Self-incompatibility in the Hawaiian Madiinae (Compositae): an exception to Baker's Rule. *Evolution*, **40**, 430–4.

Carr, G. D., Robichaux, R. H., Witter, M. S. & Kyhos, D. W. (1989). Adaptive radiation of the Hawaiian silversword alliance (Compositae–Madiinae): a comparison with Hawaiian picture-winged *Drosophila*. In *Genetics, Speciation, and the Founder Principle*, ed. W. Giddings, K. Y. Kaneshiro & W. W. Anderson, pp. 79–97. New York: Oxford University Press.

Chesser, R. K. & Baker, R. J. (1986). On factors affecting the fixation of chromosomal rearrangements and neutral genes: computer simulations. *Evolution*, **40**, 625–32.

Coffey, J. C. (1990). Juncaceae. In *Manual of the Flowering Plants of Hawaii*, vol. 2, ed. W. L. Wagner, D. R. Herbst & S. H. Sohmer, pp. 1451–5. Honolulu: University of Hawaii Press; Bishop Museum Press.

Conn, B. J. (1980). A taxonomic revision of *Geniostoma* subg. *Geniostoma* (Loganiaceae). *Blumea*, **26**, 245–364.

Constance, L. & Affolter, F. (1990). Apiaceae. In *Manual of the Flowering Plants of Hawaii*, vol. 1., ed. W. L. Wagner, D. R. Herbst & S. H. Sohmer, pp. 198–213. Honolulu: University of Hawaii Press; Bishop Museum Press.

Dalla-Favera, R., Bregni, M., Erikson, J., Patterson, D., Gallo, R. C. & Croce, C. M. (1982). Human *c-myc* oncogene is located on the region of chromosome 8 that is translocated in Burkitt lymphoma cells. *Proceedings of the National Academy of Sciences USA*, **79**, 7824–7.

Darwin, S. P. & Chaw, S. (1990). *Bobea* Gaud. In *Manual of the Flowering Plants of Hawaii*, vol. 2., ed. W. L. Wagner, D. R. Herbst & S. H. Sohmer, pp. 1114–8. Honolulu: University of Hawaii Press; Bishop Museum Press.

Dobzhansky, T. (1970). *Genetics of the Evolutionary Process*. New York: Columbia University Press.

Erikson, J., Finnan, J., Nowell, P. C. & Croce, C. M. (1982). Translocation of immunoglobulin *Vh* genes in Burkitt lymphoma. *Proceedings of the National Academy of Sciences USA*, **79**, 5611–15.

Fedorov, A. (ed.) (1974). *Chromosome Numbers of Flowering Plants*. Koenigstein: Otto Koeltz Publishers. [Reprint of 1969 edition.]

Fosberg, F. R. (1948). Derivation of the flora of the Hawaiian Islands. In *Insects of Hawaii*, vol. 1., ed. E. C. Zimmerman, pp. 107–19. Honolulu: University of Hawaii Press.

Ganders, F. R. & Nagata, K. M. (1984). The role of hybridization in the evolution of *Bidens* on the Hawaiian Islands. In *Plant Biosystematics*, ed. W. F. Grant, pp. 179–94. Toronto: Academic Press.

Garber, E. D. (1972). *Cytogenetics: An Introduction*. New York: McGraw-Hill, Inc.

Gardner, R. C. (1976). Evolution and adaptive radiation in *Lipochaeta* (Compositae) of the Hawaiian Islands. *Systematic Botany*, 1, 383–91.

Gardner, R. C. (1977). Chromosome numbers and their systematic implications in *Lipochaeta* (Compositae: Heliantheae). *American Journal of Botany*, 64, 810–13.

Gillett, G. W. (1966). Hybridization and its taxonomic implications in the *Scaevola gaudichaudiana* complex of the Hawaiian Islands. *Evolution*, 20, 506–16.

Gillett, G. W. (1969). The nomenclatural and taxonomic status of the Hawaiian shrub *Scaevola gaudichaudii* H. & A. *Pacific Science*, 23, 125–8.

Gillett, G. W. & Lim, E. K. S. (1970). An experimental study of the genus *Bidens* (Asteraceae) of the Hawaiian Islands. *University of California Publications in Botany*, 56, 1–63.

Givnish, T. J., Sytsma, K. J., Smith, J. F. & Hahn, W. J. (1995). Molecular evolution, adaptive radiation and geographic speciation in *Cyanea* (Campanulaceae, Lobelioideae). In *Hawaiian Biogeography: Evolution on a Hot Spot Archipelago*, ed. W. L. Wagner & V. A. Funk, pp. 288–337. Washington DC: Smithsonian Institution Press.

Goldblatt, P. (1980). Polyploidy in angiosperms: monocotyledons. In *Polyploidy, Biological Relevance*, ed. W. H. Lewis, pp. 219–39. New York: Plenum Press.

Goldblatt, P. (ed.) (1981). *Index to Plant Chromosome Numbers 1975–1978*. Monographs in Systematic Botany from the Missouri Botanical Garden, vol. 5. St Louis: Missouri Botanical Garden.

Goldblatt, P. (ed.) (1984). *Index to Plant Chromosome Numbers 1979–1981*. Monographs in Systematic Botany from the Missouri Botanical Garden, vol. 8. St Louis: Missouri Botanical Garden.

Goldblatt, P. (ed.) (1988). *Index to Plant Chromosome Numbers 1984–1985*. Monographs in Systematic Botany from the Missouri Botanical Garden, vol. 23. St Louis: Missouri Botanical Garden.

Grant, V. (1963). *The Origin of Adaptations*. New York: Columbia University Press.

Gupta, S. & Gillett, G. W. (1969). Observations on Hawaiian species of *Wikstroemia* (Angiospermae: Thymelaeaceae). *Pacific Science*, 23, 83–8.

Hall, W. P. (1983). Modes of speciation and evolution in the sceloporine iguanid lizards. I. Epistemology of the comparative approach and introduction to the problem. In *Advances in Herpetology and Evolutionary Biology*, ed. A. G. J. Rhodin. & K. Miyata, pp. 643–79. Cambridge: Harvard Museum of Comparative Zoology.

Hedrick, P. W. (1981). The establishment of chromosomal variants. *Evolution*, 35, 322–32.

Hedrick, P. W. & Levin, D. A. (1984). Kin-founding and the fixation of chromosomal variants. *American Naturalist*, 124, 789–97.

John, B. & Lewis, K. R. (1959). Selection for interchange heterozygosity in an inbred culture of *Blaberus discoidalis* (Serville). *Genetics*, 44, 251–67.

Jones, J. D. G., Dunsmuir, P. & Bedbrook, J. (1985). High level expression of introduced chimaeric genes in regenerated transformed plants. *European Molecular Biology Organization Journal*, **4**, 2411–18.

Kiehn, M. & Lorence, D. (1996). Chromosome counts on angiosperms cultivated at the National Tropical Botanical Garden, Kauai, Hawaii. *Pacific Science*, **50**, 317–23.

Kim, I. & Carr, G. D. (1990). Cytogenetics and hybridization of *Portulaca* in Hawaii. *Systematic Botany*, **15**, 370–77.

Klein, G. (1983). Specific chromosomal translocations and the genesis of b-cell derived tumors in mice and men. *Cell*, **32**, 311–15.

Kyhos, D. W. & Carr, G. D. (1994). Chromosome stability and lability in plants. *Evolutionary Theory*, **10**, 227–48.

Kyhos, D. W., Carr, G. D. & Baldwin, B. G. (1990). Biodiversity and cytogenetics of the tarweeds (Asteraceae: Heliantheae–Madiinae). *Annals of the Missouri Botanical Garden*, **77**, 84–95.

Lammers, T. G. (1988). Chromosome numbers and their systematic implications in Hawaiian Lobelioideae (Campanulaceae). *American Journal of Botany*, **75**, 1130–34.

Lammers, T. G. (1990). Campanulaceae. In *Manual of the Flowering Plants of Hawaii*, vol. 1., ed. W. L. Wagner, D. R. Herbst & S. H. Sohmer, pp. 420–88. Honolulu: University of Hawaii Press; Bishop Museum Press.

Lammers, T. G. (1995). Patterns of speciation and biogeography in *Clermontia* (Campanulaceae, Lobelioideae). In *Hawiian Biogeography: Evolution on a Hot Spot Archipelago*, ed. W. L. Wagner & V. A. Funk, pp. 338–62. Washington DC: Smithsonian Institution Press.

Lammers, T. G. & Lorence, D. H. (1993). A new species of *Cyanea* (Campanulaceae: Lobelioideae) from Kauai and the resurrection of *C. remyi*. *Novon*, **3**, 431–6.

Lande, R. (1979). Effective deme sizes during long-term evolution estimated from rates of chromosomal rearrangements. *Evolution*, **33**, 234–51.

Lande, R. (1984). The expected fixation rate of chromosomal inversions. *Evolution*, **38**, 743–52.

Larson, A., Prager, E. M. & Wilson, A. C. (1984). Chromosomal evolution, speciation and morphological change in vertebrates: the role of social behavior. *Chromosomes Today*, **8**, 215–28.

Lewis, H. (1973). The origin of diploid neospecies in *Clarkia*. *American Naturalist*, **107**, 161–70.

Linney, G. K. (1988). Evolution and systematics of *Alectryon* (Sapindaceae) with special reference to the large-fruited species from Hawaii. Ph.D. dissertation, University of Hawaii.

Lowrey, T. K. (1986). A biosystematic revision of Hawaiian *Tetramolopium* (Compositae: Astereae). *Allertonia*, **4**, 203–65.

Luegmayr, E. (1993). Pollen of Hawaiian *Cyrtandra* (Gesneriaceae) including notes on southeast Asian taxa. *Blumea*, **38**, 25–38.

Macdonald, G. A., Abbott, A. T. & Peterson, F. L. (1983). *Volcanoes in the Sea: The Geology of Hawaii*, 2nd edn. Honolulu: University of Hawaii Press.

Manuelidis, L. (1990). A view of interphase chromosomes. *Science*, **250**, 1533–40.

Mayer, S. (1991). Artificial hybridization in Hawaiian *Wikstroemia* (Thymelaeaceae). *American Journal of Botany*, **78**, 122–30.

Moore, R. J. (ed.) (1973). Index to plant chromosome numbers 1967–1971. *Regnum Vegetabile*, **90**, 1–539.

Moore, R. J. (ed.) (1974). Index to plant chromosome numbers for 1972. *Regnum Vegetabile*, **91**, 1–108.

Moore, R. J. (ed.) (1977). Index to plant chromosome numbers for 1973–74. *Regnum Vegetabile*, **96**, 1–257.

Nagy, F., Morelli, G., Fraley, R. T., Rogers, S. G. & Chua, N. H. (1985). Photoregulated expression of a pea *rbcS* gene on leaves of transgenic plants. *European Molecular Biology. Organization Journal*, **4**, 3063–8.

Napoli, C., Lemieux, C. & Jorgensen, R. (1990). Introduction of a chimaeric chalcone synthase gene into petunia results in reversible co-suppression of homologous genes *in trans*. *Plant Cell*, **2**, 279–89.

Nicharat, S. & Gillett, G. W. (1970). A review of the taxonomy of Hawaiian *Pipturus* (Urticaceae) by anatomical and cytological evidence. *Brittonia*, **22**, 191–206.

Ornduff, R. (ed.) (1968). Index to plant chromosome numbers for 1966. *Regnum Vegetabile*, **55**, 1–126.

Patterson, R. (1984*a*). Flavonoid uniformity in diploid species of Hawaiian *Scaevola* (Goodeniaceae). *Systematic Botany*, **9**, 263–5.

Patterson, R. (1984*b*). Flavonoid diversification in the Hawaiian species of *Scaevola* (Goodeniaceae). *American Journal of Botany*, **71**, 183 (abstract).

Patterson, R. (1995). Phylogenetic analysis of Hawaiian and other Pacific species of *Scaevola* (Goodeniaceae). In *Hawiian Biogeography: Evolution on a Hot Spot Archipelago*, ed. W. L. Wagner & V. A. Funk, pp. 363–78. Washington DC: Smithsonian Institution Press.

Rabakonandrianina, E. (1980). Infrageneric relationships and the origin of the Hawaiian endemic genus *Lipochaeta* (Compositae). *Pacific Science*, **34**, 29–39.

Rabakonandrianina, E. & Carr, G. D. (1981). Intergeneric hybridization, induced polyploidy and the origin of the Hawaiian endemic *Lipochaeta* from *Wedelia* (Compositae). *American Journal of Botany*, **68**, 206–15.

Sanders, R. W., Stuessy, T. F. & Rodríguez R., R. (1983). Chromosome numbers from the flora of the Juan Fernandez Islands. *American Journal of Botany*, **70**, 799–810.

Sastrapradja, S. (1968). Chromosome study of Hawaiian *Peperomia* (Piperaceae) species. *Annales Bogorienses*, **4**, 245–51.

Skottsberg, C. (1953). Chromosome numbers in Hawaiian flowering plants. *Arkiv for Botanik*, **3**, 63–70.

Skottsberg, C. (1972). The genus *Wikstroemia* in the Hawaiian Islands. *Acta Regiae Societatis Scientiarum et Litterarum Gothoburgensis, Botanica*, **1**, 1–166.

Smith, J. F., Burke, C. C. & Wagner, W. L. (1995). Interspecific hybridization in natural populations of Hawaiian *Cyrtandra*: evidence from RAPD markers. *American Journal of Botany*, **82**, 162 (abstract).

Storey, W. B. (1966). Chromosome numbers in *Cyrtandra*. In *Monograph of Cyrtandra (Gesneriaceae) on Oahu, Hawaiian Islands. Bernice Pauahi Bishop Museum Bulletin*, **229**, 31–33.

Strother, J. L. (1991). Taxonomy of *Complaya, Elaphandra, Iogeton, Jefea, Wamalchitamia, Wedelia, Zexmenia* and *Zyzyxia* (Compositae–Heliantheae–Ecliptinae). *Systematic Botany Monographs*, **33**, 1–111.

Taub, R., Kirsch, I., Morton, C., Lenoir, G., Swan, D., Tronick, S., Aaronson, S. & Leder, P. (1982). Translocation of the *c-myc* gene into the immunoglobulin heavy chain locus in the human Burkitt lymphoma and murine plasmacytoma cells. *Proceedings of the National Academy of Sciences USA,* **79**, 7837–41.

Thompson, M. M. (1995). Chromosome numbers of *Rubus* species at the National Clonal Germplasm Repository. *Hortscience,* **30**, 1447–52.

Vander Kloet, S. P. (1993). Biosystematic studies of *Vaccinium* Section *Macropelma* (Ericaceae) in Hawaii. *Pacific Science,* **47**, 76–85.

Van der Krol, A. R., Mur, L. A., Beld, M., Mol, J. N. & Stuitje, A. R. (1990). Flavonoid genes in petunia: addition of a limited number of gene copies may lead to a suppression of gene expression. *Plant Cell,* **2**, 291–9.

Wagner, W. L. (1986). A new look at *Cyrtandra* (Gesneriaceae) in Hawaii. *American Journal of Botany,* **73**, 792–3 (abstract).

Wagner, W. L., Herbst, D. R. & Sohmer, S. H. (1990). *Manual of the Flowering Plants of Hawaii.* Honolulu: University of Hawaii Press; Bishop Museum Press.

Witter, M. S. (1990). Evolution in the Madiinae: evidence from enzyme electrophoresis. *Annals of the Missouri Botanical Garden,* **77**, 110–17.

2

Evolution in the endemic Hawaiian Compositae

BRUCE G. BALDWIN

Abstract

All but one or two indigenous species of Compositae on the Hawaiian Islands are endemic to the archipelago, the descendants of ten founder species. Five of the ten endemic lineages (the silversword alliance, *Bidens*, *Lipochaeta* sect. *Aphanopappus*, *Lipochaeta* sect. *Lipochaeta* and *Tetramolopium*) comprise 77 of the 90 endemic species of Hawaiian Compositae and have been regarded as examples of adaptive radiation. The great breadth of morphological and ecological variation in the silversword alliance and Hawaiian *Bidens*, the two largest groups, is exceptional for insular floras worldwide. Biosystematic and molecular evolutionary data have revealed high genetic similarity between species and lack of effective sterility barriers in each of the five major groups of endemic Hawaiian Compositae, consistent with the young age of the lineages and the long-persistent, woody life-forms of most species. Most species of Hawaiian Compositae are self-compatible, but exceptional breeding system diversity exists, in part associated with selection for outcrossing, with self-incompatibility in the silversword alliance, gynodioecy in Hawaiian *Bidens*, and gynomonoecy and monoecy in Hawaiian *Tetramolopium*. Phylogenetic comparisons between the silversword alliance and Hawaiian *Tetramolopium* show contrasting patterns of adaptive shifts, with recurrent transformations in life-form and habitat preferences in different lineages of the silversword alliance, and a minimal number and narrower range of such transformations in the younger Hawaiian *Tetramolopium* group. The extent to which lineage size and numbers of major evolutionary shifts in Hawaiian Compositae are attributable to lineage age, origin of key innovations or 'preadaptations' is uncertain. Some evolutionary changes in the silversword alliance (e.g., chromosome evolution, loss of ray flowers), otherwise unknown in Hawaiian Compositae, have arisen independently in the closely related, continental *Madia/Raillardiopsis* group and may be attributable in part to homologous, underlying developmental-genetic characteristics. The evolutionary importance of hybridization in Hawaiian Compositae is still debatable, but comparisons of chloroplast DNA, nuclear ribosomal DNA and cytogenetic data in the silversword alliance provide compelling evidence that ancient hybridization has influenced the genetic composition of at least five species of Kauai *Dubautia*. Large impediments to progress in understanding the evolutionary history of Hawaiian Compositae and other insular plant groups are extinction and decimation of natural populations, and lack of identified, phylogenetically interpretable gene regions that have evolved with sufficient rapidity to have recorded the pattern of diversification in explosive species radiations.

During the last 25 years, Hawaiian plant evolution has been the subject of considerable experimental and, more recently, molecular research. Most of this effort has focused on examining relationships and evolutionary processes in the five most highly diversified lineages of Hawaiian Compositae: the silversword alliance (*Argyroxiphium*, *Dubautia* and *Wilkesia*) (e.g., Carr & Kyhos, 1981, 1986; Crins, Bohm & Carr, 1988; Witter & Carr, 1988; Robichaux *et al.*, 1990; Baldwin, Kyhos & Dvořák, 1990; Baldwin *et al.*, 1991; Baldwin & Robichaux, 1995); Hawaiian *Bidens* (e.g., Gillett & Lim, 1970; Ganders & Nagata, 1984; Helenurm & Ganders, 1985; Sun & Ganders, 1986); the two major lineages of *Lipochaeta* (i.e., sect. *Aphanopappus* and *Lipochaeta*) (e.g., Gardner, 1976; Rabakonandrianina & Carr, 1981); and Hawaiian *Tetramolopium* (e.g., Lowrey & Crawford, 1985; Lowrey, 1986, 1995). Together, these groups comprise 77 of the 90 Hawaiian endemic species of Compositae, the family with the second largest number of indigenous Hawaiian plant species (Table 2.1). Relatively little systematic attention has been focused recently on the four remaining polytypic groups of endemic Compositae (Wagner, Herbst & Sohmer, 1990; Funk & Wagner, 1995*a*; Kim, Keeley & Jansen, 1996), each with three species: *Artemisia*, *Hesperomannia*, *Lagenophora* (= *Lagenifera*) and *Remya*. Although little is known about the evolutionary history of these four small groups, data from the five largest lineages of Hawaiian Compositae are sufficiently extensive to allow comparisons of major evolutionary trends and patterns. In this

Table 2.1. *Endemic and indigenous Hawaiian Compositae*

Hawaiian Compositae lineage	No. of species[1]	One-island endemics[2]	Distribution of entire group[3]	Closest relatives (source of founder) outside Hawaiian Isl.[4]	Subtribe, Tribe[5]
Lineages endemic to Hawaiian Islands:					
Silversword alliance: *Argyroxiphium, Dubautia, Wilkesia*	28	23	Kau, O, Mo, Lan, M,H	*Madia/Raillardiopsis* (west. North America)	Madiinae, Heliantheae s.l.
Bidens	19	12	N, Kau, O, Mo, Lan, Kah, M, H	*Bidens* (tropical America)	Coreopsidinae, Heliantheae s.l.
Lipochaeta sect. Aphanopappus	13	9	KA, Lay, N, Kau, O, Mo, Lan, Kah, M, N, H	*Wollastonia (Wedelia) biflora* (Micronesia)	Verbesininae, Heliantheae s.l.
Lipochaeta sect. Lipochaeta	6	2	N, Kau, O, Mo, Lan, Kah, M, H	*Wedelia* s.l.	Verbesininae, Heliantheae s.l.
Tetramolopium	11	6–7	Kau, O, Mo, Lan, M, H	*Tetramolopium* (New Guinea)	Asterinae, Astereae
Artemisia	3	2	N, Kau, O, Mo, Lan, Kah, M, H	*Artemisia*	Artemisiinae, Anthemideae
Hesperomannia	3	1	Kau, O, Mo, Lan, M	*Vernonia* (Africa)	Vernoniinae, Vernonieae
Lagenophora (= *Lagenifera*)	3	3	Kau, Mo, M	*Lagenophora*	Asterinae, Astereae
Remya	3	3	Kau, M	*Olearia* (Australasia)	Asterinae, Astereae
Gnaphalium	1	0	KA,Mid, N, Kau, O, Mo, Lan, M, H	*Gnaphalium*	Gnaphaliinae, Gnaphalieae

Table 2.1 (*cont.*)

Hawaiian Compositae lineage	No. of species[1]	Distribution of entire group[3]	Closest relatives (source of founder) outside Hawaiian Isl.[4]	Subtribe, Tribe[5]
Species indigenous but not endemic to Hawaiian Islands:				
Adenostemma lavenia	1	Kau, O, Mo, Lan, M, H	*Adenostemma lavenia*	Adenostemmatinae, Eupatorieae
Madia sativa[6]	1	M	*Madia sativa* (North or South America)	Madiinae, Heliantheae s.l.

[1]From Wagner, Herbst & Sohmer, 1990, with subtraction of *Lipochaeta ovata* (= *Wollastonia biflora*, W. L. Wagner, personal communication).

[2]Kahoolawe, Lanai, Maui and Molokai are treated as one island because of geographic continuity during the Pleistocene as Maui Nui (Carson & Clague, 1995).

[3]H = Hawaii or Big Island, Kah = Kahoolawe, Kau = Kahoolawe, Kau = Kauai, KA = Kure Atoll, Lan = Lanai, Lay = Laysan, Mid = Midway Atoll, Mo = Molokai, M = Maui, N = Niihau, O = Oahu.

[4]See Baldwin *et al.*, 1991; Baldwin, 1992; Funk & Wagner, 1995a; Ganders, 1992; Kim, Keeley & Jansen, 1996; Lowrey, 1995; Rabakonandrianina & Carr, 1981. Source of founder unknown if not indicated.

[5]See Bremer (1994).

[6]Indigenous status uncertain; may be a human introduction (Wagner, Herbst & Sohmer, 1990).

paper, I provide a general summary and systematic synthesis of comparative data on Hawaiian Compositae, which comprises some of the most spectacular examples of adaptive radiation among island angiosperms.

The Hawaiian setting has been conducive to organismal diversification marked by extreme morphological and ecological shifts, i.e., adaptive radiation (Carlquist, 1980, 1995). Rarity of successful plant colonization of the Hawaiian archipelago, as reflected by the disharmonic flora (Fosberg, 1948; Wagner, Herbst & Sohmer, 1990; Wagner, 1991) and the extreme geographic isolation of the archipelago, suggests that immigrants to young islands probably encountered little interference from other species, and, consequently, experienced relaxation of biotically imposed selection. The array of possible phenotypes descended from a founder may have been enhanced by the release of variation associated with population bottlenecks (Carson & Wisotzkey, 1989). Extraordinary diversity and close proximity of contrasting environmental conditions on islands of the Hawaiian chain (Carlquist, 1980; Gagne & Cuddihy, 1990) probably offered diverse key opportunities to particular variants and descendant lineages, which were further molded by selection. Recurrent opportunities for diversification of Hawaiian plants with inter-island dispersal capabilities arose with the successive origin of new islands from the Hawaiian hot spot during the last 29 million years (Carson & Clague, 1995), i.e., throughout most of the known history of Compositae (see Bremer, 1994; DeVore & Stuessy, 1995). Rarity of inter-island dispersal, e.g., in the silversword alliance (Baldwin & Robichaux, 1995) and *Bidens* (Ganders & Nagata, 1984), may have also served to promote reproductive isolation and divergent evolution. The above factors may explain much of the exceptional diversity found in the endemic lineages of Hawaiian Compositae, discussed below.

Phenotypic diversity within endemic Hawaiian Compositae

Pronounced diversity in general morphological and ecological characteristics are found within each of the five major lineages of endemic Hawaiian Compositae. Life-form diversity and ecological breadth in the Hawaiian silversword alliance and Hawaiian *Bidens* are greater than those found among the continental relatives of each group or in the other three major Hawaiian endemic lineages of Compositae (also see Carr, 1987). In the silversword alliance, the widest array of semi-woody and woody life-forms known in Hawaiian plants is found in association with an extreme range of habitat preferences (Carr, 1985): acaulescent or

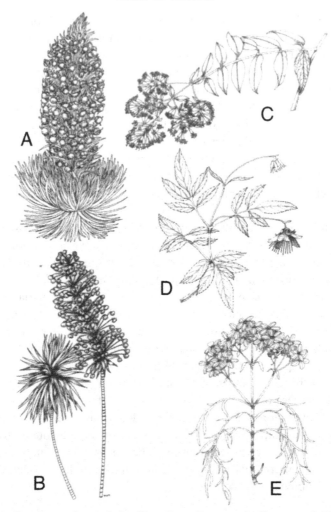

Fig. 2.1. Representatives of the Hawaiian silversword alliance and Hawaiian *Bidens*. A, the Haleakala silversword, *Argyroxiphium sandwicense* subsp. *macrocephalum* (×0.1), an acaulescent monocarpic rosette-plant of alpine cinder slopes; B, *Wilkesia gymnoxiphium* (×0.05), a caulescent monocarpic rosette-plant of dry ridges; C, *Dubautia latifolia* (×0.2), a vine of mesic forest; D, *Bidens cosmoides* (×0.15), a clambering, suffruticose perennial herb of mesic forest; E, *B. menziesii* subsp. *menziesii* (×0.2), an erect shrub of dry open habitats and subalpine forest. [Reproduced from Wagner, Herbst & Sohmer (1990) with permission from Bishop Museum Press, Bishop Museum, Honolulu, Hawaii.]

short-stemmed, thick-leaved rosette plants of high-elevation cinder slopes and bogs (*Argyroxiphium*, five species; Fig. 2.1A); caulescent, fibrous-leaved rosette plants of low to mid-elevation xeric sites (*Wilkesia*, two species; Fig. 2.1B); and trees, shrubs, mat-plants, cushion plants and vines that, as a group, span the extremes in elevation and precipitation found on the islands (*Dubautia*, 21 species; Fig. 2.1C). Woody and herbaceous perennial forms that have arisen in Hawaiian *Bidens* (19 species; Fig. 2.1D, E) include small trees, shrubs, and erect and prostrate perennial herbs that collectively occur in dry, wet and bog habitats from near sea level to over 2200 m in elevation (Ganders, 1989; Ganders & Nagata, 1990). Each of the two sections of *Lipochaeta* (Fig. 2.2) is habitally conservative, with subshrubs accounting for 18 of the 19 species, but habitats occupied by one or more of these species range from coastal to high elevation situations across a broad xeric to mesic moisture gradient (Gardner, 1976). One species in *L.* sect. *Lipochaeta*, *L. remyi* (Fig. 2.2A), is an obligately annual herb, an unusual life-form in the indigenous Hawaiian flora. Hawaiian *Tetramolopium* (11 species; Fig. 2.3) includes prostrate, cespitose and erect shrubs that taken together occur from near sea level to alpine slopes, a broader array of habits and habitats than is characteristic for the rest of the genus (Lowrey, 1986, 1995).

Marked morphological diversity is also found at a finer scale in Hawaiian Compositae, particularly in the silversword alliance and Hawaiian *Bidens*. In the silversword alliance, leaf forms range from strap-shaped with parallel venation and few cross-connecting veinlets in the yucca-like genus *Wilkesia* (Fig. 2.1B) to broadly elliptic with highly reticulate venation in the mesophytic vine *Dubautia latifolia* (Fig. 2.1C; Carr, 1985). Capitulescence types in the silversword alliance include massive racemiform to paniculiform arrangements that can reach 2 m in height with up to 600 large, rayed heads (each with up to 600 disc florets) in *Argyroxiphium sandwicense* (Fig. 2.1A); similarly large forms with whorled peduncles and discoid heads in species of *Wilkesia*; and a wide assortment of different racemiform and paniculiform types in *Dubautia*, all with discoid heads and with as few as two florets per capitulum (in *D. pauciflorula*) (Carr, 1985). In Hawaiian *Bidens* (Fig. 2.1D, E), leaf lobing among species ranges from shallowly serrate to tripinnately compound (Ganders, 1989; Ganders & Nagata, 1990). Capitulescence types in Hawaiian *Bidens* range from unbranched forms terminating with large, solitary heads to compound cymiform arrangements with more than 50 heads (Ganders & Nagata, 1990). Capitula are radiate in endemic Hawaiian *Bidens* and show diversity in fruit and pappus morphology,

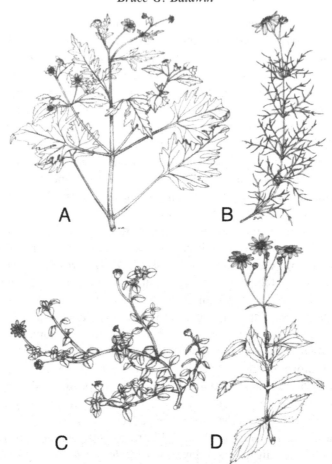

Fig. 2.2. Representative species of *Lipochaeta* sect. *Aphanopappus* and *L.* sect. *Lipochaeta*. A, *L. remyi* sect. *Aphanopappus* (×0.45), an annual herb of dry forest and shrubland; B, *L. tenuifolia* sect. *Aphanopappus* (×0.3), a suffruticose perennial herb of mesic forest; C, *L. integrifolia* sect. *Lipochaeta* (×0.3), a fleshy-leaved perennial herb of coastal habitats; D, *L. connata* sect. *Lipochaeta* (×0.2), a suffruticose perennial herb of dry forest. [Reproduced from Wagner, Herbst & Sohmer (1990) with permission from Bishop Museum Press, Bishop Museum, Honolulu, Hawaii.]

apparently associated with a shift from biotic to abiotic dispersal, that is exceptional in the genus (Ganders, 1989). *Bidens cosmoides* (Fig. 2.1D), a Kauai endemic, possesses extraordinarily elongate reddish-orange disc corolla tubes that extend up to 30 mm past the anthers, an innovation that appears to be correlated with a shift to bird pollination (Ganders & Nagata, 1983).

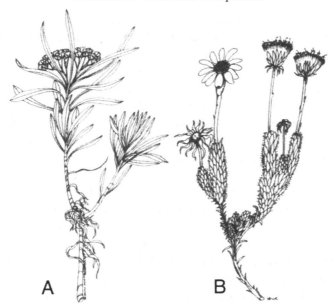

Fig. 2.3. Representative endemic species of Hawaiian *Tetramolopium*. A, *T. consanguineum* subsp. *consanguineum* (×0.45), an erect shrub of mesic forest; B, *T. humile* subsp. *haleakalae* (×0.85), a cespitose shrublet of dry, open subalpine to alpine habitats. [Reproduced from Wagner, Herbst & Sohmer (1990) with permission from Bishop Museum Press, Bishop Museum, Honolulu, Hawaii.]

Evidence for origin and monophyly of Hawaiian Compositae groups

A fundamental assumption of hypotheses of adaptive radiation in the Hawaiian archipelago is descent from a single ancestral species that arrived on the islands via long-distance dispersal. A common insular ancestry of each of the endemic Hawaiian Compositae groups is indicated by multiple lines of evidence, some of which offer insights into their extra-Hawaiian sources.

Monophyly of the Hawaiian silversword alliance is strongly supported by results from phylogenetic analyses based on chloroplast DNA (cpDNA) restriction sites (Baldwin, Kyhos & Dvořák, 1990; Baldwin *et al.*, 1991) and internal transcribed spacer (ITS) sequences of nuclear ribosomal DNA (rDNA) (Baldwin & Robichaux, 1995). Inclusion of western American tarweeds (subtribe Madiinae) and other Heliantheae s.l. in the molecular studies allowed testing of Carlquist's (1959*a, b*) hypothesis that the silversword alliance was founded by an American tarweed ancestor. Results from independent analyses of cpDNA

(Baldwin *et al.*, 1991) and the nuclear rDNA ITS region (Baldwin, 1992, 1996; Baldwin & Robichaux, 1995) uphold Carlquist's (1959*a, b*) hypothesis and demonstrate that the silversword alliance is not the sister-group of American tarweeds, as might be expected if the origin of the Hawaiian group preceded diversification of modern tarweed species, but instead arose following major diversification of the continental lineage. Both lines of evidence indicate that the silversword alliance is part of a lineage that includes all species in the primarily Californian genera *Madia* (19 species) and *Raillardiopsis* (two species), exclusive of the more distantly related 65 remaining species of continental tarweeds. Reconstructions of character evolution and historical biogeography based on parsimony indicate that the ancestor of the Hawaiian taxa was an herbaceous, Californian species that must have been externally bird-dispersed to the archipelago across *c*. 3900 km of the Pacific Ocean. Prior to diversification of the modern Hawaiian species, aerial (semi-) woodiness was acquired, a prevalent transformation in island plants (Carlquist, 1965, 1974, 1980) that is resolved unequivocally in the silversword alliance from phylogenetic evidence.

Radiation of Hawaiian *Bidens* from a single founder species is supported by phylogenetic analyses based on morphological (Ganders, 1992) and cpDNA (S. O'Kane, personal communication) data. Morphological innovations of the fruit and pappus are conspicuous features of phylogenetic significance in the Hawaiian assemblage (Ganders, 1989). The precise relationship of Hawaiian *Bidens* to the other *c*. 200 species of the genus has been a more complex phylogenetic and biogeographic problem than the determination of extra-Hawaiian relationships of the silversword alliance. Gillett & Lim (1970) envisioned an origin of Hawaiian *Bidens* from southwestern North America, based in part on the geographic proximity to the Hawaiian archipelago. Gillett (1972) proposed that southeastern Polynesian *Bidens* was, in turn, derived from Hawaiian *Bidens*, based in part on interfertility of the two groups. Ganders (1992) also proposed an American origin of Hawaiian *Bidens*, but, contrary to Gillett (1972), endorsed a sister-group relationship between the Hawaiian species and a southerly assemblage of American *Bidens* species, *B.* sect. *Greenmania*, which occurs from Brazil to Mexico.

In *Tetramolopium*, the precise relationships of the Hawaiian species to the New Guinean taxa and a recently discovered taxon from the Cook Islands (see Lowrey, 1995) are somewhat uncertain, although data from phylogenetic analysis of rDNA ITS sequences (Chan, 1994; Chan *et al.*, 1994) and cluster analysis of low-copy number RFLPs (Lowrey *et al.*,

1995; Okada, Whitkus & Lowrey, 1995) support monophyly of the Hawaiian assemblage. Based on biosystematic and geological considerations, a New Guinean origin of the ancestors of Hawaiian and Cook Island *Tetramolopium* was favoured by Fosberg (1948, with reference to the Hawaiian species only) and Lowrey (1995), and is consistent with molecular data (Chan 1994; Chan *et al.*, 1994). Strong morphological similarities between the *c.* 25 species of *Tetramolopium* in New Guinea (especially *T. alinae*) and *T. humile* (Fig. 2.3B) on the Hawaiian Islands, which share a cespitose subshrub habit and occur in alpine habitats (unlike the other Hawaiian species), led Lowrey (1986) to group New Guinean *Tetramolopium* and *T. humile* into *T.* sect. *Alpinum*, and to conclude that *T. humile* may be morphologically similar to the common ancestor of all species of Hawaiian *Tetramolopium* (Lowrey, 1995). A phylogenetic analysis based on morphological data (Lowrey, 1995) placed *T. humile*, an alpine Hawaiian species, as sister to the other Hawaiian taxa, in accord with Lowrey's earlier predictions. The newly discovered unnamed species from the Cook Islands (see Lowrey, 1995) is resolved as sister to the Hawaiian group in the ITS tree (Chan, 1994; Chan *et al.*, 1994). Whether dispersal occurred from New Guinea to each of the two island groups independently or involved successive dispersal from one archipelago to the other is unclear from the phylogenetic pattern alone (Chan, 1994; Chan *et al.*, 1994).

In *Lipochaeta*, cytogenetic work by Rabakonandrianina (1980) and Rabakonandrianina & Carr (1981) has shown that the genus is polyphyletic (i.e., biphyletic), contrary to Gardner's (1977) interpretation that *Lipochaeta* is monophyletic. Artificial hybrids between members of sect. *Aphanopappus* ($n = 15$; Fig. 2.2A, B) and sect. *Lipochaeta* ($n = 26$; Fig. 2.2C, D) were highly sterile, with 15 pairs of chromosomes and 11 univalents at diakinesis and meiotic metaphase I, whereas hybrids between members of sect. *Aphanopappus* and *Wedelia* (*Wollastonia*) *biflora* ($n = 15$), an extra-Hawaiian species of otherwise broad distribution in the Pacific, displayed normal meiosis, with 15 pairs of chromosomes at diakinesis and metaphase I (Rabakonandrianina, 1980; Rabakonandrianina & Carr, 1981). The chromosomal data indicate that the members of sect. *Lipochaeta* are allopolyploids derived from hybridization between a 15-paired taxon, with a genomic arrangement similar to that found in members of sect. *Aphanopappus*, and an 11-paired taxon. Absence of species of *Wedelia* s.l., including any species with 11 pairs of chromosomes, from the Hawaiian Islands led Rabakonandrianina (1980) and Rabakonandrianina & Carr (1981) to conclude that the two sections of

Lipochaeta are not sister-groups and probably represent lineages that
have descended from different founders in the archipelago, as proposed
earlier by Carr (1978). Available evidence is consistent with the interpretation that each of the two sections is monophyletic.

Evidence of high genetic similarity within lineages of species

Multiple lines of evidence suggest that levels of genetic divergence within
groups of endemic Hawaiian Compositae are lower than would be
expected in continental taxa that displayed such extensive morphological
and ecological variation. Crossability barriers are absent within the silversword alliance (Carr, 1985), Hawaiian *Bidens* (Gillett & Lim, 1970;
Ganders & Nagata, 1984), *Lipochaeta* sect. *Aphanopappus*
(Rabakonandrianina, 1980), *L.* sect. *Lipochaeta* (Rabakonandrianina,
1980) and Hawaiian *Tetramolopium* (Lowrey, 1986), with vigorous
hybrids documented from all of many crossing combinations attempted
in each group and between members of different sections in *Lipochaeta*
(Rabakonandrianina, 1980). Furthermore, interspecific hybrids in all five
groups are of normal fertility, except in some crosses within the silversword alliance wherein the parental species are differentiated by one or
more reciprocal chromosomal translocation(s) (Carr & Kyhos, 1986).

Genetic identity estimates based on allozymes in Hawaiian *Bidens*
(0.89–1.0; Helenurm & Ganders, 1985), Hawaiian *Tetramolopium*
(0.87–1.0; Lowrey & Crawford, 1985) and the Maui Nui/Hawaii lineages
of *Dubautia* sect. *Railliardia* (0.84–1.0; Witter & Carr, 1988) are mostly
within the range typical of infraspecific comparisons in mainland species
(Gottlieb, 1981; Crawford, 1983), including comparisons between conspecific populations of *Layia* (Warwick & Gottlieb, 1985; Witter, 1990), a
continental genus that is closely related to the silversword alliance (see
Baldwin, 1996). Genetic identities based on allozymes between Kauai
species of the silversword alliance (0.43–0.98) are lower than those of
other Hawaiian Compositae (Witter & Carr, 1988), mostly in the range
of interspecific identities between closely related mainland species
(Gottlieb, 1981; Crawford, 1983). Such high divergence between Kauai
species of the silversword alliance is consistent with the greater age of
the Kauai lineages (≤ 5–6 million years) than those of *Dubautia* sect.
Railliardia on Maui Nui and Hawaii or Hawaiian *Tetramolopium*
(≤ 1.5 million years) (Lowrey & Crawford, 1985; Witter & Carr, 1988;
Baldwin & Robichaux, 1995; Lowrey, 1995). As pointed out by Witter &
Carr (1988), absence of allozyme data from the distinctive Kauai *Bidens*

species, *B. cosmoides*, hampers comparisons of genetic distances within Hawaiian *Bidens* with those in the silversword alliance.

Similarly, genetic distance estimates based on cpDNA restriction sites and DNA sequences of the ITS region of nuclear rDNA suggest low levels of genetic divergence among endemic Hawaiian Compositae. In the silversword alliance, distance values based on cpDNA variation ranged from 0 to 0.46% (Baldwin, Kyhos & Dvořák, 1990), within the ranges estimated from studies of close congeners in continental floras. Variation found in cpDNA in Hawaiian *Bidens* was even lower, too low to offer any resolution of phylogenetic relationships (S. O'Kane, personal communication). Preliminary investigations of cpDNA restriction sites in *Tetramolopium* (T. Lowrey & D. Crawford, personal communication) and both sections of *Lipochaeta* (S. Keeley, personal communication) revealed minimal variation. Sequence divergence values for ITS (ITS 1 + ITS 2) in the silversword alliance (0–5.3%) and in *Tetramolopium* (0–0.9%; Chan, 1994) are, as in many plant groups, considerably higher than those estimated from cpDNA of the same plants, but are again within the range reported for closely related congeners in mainland plant groups (see Baldwin *et al.*, 1995). Negligible ITS sequence variation has been detected in *Lipochaeta*, even in comparisons between members of the two sections (S. Keeley, personal communication).

Multiple lines of evidence of high genetic similarity among taxa in all five major groups of Hawaiian Compositae support the interpretation that the crown lineages (i.e., the groups that include the most recent common ancestor of extant species and all descendants) are young, consistent with considerations of geology and population biology. Geological reconstructions allow for the possibility that Hawaiian lineages could have originated on extinct or nearly extinct islands dating back *c.* 29 million years, i.e., the age of Kure Atoll, with subsequent island-hopping from west to east as new islands emerged from the Hawaiian hot spot (Carson & Clague, 1995). The group that shows the most evidence of genetic and morphological divergence, the silversword alliance, may have originated on an island older than Kauai, the oldest modern high island, but biogeographic reconstructions based on a phylogeny from rDNA ITS sequences support a Kauai (or younger island) origin of the most recent common ancestor of modern species (Baldwin & Robichaux, 1995). Minor equivocality about a Kauai origin of the crown lineage of the silversword alliance is attributable to restriction of *Argyroxiphium*, the apparent sister group of *Dubautia* and *Wilkesia*, to the two youngest islands, Maui and Hawaii. Restriction of species of

Argyroxiphium to high elevation bogs and cinder slopes on Maui and Hawaii is consistent with the hypothesis that *Argyroxiphium* once occurred on the older islands of Kauai and, perhaps, Oahu, but became extinct there following loss of sufficient suitable habitat through erosion and island subsidence (Baldwin & Robichaux, 1995; Carlquist, 1995). Biogeographic reconstructions based on phylogenetic analyses of Hawaiian *Bidens* (Ganders, 1992), *Hesperomannia* (Funk & Wagner, 1995a), and *Remya* (Funk & Wagner, 1995a) also suggest that modern species of the three genera descended from Kauai ancestors, although the *Hesperomannia* data must be reanalysed with a *Vernonia*, rather than Mutisieae, outgroup (Kim, Keeley, & Jansen, 1996). Similar analyses of *Tetramolopium* support more recent arrival to the Hawaiian Islands, probably to Maui Nui (< 1.5 Ma) (Lowrey, 1995).

A contributing factor to lack of genetic divergence within Hawaiian Compositae groups could be acquisition of woody or perennial habits, which are generally associated with slower molecular and chromosomal evolution than is the annual habit (see Baldwin *et al.*, 1995). Nevertheless, reduced population sizes and founder effects in the Hawaiian setting might be expected to accelerate molecular and chromosomal evolution relative to rates experienced by plants in more typical continental situations. Evidence of genetic bottlenecks and fixation of alleles associated with speciation in the silversword alliance (Witter & Carr, 1988) conforms to this expectation, whereas high levels of allozymic variation within populations in Hawaiian *Bidens* (Helenurm & Ganders, 1985), as high as in comparisons between taxa, are inconsistent with the expected outcome of founder effects.

Breeding system evolution

As expected (Baker, 1955), members of most Hawaiian Compositae lineages are self-compatible, with the notable exception of species in the Hawaiian silversword alliance (Carr, Powell & Kyhos, 1986). Carr, Powell & Kyhos (1986) demonstrated that some members of all three silversword alliance genera are self-incompatible, presumably the plesiomorphic or primitive condition in the group. If the ancestor of the silversword alliance was indeed self-incompatible, establishment on the Hawaiian Islands would have necessitated the introduction of two or more plants of different parentage, or the introduction of a plant capable of vegetative reproduction or limited selfing until sufficient S-allele diversity arose, through mutation, to allow efficient sexual reproduction (Carr,

Powell & Kyhos, 1986). Maintenance of self-incompatibility in much of the silversword alliance has probably been the result of strong selection for outcrossing in the face of insular conditions inimical to retention of genetic variation.

Similarly, unusual diversity of sexual expression in Hawaiian *Bidens* may reflect selective advantages for outcross progeny. Gynodioecy, i.e., the condition wherein some individuals possess normal heterogamous capitula whereas others are functionally pistillate, has been documented in nine of 19 species of Hawaiian *Bidens* (Sun & Ganders, 1986). Sun & Ganders (1986) showed that the frequency of pistillate individuals in populations was positively correlated with the incidence of selfing in hermaphrodites, as expected if inbreeding depression in bisexual plants allows persistence of pistillate plants (Lloyd, 1975). Subsequently, Schultz & Ganders (1996) demonstrated that inbreeding depression and levels of selfing in hermaphrodites of gynodioecious *B. sandvicensis* were sufficiently high to explain the loss of staminate function and maintenance of females in the species. Developmental similarities of microsporogenesis failure in pistillate plants of different *Bidens* species led Sun & Ganders (1987) to conclude that the widespread occurrence of gynodioecy in Hawaiian *Bidens* is attributable to descent from a common gynodioecious ancestor in Hawaii.

In *Tetramolopium*, gynomonoecy, the condition wherein all plants possess pistillate ray florets and perfect disc florets, is found in only four Hawaiian species, i.e., those of *T.* sect. *Sandwicense* (Lowrey, 1986). Lowrey's (1995) phylogenetic considerations led him to conclude that monoecy, the condition wherein all plants produce pistillate ray florets and functionally staminate disc florets, is the plesiomorphic or primitive type of sex expression in the genus. Origin of gynomonoecy from monoecy is the reverse of what would be expected if selection for increased outcrossing was responsible for differential sex expression in Hawaiian *Tetramolopium*. Based on Lowrey's morphology-based phylogeny (1995), gynomonoecy arose once in Hawaiian *Tetramolopium* and is associated with a suite of morphological characteristics suggestive of an outcrossing breeding system. Lowrey (1986) raised the possibilities that the shift to gynomonoecy in Hawaiian *Tetramolopium* may reflect a shift to different pollinators or an adjustment toward numerical optimality of pollen and seed-bearing florets.

Evidence for major morphological and ecological shifts

Overview of morphological and ecological diversity in each of the major Hawaiian Compositae lineages provides evidence of striking changes during evolution in each group. Understanding of phylogenetic relationships is needed, however, to resolve the number and types of transformations that occurred in each lineage. Detailed phylogenetic data are now available from the silversword alliance and Hawaiian *Tetramolopium*.

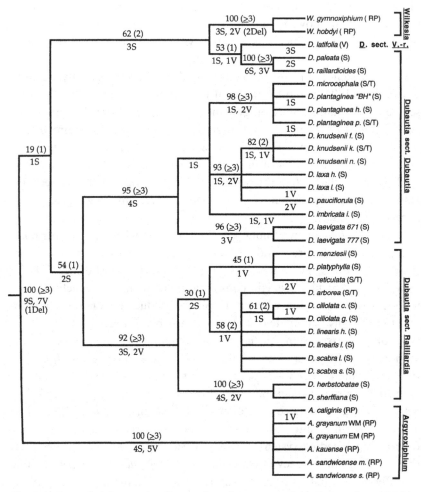

Fig. 2.4. One of eight maximally parsimonious trees of 18–26S nuclear rDNA sequence variation in Madiinae, pruned to show only the Hawaiian silversword alliance. Numbers of transitions (S) and transversions (V) mapped to each branch by ACCTRAN optimization (PAUP 3.1.1) appear below branches.

In the silversword alliance, phylogenetic data from rDNA ITS sequences (Baldwin & Robichaux, 1995; Fig. 2.4) indicate that at least four habital shifts from shrubs to trees have occurred in *Dubautia*, with two shifts on Kauai (*D. knudsenii* and *D. plantaginea/D. microcephala*), one on east Maui (*D. reticulata*) and one on Hawaii (*D. arborea*). In fact, all five species with tree habits also include shrubby individuals. The same phylogeny dictates two origins of either the rosette-plant or shrub habit, depending on which of the two conditions is primitive in the alliance (the apomorphic state being derived twice). Ecologically, shifts between wet and dry habitats have occurred in each of the four primary silversword alliance lineages and on each of three, and possibly all four, major islands or island groups (Baldwin & Robichaux, 1995). Association of such major ecological changes with diversification on different islands underscores the importance of adaptive radiation in the evolution of the silversword alliance.

Phylogenetic analysis of *Tetramolopium* by Lowrey (1995) yielded results consistent with single major shifts between life-forms, ecological preferences and sex expressions. As pointed out by Lowrey (1995), the conservative pattern of evolutionary change and near-absence of wet-forest species in *Tetramolopium* is highly consistent with the compelling hypothesis that *Tetramolopium* is a relatively recent arrival to the Hawaiian Islands compared to the silversword alliance. The area cladogram of *Tetramolopium* is, in fact, highly unusual among those of Hawaiian organisms in general for showing unequivocal evidence of origin on a young island group (Maui Nui) followed by dispersal to older and younger high islands (Lowrey, 1995; Funk & Wagner, 1995*b*). Confinement of most Hawaiian *Tetramolopium* species to open, dry habitats has been suggested to be the possible outcome of competitive exclusion from wet habitats by species from older Hawaiian plant lineages (Lowrey, 1995).

Caption for fig. 2.4 (*cont.*)
Deletion mutations (Del), not included in the phylogenetic analysis, appear below branches. Bootstrap values, based on 100 replicate analyses of the resampled data set, appear above branches outside parentheses. Decay index values appear above branches in parentheses. See Baldwin & Robichaux (1995) for additional details about the analysis and populations/taxa sampled. Letters in parentheses following species names indicate life-form or habit of the taxon. Abbreviations: RP = rosette plant, S = shrub or subshrub, T = tree, V = vine, *D.* sect. *V.-r.* = *Dubautia* sect. *Venoso-reticulatae.*

Accelerated radiation and phenotypic disparity among lineages

The extent, if any, to which rates of diversification in Hawaiian Compositae have exceeded those of closely related lineages is uncertain. Phylogenetic data are needed that encompass each Hawaiian group and its closest extra-Hawaiian relatives to address this question. At present, such data are available only from the silversword alliance and the closely related *Madia/Raillardiopsis* group, a primarily Californian assemblage (Baldwin, 1993, unpublished). Comparison of relative lineage size between the Californian and Hawaiian groups does not support the hypothesis of accelerated rates of speciation in the Hawaiian silversword alliance relative to the continental tarweeds (see Baldwin, 1997). The necessary assumption of equal rates of extinction in California and Hawaii is, however, highly questionable, given the more geologically unstable, spatially restricted habitats of the Hawaiian Islands relative to those of California. Rates of both speciation and extinction in the silversword alliance may be significantly higher than in *Madia/ Raillardiopsis*.

Improved understanding of relative phenotypic disparity between insular and continental groups is perhaps more important for testing insular adaptive radiation in the classic sense (see Carlquist, 1980) than an assessment of relative diversification rates. For example, shifts in chromosome structure, ray flower presence, elevational occurrence and habitat preference in the silversword alliance are paralleled to varying degrees in the closely related, continental *Madia/Raillardiopsis* group, changes that may in part reflect a similar lack of evolutionary constraints attributable to homologous, developmental-genetic characteristics (Baldwin, 1993, unpublished). More in-depth data on phenotypic disparity and developmental genetics of Hawaiian plants are badly needed.

Importance of hybridization in evolution

As discussed above, sterility barriers are absent throughout four of the five major groups of Hawaiian Compositae and in other Hawaiian plant lineages that have been studied biosystematically. Gillett & Lim (1970) noted that natural hybridization in Hawaiian angiosperms 'probably is most conspicuous in the Asteraceae'. Potential for natural hybridization is apparently more limited in Hawaiian *Bidens*, both sections of *Lipochaeta* and Hawaiian *Tetramolopium* than in the silversword alliance. Ganders & Nagata (1984) estimated that allopatry and ecological isolation prevent natural hybridization in 93% of all possible interspecific combinations in Hawaiian *Bidens*. Ganders & Nagata (1984) presented evidence for a maximum of nine possible natural hybrid combinations in Hawaiian *Bidens*, one of which was noted to form hybrid swarms over a large area on Oahu. Geographical and ecological isolation also have been invoked to explain lack of documented natural interspecific hybridization in Hawaiian *Tetramolopium* (Lowrey, 1986). Natural hybridization in *Lipochaeta* has rarely been reported. Only three natural hybrid combinations, all involving the most widespread species (*L. integrifolia*, Fig. 2.2C), are documented (Wagner, Herbst & Sohmer, 1986; W. L. Wagner, personal communication). In contrast, natural hybrids of 35 interspecific combinations, some of them intergeneric, have been documented in the silversword alliance, a small proportion of which are known to form hybrid swarms (Carr, 1985). Partial chromosomal sterility barriers between species with different genomic arrangements have been shown by Carr (1995) to be ineffective at preventing gene flow between even the most intersterile species. Highly fertile, first-generation backcross progeny can be recovered readily from F_1 hybrids of low fertility (9% pollen stainability) between *Argyroxiphium sandwicense* subsp. *macrocephalum* (Fig. 2.1A) and *Dubautia menziesii* (Carr, 1995).

Although natural hybridization has been well documented in Hawaiian Compositae, the evolutionary significance of such hybridization has been more difficult to ascertain. Hybridization cannot be discounted as an evolutionary factor in Hawaiian plants on the sole basis of perceived rarity of the phenomenon today. Geoclimatic and human-induced changes on the Hawaiian Islands (Cuddihy & Stone, 1990; Carson & Clague, 1995) have been sufficiently extreme through time to demand caution in extrapolation from present conditions to those of the past. Hybridization may be of episodic importance in Hawaiian plant evolution, e.g., following rapid environmental changes associated with volcanism (Carr, 1987, 1995). Close proximity of ecologically divergent, interfertile species separated by sharp ecotones, a common situation on the Hawaiian Islands, may allow for extensive interspecific hybridization and the origin of new stable, recombinant entities after environmental disruption. On a more contemporary scale, range fragmentation of Hawaiian species by human activities limits confidence in the evolutionary significance of allopatry as a natural barrier to hybridization. Genetic evidence of ancient hybridization is needed to assess confidently the role of reticulation in the evolution of Hawaiian plants.

Comparisons of cpDNA and nuclear rDNA ITS trees of the silver-sword alliance (Baldwin *et al.*, 1995; Baldwin, 1997) provide compelling evidence that ancient hybridization has had a lasting impact on the genomic constitution of some modern species. Relationships of Kauai species of *Dubautia* inferred from cpDNA data (Baldwin, Kyhos & Dvořák, 1990) conflict strongly with those inferred from nuclear rDNA (Baldwin & Robichaux, 1995) and cytogenetic (Carr & Kyhos 1981, 1986) evidence. Complete congruence between cytogenetic and rDNA relationships of Kauai *Dubautia* reinforces confidence that nuclear DNA evolution has, at least in part, been reconstructed accurately from both lines of evidence. Strong support for relationships in the cpDNA trees that are in conflict with those of the rDNA and cytogenetic reconstructions is evidence that homoplasy, or evolutionary noise, is not responsible for the discordant cpDNA results, especially in light of the low levels of sequence divergence between taxa. The remaining, alternative interpretations for the cpDNA results are cpDNA lineage sorting or cpDNA transfer through introgressive hybridization. Lineage sorting (see Pamilo & Nei, 1988; Doyle, 1992) presupposes a diversity of different cpDNA types in an ancestral taxon with subsequent differential transmission and fixation of cpDNA types in derivative lineages. The lineage sorting scenario is difficult to reconcile with expectations of genetic bottlenecks during insular speciation, as reflected in allozymic data from the silversword alliance (Witter & Carr, 1988), particularly in light of the extensive ancestral cpDNA diversity that would need to be postulated in this example. Multiple instances of ancient cpDNA acquisition through introgressive hybridization appear to be the only tenable explanations for the conflicting cpDNA and nuclear DNA relationships in Kauai *Dubautia*. Comparisons of cpDNA (Baldwin, Kyhos & Dvořák, 1990), nuclear rDNA (Baldwin & Robichaux, 1995) and allozymic (Witter & Carr, 1988) evidence point to the possibility that *D. knudsenii* and *D. pauciflorula* may have descended from a common ancestor influenced by hybridization (Baldwin, 1997), i.e., that plants of hybrid constitution were involved in diversification and were, consequently, of long-term evolutionary importance.

Future prospects for evolutionary studies

A critical challenge facing Hawaiian botanists interested in studying patterns of evolution is the identification of phylogenetically interpretable gene regions that have evolved with sufficient rapidity to allow robust resolution of species relationships by DNA analysis. In order to include extinct Hawaiian species in phylogenetic studies, a lamentable need in most Hawaiian groups, it is especially critical that the DNA region(s) under consideration be small and, preferably, highly repeated to allow PCR amplification from herbarium material. Nuclear rDNA ITS sequences conform to these requirements and have proven useful for investigating relationships in the silversword alliance (Baldwin & Robichaux, 1995; Baldwin *et al.*, 1995), but insufficient numbers of variable ITS sites were found in studies of *Lipochaeta* (S. Keeley, personal communication) and *Tetramolopium* (Chan, 1994) to allow robust phylogenetic reconstruction. As shown in the silversword alliance (Baldwin *et al.*, 1995; Baldwin, 1997), multiple independent sets of phylogenetically useful genetic markers are, in fact, needed to assess confidently relationships if hybridization has played a role in evolution, a possibility that cannot be dismissed for the Hawaiian flora in general. More basic investigation of rapidly evolving gene regions in plants will be necessary before much of the hidden history of Hawaiian Compositae can be revealed.

Acknowledgements

The research reported here was supported in part by the US National Science Foundation (DEB-9458237), and by generous gifts from Roderick B. Park, other Friends of the Jepson Herbarium and the California Native Plant Society. I thank Timothy K. Lowrey, Robert H. Robichaux, John L. Strother, Tod F. Stuessy and Warren L. Wagner for helpful comments on the manuscript. I also thank Tod F. Stuessy and Mikio Ono for inviting me to participate in the Symposium on Speciation of Vascular Plants on Pacific Islands at the XV International Botanical Congress in Yokohama, Japan, where the results herein were first presented.

Literature cited

Baker, H. G. (1955). Self-compatibility and establishment after 'long-distance' dispersal. *Evolution*, **9**, 347–9.

Baldwin, B. G. (1992). Phylogenetic utility of the internal transcribed spacers of nuclear ribosomal DNA in plants: an example from the Compositae. *Molecular Phylogenetics and Evolution*, **1**, 3–16.

Baldwin, B. G. (1993). Molecular phylogenetics of *Madia* (Compositae–Madiinae) based on ITS sequences of 18–26S nuclear ribosomal DNA. *American Journal of Botany* **80** (supplement, abstracts), 130.

Baldwin, B. G. (1996). Phylogenetics of the California tarweeds and the Hawaiian silversword alliance (Madiinae; Heliantheae *sensu lato*). In *Proceedings of the*

International Compositae Conference, Kew, 1994, vol. 1, _Systematics_, ed. D. J. N. Hind, H. Beentje & G. V. Pope, pp. 377–91. London: Royal Botanic Gardens, Kew.

Baldwin, B. G. (1997). Adaptive radiation of the Hawaiian silversword alliance: congruence and conflict of phylogenetic evidence from molecular and non-molecular investigations. In _Molecular Evolution and Adaptive Radiation_, ed. T. J. Givnish & K. J. Sytsma, pp. 103–28. Cambridge: Cambridge University Press.

Baldwin, B. G., Kyhos, D. W. & Dvořák, J. (1990). Chloroplast DNA evolution and adaptive radiation in the Hawaiian silversword alliance (Asteraceae–Madiinae). _Annals of the Missouri Botanical Garden_, **77**, 96–109.

Baldwin, B. G., Kyhos, D. W., Dvořák, J. & Carr, G. D. (1991). Chloroplast DNA evidence for a North American origin of the Hawaiian silversword alliance (Asteraceae). _Proceedings of the National Academy of Sciences USA_, **88**, 1840–3.

Baldwin, B. G. & Robichaux, R. H. (1995). Historical biogeography and ecology of the Hawaiian silversword alliance (Asteraceae): new molecular phylogenetic perspectives. In _Hawaiian Biogeography: Evolution on a Hot Spot Archipelago_, ed. W. L. Wagner & V. A. Funk, pp. 259–87. Washington DC: Smithsonian Institution Press.

Baldwin, B. G., Sanderson, M. J., Porter, J. M., Wojciechowski, M. F., Campbell, C. S. & Donoghue, M. J. (1995). The ITS region of nuclear ribosomal DNA: a valuable source of evidence on angiosperm phylogeny. _Annals of the Missouri Botanical Garden_, **82**, 247–77.

Bremer, K. (1994). _Asteraceae: Cladistics and Classification_. Portland: Timber Press.

Carlquist, S. (1959_a_). Vegetative anatomy of _Dubautia_, _Argyroxiphium_ and _Wilkesia_ (Compositae). _Pacific Science_, **13**, 195–210.

Carlquist, S. (1959_b_). Studies on Madiinae: anatomy, cytology and evolutionary relationships. _Aliso_, **4**, 171–236.

Carlquist, S. (1965). _Island Life: A Natural History of Islands of the World_. New York: Natural History Press.

Carlquist, S. (1974). _Island Biology_. New York: Columbia University Press.

Carlquist, S. (1980). _Hawaii, A Natural History. Geology, Climate, Native Flora and Fauna Above the Shoreline_, 2nd edn. Honolulu: Pacific Tropical Botanical Garden.

Carlquist, S. (1995). Introduction. In _Hawaiian Biogeography: Evolution on a Hot Spot Archipelago_, ed. W. L. Wagner & V. A. Funk, pp. 1–13. Washington DC: Smithsonian Institution Press.

Carr, G. D. (1978). Chromosome numbers of Hawaiian flowering plants and the significance of cytology in selected taxa. _American Journal of Botany_, **65**, 236–42.

Carr, G. D. (1985). Monograph of the Hawaiian Madiinae (Asteraceae): _Argyroxiphium_, _Dubautia_ and _Wilkesia_. _Allertonia_, **4**, 1–123.

Carr, G. D. (1987). Beggar's ticks and tarweeds: masters of adaptive radiation. _Trends in Ecology and Evolution_, **2**, 192-5.

Carr, G. D. (1995). A fully fertile intergeneric hybrid derivative from _Argyroxiphium sandwicense_ subsp. _macrocephalum_ × _Dubautia menziesii_ (Asteraceae) and its relevance to plant evolution in the Hawaiian Islands. _American Journal of Botany_, **82**, 1574–81.

Carr, G. D. & Kyhos, D. W. (1981). Adaptive radiation in the Hawaiian silversword alliance (Compositae–Madiinae). I. Cytogenetics of spontaneous hybrids. *Evolution*, **35**, 543–56.

Carr, G. D. & Kyhos, D. W. (1986). Adaptive radiation in the Hawaiian silversword alliance (Compositae–Madiinae). II. Cytogenetics of artificial and natural hybrids. *Evolution*, **40**, 959–76.

Carr, G. D., Powell, E. A. & Kyhos, D. W. (1986). Self-incompatibility in the Hawaiian Madiinae (Compositae): an exception to Baker's Rule. *Evolution*, **40**, 430–4.

Carson, H. L. & Clague, D. A. (1995). Geology and biogeography of the Hawaiian Islands. In *Hawaiian Biogeography: Evolution on a Hot Spot Archipelago*, ed. W. L. Wagner & V. A. Funk, pp. 14–29. Washington DC: Smithsonian Institution Press.

Carson, H. L. & Wisotzkey, R. G. (1989). Increase in genetic variance following a population bottleneck. *American Naturalist*, **134**, 668–73.

Chan, R. K. G. (1994). Molecular Phylogenetic Relationship of Hawaiian *Tetramolopium* Species (Compositae) using DNA Sequence Data from the Internal Transcribed Spacer (ITS) Regions. Ph.D. dissertation, University of New Mexico.

Chan, R. K. G., Lowrey, T. K., Natvig, D. & Whitkus, R. (1994). Phylogenetic analysis of internal transcribed spacer (ITS) sequences from nuclear ribosomal DNA of Hawaiian, Cook Island and New Guinea *Tetramolopium* (Compositae; Astereae). *American Journal of Botany*, **82** (supplement, abstracts), 118–19.

Crawford, D. J. (1983). Phylogenetic and systematic inferences from electrophoretic studies. In *Isozymes in Plant Genetics and Breeding*, part A, ed. S. D. Tanksley & T. J. Orton, pp. 257–87. New York: Elsevier.

Crins, W. J., Bohm, B. A. & Carr, G. D. (1988). Flavonoids as indicators of hybridization in a mixed population of lava-colonizing Hawaiian tarweeds (Asteraceae: Heliantheae: Madiinae). *Systematic Botany*, **13**, 567–71.

Cuddihy, L. W. & Stone, C. P. (1990). *Alteration of Native Hawaiian Vegetation: Effects of Humans, their Activities and Introductions*. University of Hawaii Cooperative National Park Resources Studies Unit. Honolulu: University of Hawaii Press.

DeVore, M. L. & Stuessy, T. F. (1995). The place and time of origin of the Asteraceae, with additional comments on the Calyceraceae and Goodeniaceae. In *Advances in Compositae Systematics*, ed. D. J. N. Hind, C. Jeffrey & G. V. Pope, pp. 23–40. London: Royal Botanic Gardens, Kew.

Doyle, J. J. (1992). Gene trees and species trees: Molecular systematics as one-character taxonomy. *Systematic Botany*, **17**, 144–63.

Fosberg, F. R. (1948). Derivation of the flora of the Hawaiian Islands. In *Insects of Hawaii*, vol. 1, ed. E. C. Zimmerman, pp. 107–19. Honolulu: University of Hawaii Press.

Funk, V. A. & Wagner, W. L. (1995a). Biogeography of seven ancient Hawaiian plant lineages. In *Hawaiian Biogeography: Evolution on a Hot Spot Archipelago*, ed. W. L. Wagner & V. A. Funk, pp. 160–94. Washington DC: Smithsonian Institution Press.

Funk, V. A. & Wagner, W. L. (1995b). Biogeographic patterns in the Hawaiian Islands. In *Hawaiian Biogeography: Evolution on a Hot Spot Archipelago*, ed. W. L. Wagner & V. A. Funk, pp. 379–419. Washington DC: Smithsonian Institution Press.

Gagné, W. C. & Cuddihy, L. W. (1990). Vegetation. In *Manual of the Flowering Plants of Hawaii*, ed. W. L. Wagner, D. R. Herbst & S. H. Sohmer, pp. 45–114. Honolulu: University of Hawaii Press; Bishop Museum Press.

Ganders, F. R. (1989). Adaptive radiation in Hawaiian *Bidens*. In *Genetics, Speciation and the Founder Principle*, ed. L. V. Giddings, K. Y. Kaneshiro & W. W. Anderson, pp. 99–112. New York: Oxford University Press.

Ganders, F. R. (1992). Phylogeny, biogeography and evolution of Hawaiian *Bidens*. *American Journal of Botany*, **79** (supplement, abstracts), 126–7.

Ganders, F. R. & Nagata, K. M. (1983). Relationships and floral biology of *Bidens cosmoides* (Asteraceae). *Lyonia*, **2**, 23–31.

Ganders, F. R. & Nagata, K. M. (1984). The role of hybridization in the evolution of *Bidens* on the Hawaiian Islands. In *Plant Biosystematics*, ed. W. F. Grant, pp. 179–194. Toronto: Academic Press.

Ganders, F. R. & Nagata, K. M. (1990). *Bidens*. In *Manual of the Flowering Plants of Hawaii*, ed. W. L. Wagner, D. R. Herbst & S. H. Sohmer, pp. 267–83. Honolulu: University of Hawaii Press; Bishop Museum Press.

Gardner, R. C. (1976). Evolution and adaptive radiation in *Lipochaeta* (Compositae: Heliantheae) of the Hawaiian Islands. *Systematic Botany*, **1**, 383–91.

Gardner, R. C. (1977). Chromosome numbers and their systematic implications in *Lipochaeta* (Compositae: Heliantheae). *American Journal of Botany*, **64**, 810–13.

Gillett, G. W. (1972). Genetic affinities between Hawaiian and Marquesan *Bidens* (Asteraceae). *Taxon*, **21**, 479–83.

Gillett, G. W. & Lim, E. K. S. (1970). An experimental study of the genus *Bidens* (Asteraceae) of the Hawaiian Islands. *University of California Publications in Botany*, **56**, 1–63.

Gottlieb, L. D. (1981). Electrophoretic evidence and plant populations. *Progress in Phytochemistry*, **7**, 1–45.

Helenurm, K. & Ganders, F. R. (1985). Adaptive radiation and genetic differentiation in Hawaiian *Bidens*. *Evolution*, **39**, 753–65.

Kim, H. G., Keeley, S. C. & Jansen, R. K. (1996). Phylogenetic position of the Hawaiian endemic *Hesperomannia* (Mutisieae) based on *ndh*F sequence data. *American Journal of Botany*, **83** (supplement, abstracts), 167.

Lloyd, D. G. (1975). The maintenance of gynodioecy and androdioecy in angiosperms. *Genetica*, **45**, 325–39.

Lowrey, T. K. (1986). A biosystematic revision of Hawaiian *Tetramolopium* (Compositae: Astereae). *Allertonia*, **4**, 203–65.

Lowrey, T. K. (1995). Phylogeny, adaptive radiation and biogeography of Hawaiian *Tetramolopium* (Asteraceae: Astereae). In *Hawaiian Biogeography: Evolution on a Hot Spot Archipelago*, ed. W. L. Wagner & V. A. Funk, pp. 195–220. Washington DC: Smithsonian Institution Press.

Lowrey, T. K., Chan, R., Daniels, D. & Whitkus, R. (1995). Phylogenetic relationships and taxonomic status of Cook Islands *Tetramolopium* (Compositae) based on artificial hybridization and molecular evidence. *American Journal of Botany*, **82** (supplement, abstracts), 147–8.

Lowrey, T. K. & Crawford, D. J. (1985). Allozyme divergence and evolution in *Tetramolopium* (Compositae: Astereae) on the Hawaiian Islands. *Systematic Botany*, **10**, 64–72.

Okada, M., Whitkus, R. & Lowrey, T. K. (1995). RFLP diversity and species relationships in *Tetramolopium* (Asteraceae) in Hawaii and the Cook Islands. *American Journal of Botany*, **82** (supplement, abstracts), 153.

Pamilo, P. & Nei, M. (1988). Relationships between gene trees and species trees. *Molecular Biology and Evolution*, **5**, 568–83.

Rabakonandrianina, E. (1980). Infrageneric relationships and the origin of the Hawaiian endemic genus *Lipochaeta* (Compositae). *Pacific Science*, **34**, 29–39.

Rabakonandrianina, E. & Carr, G. D. (1981). Intergeneric hybridization, induced polyploidy and the origin of the Hawaiian endemic *Lipochaeta* from *Wedelia* (Compositae). *American Journal of Botany*, **68**, 206–15.

Robichaux, R. H., Carr, G. D., Liebman, M. & Pearcy, R. W. (1990). Adaptive radiation of the Hawaiian silversword alliance (Compositae–Madiinae): ecological, morphological and physiological diversity. *Annals of the Missouri Botanical Garden*, **77**, 64–72.

Schultz, S. T. & Ganders, F. R. (1996). Evolution of unisexuality in the Hawaiian flora: a test of microevolutionary theory. *Evolution*, **50**, 842-55.

Sun, M. & Ganders, F. R. (1986). Female frequencies in gynodioecious populations correlated with selfing rates in hermaphrodites. *American Journal of Botany*, **73**, 1645–8.

Sun, M. & Ganders, F. R. (1987). Microsporogenesis in male-sterile and hermaphroditic plants of nine gynodioecious taxa of Hawaiian *Bidens* (Asteraceae). *American Journal of Botany*, **74**, 209–17.

Wagner, W. L. (1991). Evolution of waif floras: a comparison of the Hawaiian and Marquesan archipelagos. In *The Unity of Evolutionary Biology, the Proceedings of the Fourth International Congress of Systematics and Evolutionary Biology*, vol. 1, ed. E. C. Dudley, pp. 267–84. Portland: Dioscorides Press.

Wagner, W. L., Herbst, D. R. & Sohmer, S. H. (1986). Contributions to the flora of Hawaii I. Acanthaceae–Asteraceae. *Bishop Museum Occassional Papers*, **26**, 102–22.

Wagner, W. L., Herbst, D. R. & Sohmer, S. H. (1990). *Manual of the Flowering Plants of Hawaii*. Honolulu: University of Hawaii Press; Bishop Museum Press.

Warwick, S. I. & Gottlieb, L. D. (1985). Genetic divergence and geographic speciation in *Layia* (Compositae). *Evolution*, **39**, 1236–41.

Witter, M. S. (1990). Evolution in the Madiinae: evidence from enzyme electrophoresis. *Annals of the Missouri Botanical Garden*, **77**, 110–17.

Witter, M. S. & Carr, G. D. (1988). Adaptive radiation and genetic differentiation in the Hawaiian silversword alliance (Compositae: Madiinae). *Evolution*, **42**, 1278–87.

Part two
Juan Fernandez Islands

Introduction

Few young students in the western world have failed to read the famous novel, *Robinson Crusoe*, by Daniel Defoe (1719). Although set in the Caribbean region, the story does relate to the real Robinson Crusoe Islands in the Pacific Ocean 667 km off the coast of continental Chile. These islands, also known as the Juan Fernandez Islands after the Spanish navigator who first discovered them in 1574 (Medina, 1974), harboured a Robinson Crusoe-type sailor for five years during 1704–9. After having suffered passage around Cape Horn, Alexander Selkirk had a dispute with Captain Thomas Stradling of the ship, *Cinque Ports*, and demanded to be set off at the next available land, which to his misfortune happened to be the Juan Fernandez Islands. In this solitude he endured a quiet and eventually rewarding experience that created genuine interest in journalists of the day (including Defoe) when he finally returned to England in 1711.

The Juan Fernandez Islands have been of even greater interest for economic and strategic reasons to European countries for over four centuries. The two greatest considerations were as a location for refitting boats and recuperating crews after the long and arduous trip around the tip of South America, and as a place to mount raids against ships and coastal towns throughout the colonial Spanish empire. Ships coming from Europe had an initial long journey across the Atlantic Ocean and down the eastern coast of South America, generally touching Brazil and Argentina. The trip around the tip of the continent was especially difficult with cold temperatures, high seas and strong winds. Many lives were often lost. The Juan Fernandez Islands were well situated for resting and rebuilding after this long and trying part of the voyage. The islands, with the one protected bay, Bahia Cumberland, provided an excellent place from which to attack selected targets along the coast of colonial

Spain. British, Dutch and French buccaneering vessels proved quite adept in striking and running to and from the coast with remarkable success against the Spaniards, until the latter took control of the easternmost island, Masatierra, and fortified it with cannon in the mid-18th century (Woodward, 1969).

Botanical interest in the Juan Fernandez Islands began with the first collections by Mary Graham in 1823 who accompanied Lord Cochrane on a visit to Masatierra. This visit was followed in 1830 by Carlos Bertero who collected more extensively. These investigations were extended by collections of Moseley on the Challenger Expedition, studied and written up by W. B. Hemsley (1884). The first comprehensive flora of the archipelago was produced by Federico Johow (1896), and it served as the basis for all future work. The last broad inventory of the vascular flora was done by Carl Skottsberg (1921) based on collections during 1908 and 1916–17.

In 1980, a series of evolutionary studies were begun on the vascular flora of the Juan Fernandez Islands by personnel from the Ohio State University and Universidad de Concepción, Chile. The objectives were to extend our understanding of patterns and processes of plant evolution in this archipelago. Since that time, eight expeditions have been completed and more than 50 papers have been published (see literature citations at the end of the Chapters 3, 4 and 5).

The Juan Fernandez archipelago is extremely well suited for addressing evolutionary questions due to the following factors: only two major islands, closeness to the principal source of new immigrants from southern South America, east-west orientation of the two islands relative to the principal source area, known ages of the islands from K–Ar dating (Stuessy *et al.*, 1984), and a small endemic flora of only 127 vascular plant species (Stuessy *et al.*, 1992).

This part of the book focuses on aspects of evolutionary phenomena on the Juan Fernandez Islands. Chapter 3 deals with general patterns of isolating mechanisms and modes of speciation. Chapter 4 deals with the two largest endemic genera, *Dendroseris* and *Robinsonia* (Asteraceae), summarizing previous work, comparing and contrasting their evolution, and adding new interpretations. Chapter 5 examines factors determining species diversity on the Juan Fernandez Islands. In addition to the size of island and distance from source area, ecological and historical factors are shown to be equally important in dictating levels of specific diversity.

Literature cited

Defoe, D. (1719). *The Life and Strange Surprising Adventures of Robinson Crusoe.* London.

Hemsley, W. B. (1884). Report on the botany of Juan Fernandez, the southeastern Moluccas and the Admiralty Islands. In *Report on the Scientific Results of the Voyage of HMS Challenger During the Years 1873–76,* vol. 1, part 3, *Botany,* ed. C. W. Thomson & J. Murray, pp. 1–96. London: HMSO.

Johow, F. (1896). *Estudios sobre la Flora de las Islas de Juan Fernandez.* Santiago: Gobierno de Chile.

Medina, J. T. (1974). *El Piloto Juan Fernandez, Descrubridor de las Islas que Llevan Su Nombre, y Juan Jufre, Armador de la Expedición que Hizo en Busca de Otras en el Mar del Sur,* 2nd edn. Santiago: Gabriela Mistral.

Skottsberg, C. (1921). The phanerogams of the Juan Fernandez Islands. In *The Natural History of Juan Fernandez and Easter Island,* vol. 2, *Botany,* ed. C. Skottsberg, pp. 95–240. Stockholm: Almqvist & Wiksells.

Stuessy, T. F., Crawford, D. J. & Marticorena, C. (1990). Patterns of phylogeny in the endemic vascular flora of the Juan Fernandez Islands, Chile. *Systematic Botany,* **15,** 338–46.

Stuessy, T. F., Foland, K. A., Sutter, J. F., Sanders, R. W. & Silva O., M. (1984). Botanical and geological significance of potassium-argon dates from the Juan Fernandez Islands. *Science,* **225,** 49–51.

Stuessy, T. F., Marticorena, C., Rodríguez R., R., Crawford, D. J. & Silva O., M. (1992). Endemism in the vascular flora of the Juan Fernandez Islands. *Aliso,* **13,** 297–307.

Woodward, R. L. (1969). *Robinson Crusoe's Island: A History of the Juan Fernandez Islands.* Chapel Hill: University of North Carolina Press.

3

Isolating mechanisms and modes of speciation in endemic angiosperms of the Juan Fernandez Islands

TOD F. STUESSY, DANIEL J. CRAWFORD,
CLODOMIRO MARTICORENA AND
MARIO SILVA O.

Abstract

The Juan Fernandez Islands are located 667 km west of continental Chile at 33° S latitude. They contain 104 endemic angiosperms. Although numerous isolating mechanisms and modes of speciation are known for plants of continental regions, few have been documented for those on Juan Fernandez. A survey of endemic angiosperms reveals spatial isolation (between islands) within the archipelago as the most important primary factor accounting for 70% of closely related species pairs. Environmental (habitat) isolation occurs within individual islands and is known in 18% of cases. External reproductive isolation is less common, with temporal isolation accounting for 7% of cases, mechanical isolation only 3% and autogamy known in only two situations (3%). There is no indication of internal reproductive isolation between close congeners, although direct experimental evidence is lacking. Almost no changes in chromosome number or reticulation during speciation are known. Isozyme and DNA comparisons show low levels of genetic divergence among endemic congeners, yet conspicuous morphological differences often exist. The eroded environment of the older island, Masatierra, gives fewer clues to geological and ecological components of speciation. The younger island, Masafuera, has stronger ecological zonation and species of some genera (e.g., *Erigeron*) correlate with particular environments. The overall pattern of speciation in the archipelago, therefore, is divergence at the diploid level, initiated by geographic isolation especially between islands, accelerated by ecological partitioning within islands, accompanied by marked morphological change in response to changing environments, but with relatively minor genetic modification. Based on geological age of the archipelago and molecular divergence data, new species originate within 100 000–320 000 years.

The Juan Fernandez Islands are located in the eastern Pacific Ocean 667 km off the coast of continental Chile at a latitude 33°40′ S (approximately the same latitude as Valparaiso and Santiago; see Chapter 4, Fig. 4.1). The archipelago is small, with two major islands: Masatierra (also known as Isla Robinson Crusoe) and Masafuera (or Isla Alejandro Selkirk), the former 48 km² and the latter 50 km². Off the western end of Masatierra is another small island of 2 km², Santa Clara. Radiometric dates from these islands (Stuessy *et al.*, 1984) reveal Masafuera to be one to two million years old and Masatierra (as well as Santa Clara) approximately four million years old.

Several major previous floristic studies have been done on the Juan Fernandez archipelago. The first reasonably comprehensive inventory of the vascular flora was by Hemsley (1884) based on collections by Moseley in 1875 on the Challenger expedition. The first (and only) complete flora of the islands was done by Johow (1896), a German botanist residing in Chile. This excellent work was added to and greatly extended by Skottsberg in a series of papers on the flora mostly summarized for the angiosperms in 1921 (and supplement in 1951). Since that time numerous papers have been published on particular taxa within the archipelago (e.g., on *Erigeron* and *Peperomia*; Valdebenito *et al.*, 1992a, b), but a modern comprehensive flora awaits completion (T. F. Stuessy, C. Marticorena, R. Rodríguez R. & O. Matthei, personal communication).

In addition to floristic interest in the Juan Fernandez Islands, especially concerning changes due to human influence (e.g., Sanders, Stuessy & Marticorena, 1982; Matthei, Marticorena & Stuessy, 1993), a major research initiative has focused over the past 15 years on understanding evolution of the endemic taxa. Collaborative studies involving personnel from Ohio State University and the Universidad de Concepción have resulted in many papers dealing with patterns and processes of evolution of vascular plants on the islands, especially regarding the endemic Compositae (e.g., Pacheco *et al.*, 1985, 1991; Crawford *et al.*, 1992a, b; Sanders *et al.*, 1987; Valdebenito *et al.*, 1992b; Sang *et al.*, 1994). These and other studies on different angiosperm groups have begun to give us an understanding of the overall processes of evolution that have transpired in the archipelago. Previous attempts to synthesize evolutionary information for the entire flora have resulted in a study on endemism (Stuessy *et al.*, 1992) and patterns of phylogeny (Stuessy, Crawford & Marticorena, 1990). What has not been addressed carefully is an interpretation of generalized isolating mechanisms and modes of speciation within the endemic flora.

The purposes of this paper, therefore, are to: (1) briefly provide statistics on the endemic angiospermous flora; (2) outline types of isolating mechanisms operative within angiosperms in general; (3) present data on isolating mechanisms documented within the endemic angiosperms of Juan Fernandez; (4) review modes of speciation commonly encountered among angiosperms; (5) suggest modes of speciation within the endemic angiosperms of the archipelago; and (6) compare detected isolating mechanisms with inferred modes of speciation. The flowering plants are treated here separately from the ferns because the latter tend to distribute more broadly within the archipelago (Stuessy, Crawford & Marticorena, 1990) and their reproductive systems are totally different from those of angiosperms.

It needs to be emphasized that the best way to gain an understanding of isolating mechanisms and modes of speciation between and among closely related species involves detailed populational-level studies of many different types (Crawford, 1985), such as examining pollination systems (e.g., Suzuki, 1992, 1993, 1994), and making synthetic hybrids to determine levels of cross incompatibility and intersterility (Lowrey, 1986). Such studies have yet to be accomplished on the Juan Fernandez Islands. Nonetheless, we believe that sufficient information does exist to permit hypothesizing regarding these two aspects and we believe this useful as a further stimulus for evolutionary investigations in the archipelago.

The angiospermous flora

The flowering plants on the Juan Fernandez Islands number 73 families, 234 genera and 383 species. Of these, there is one endemic family (Lactoridaceae), ten endemic genera (13% of non-introduced genera) and 104 endemic species (65%; 90 dicots and 14 monocots). It is this latter group of endemic species that we intend to examine for the purposes of understanding isolating mechanisms and modes of speciation.

To reveal isolating mechanisms and avenues of speciation *within* the Juan Fernandez archipelago, it is useful to examine numbers of endemic species within genera (Table 3.1). First are those genera that have only

Table 3.1. *Numbers of endemic species in angiosperm genera of the Juan Fernandez Islands*

Number of endemic species	Number of genera with these species	Total number of species	Per cent of total endemic species in flora
1	37	27	35.6
2	11	22	21.2
3	4	12	11.5
4	1	4	3.8
5	1	5	4.8
6	1	6	5.8
7	1	7	6.7
11	1	11	10.6
TOTALS		104	100.0

one endemic species on the islands, and which, therefore, have been derived via anagenesis from continental progenitors (Stuessy, Crawford & Marticorena, 1990). Some 37 species (in the same number of genera) are of this type. Obviously, these derivative species are clearly geographically isolated from their continental progenitors. The other endemic species are found in genera with two or more endemic species each. These genera contain the majority of information regarding origin of isolating mechanisms and modes of speciation *within* the archipelago. Hence, special attention will given to their characteristics. Genera with three or more species, making 45 species total, or 43% of angiosperms, contain especially high evolutionary information regarding events in the archipelago. The largest genus, with 11 species, is *Dendroseris* (Compositae) followed by *Robinsonia* (Compositae; seven species; see Chapter 4) and *Erigeron* (also Compositae; six species).

Isolating mechanisms in flowering plants

Different classifications of isolating mechanisms exist for flowering plants (e.g., Levin, 1978; Grant 1981, 1985; White 1978; Littlejohn, 1981; King, 1993), but we have elected to work from the perspective of Grant (1981). In our opinion, this classification is most suited for plants in general, as well as for comparison with available taxa and data for the Juan Fernandez Islands. We have modified this scheme even further, however, as follows (Table 3.2): spatial (geographic), environmental (ecological),

Table 3.2. *Isolating mechanisms in endemic angiosperms of the Juan Fernandez Islands*

Spatial (Geographic)
Environmental (Ecological–habitat differences)
Reproductive
External
Temporal (Phenological)
Mechanical (Floral features)
Autogamy
Internal
Incompatibility barriers
Hybrid problems

Modified from Grant (1981)

and reproductive (external and internal) factors. The external reproductive aspects are divided into temporal, mechanical and autogamy; the internal reproductive factors include incompatibility barriers and information regarding hybridization.

Isolating mechanisms in Juan Fernandez angiosperms

A point that should be stressed when surveying isolating mechanisms within the Juan Fernandez flora is that in some groups, as with many angiosperms (Levin, 1978), more than one mechanism may be operative. Our efforts have been directed, however, toward inferring the *primary mode* of isolation that separates related taxa. For example, two congeners distributed one on each of the islands will be regarded as isolated principally by geography (spatial) even though other factors could possibly also be involved (e.g., incompatibility barriers).

As might be anticipated in an oceanic archipelago, geography is very important as a primary mode of isolation between related taxa. Obviously the 37 species that represent single endemic species on the islands (Table 3.1) are spatially isolated from their progenitors in continental source areas. These represent 35.6%, or slightly more than one-third, of the endemic angiosperm flora of Juan Fernandez Islands. In addition, five genera (*Berberis*, *Fagara*, *Haloragis*, *Myrceugenia* and *Sophora*) have two endemic species on the islands, one each on Masatierra and Masafuera, which is also a pattern of primary geographic isolation. Taken together these two groups represent 47 endemic species or 45% of the total angiospermous endemic flora. Further, in genera with three or more species, within closely related parts of these genera (e.g., subgenera within *Dendroseris*) additional species pairs occur that are distributed one on each island. For example, in *Dendroseris*, three species pairs are known (Masatierra species listed first): *D. berteroana* (Decne.) Hook. and *D. regia* Skottsb.; *D. macrantha* (Bertero ex Decne.) Skottsb. and *D. macrophylla* D. Don; *D. neriifolia* (Decne.) Hook. et Arn. and *D. gigantea* Johow. An additional nine genera represent this same pattern which yields 50 genera or subgeneric units totalling 70% of the total endemic angiosperm species (Table 3.1). Quite obviously, geographic isolation appears to have been fundamental as a stimulus for speciation within the archipelago.

That geography appears fundamental as an isolating mechanism can be further examined by looking at the five genera having only two closely related endemic species on the Juan Fernandez Islands, one restricted to each of the major islands (Table 3.3). If the spatial isolation by itself were sufficient as a stimulus for speciation, then no additional ecological differentiation should be seen when the habitats of the species pairs are compared. As predicted, the species essentially fill the same ecological positions on each of the islands, with a slight shift toward higher elevations on Masafuera (which is 1300 m tall in contrast to 950 m on Masatierra). The only possible exception might be between *Berberis corymbosa* Hook. et Arn. (MT) and *B. masafuerana* Skottsb. (MF), with the former from dry forests and the latter registered as from the quebradas (or deep ravines), which would be a more moist and cool environment. This latter species is so rare (Skottsberg, 1921), however, that it is difficult to know how meaningful the ecological data really are in this case. In general, therefore, it appears that between these closely related species pairs, geographic isolation is fundamental.

Ecological (habitat) isolation is also a factor among species within the same island, seen especially within subgenera of the larger taxa. Genera in which this pattern occurs are: *Dendroseris, Erigeron, Gunnera, Megalachne, Peperomia, Robinsonia* and *Wahlenbergia*. For example, in *Dendroseris* subg. *Dendroseris*, *D. litoralis* Skottsb. is a coastal species and its close relative, *D. marginata* (Bertero ex Decne.) Hook. et Arn., is confined to exposed cliffs at higher elevations. In subg. *Rea*, *D. micrantha* (Bertero ex Decne.) Hook. et Arn. occurs in mid-elevation forest, and *D. pruinata* (Johow) Skottsb. is found near the coast or on sea-swept cliffs at higher elevations. In subg. *Phoenicoseris*, *D. pinnata* (Bertero ex Decne.) Hook. et Arn. grows on windy ridges at higher elevations and *D. berteroana* is restricted to the uppermost tree-fern forests. One general problem

Table 3.3. *Ecological tolerances of closely related angiosperm species in the Juan Fernandez archipelago, in genera with only two endemic species, one on each of the major islands*

Species pair	Island	Elevation (m)	Habitat
OCCUPYING THE SAME ECOLOGICAL ZONE			
Haloragis masatierrana	MT	240–570	Dry, open slope
H. masafuerana	MF	440–1200	Dry, open slope
Sophora fernandeziana	MT	380–350	Open ridges and cliffs
S. masafuerana	MF	200–800	Open ridges and cliffs
Myrceugenia fernandeziana	MT	300–500	Dry to humid forest
M. schulzei	MF	200–800	Dry to humid forest
Fagara mayu	MT	200–600	Dry forest
F. externa	MF	280–515	Dry forest
OCCUPYING DIFFERENT ECOLOGICAL ZONES			
Berberis corymbosa	MT	200–600	Dry forest
B. masafuerana	MF	215	Quebradas (ravines)

MT = Masatierra
MF = Masafuera
Source: Ecological data from Skottsberg (1921)

with interpreting ecological differentiation within Masatierra is the substantial habitat alteration both naturally over the four million years of existence of the island and due to human-induced causes since the island was discovered over 400 years ago (Woodward, 1969; Sanders, Stuessy & Marticorena, 1982). Dramatic habitat differentiation simply does not exist now on this geologically older island (Sanders *et al.*, 1987) in the same fashion as is seen so clearly in the Hawaiian archipelago (Carlquist, 1980). The younger island, Masafuera, shows greater ecological diversity and the only genus of two or more species on this island is *Erigeron* [six species, five endemic to Masafuera; the sixth, *E. fernandezianus* (Colla) Solbrig, is known also from Masatierra]. This genus does, indeed, reflect ecological differentiation between closely related species pairs, i.e., between *E. rupicola* Skottsb. and *E. sp. nov.* (H. Valdebenito, personal communication), *E. fernandezianus* and *E. luteoviridis* Skottsb., and between *E. ingae* Skottsb. and *E. turricola* Skottsb. (Valdebenito *et al.*, 1992b). There are 13 such examples within the Juan Fernandez flora, or representing 18.1% of cases of isolating mechanisms (Table 3.4).

One external reproductive isolating mechanism that is relatively easy to assess is temporal isolation. Skottsberg (1928) tabulated, based on known herbarium materials and personal field observations, the flowering times of all endemic angiosperms. We extrapolate from these data regarding closely related species pairs and whether or not they are temporally isolated. In general there is considerable overlap in flowering times, but this is often in taxa that either are distributed on different islands or ecologically separated within the same island. Five species pairs seem at least mostly distinguishable in flowering time: *Uncinia brevicaulis* Thouars and *U. douglasii* Boott (Cyperaceae); *Chenopodium sanctae-clarae* Johow and *C. crusoeanum* Skottsb. (Chenopodiaceae; this pair is also distinct geographically with the former only on Santa Clara and the other on Masatierra); *Centaurodendron dracaenoides* Johow and *C. palmiforme* Skottsb. (Compositae; see Skottsberg, 1957); *Robinsonia gayana* Decne. and *R. thurifera* Decne. (Compositae); and *Robinsonia macrocephala* Decne. from the rest of the genus. These make up five of 72 total estimated cases of isolating mechanism relationships, or 6.9%. Temporal

Table 3.4. *Summary of isolating mechanisms between pairs of closely related endemic angiosperms of the Juan Fernandez Islands*

Isolating mechanism	Occurrence within genera or subgeneric units	Per cent occurrence
Spatial (between islands)	50	69.4
Environmental (habitat)	13	18.1
External Reproductive		
Temporal	5	6.9
Mechanical	2	2.8
Autogamy	2	2.8
TOTALS	72	100.0

isolation, therefore, is not a major factor in keeping taxa distinct within the archipelago.

Mechanical isolation involving differences in pollinators is even less important within the endemic species of the Juan Fernandez Islands. Only two cases are known: between *Wahlenbergia fernandeziana* A.DC. and *W. grahamae* Hemsl. (Campanulaceae; shape of corolla); and *Cuminia eriantha* Benth. and *C. fernandezia* Colla (Labiatae; colour of corolla). The question of morphological shifts in response to pollinator adaptations is a difficult one on the Juan Fernandez Islands because our knowledge of the insect fauna is incomplete. Further, pollination biology studies have been few (see Sun *et al.*, 1996, for an exception in *Rhaphithamnus* (Verbenaceae)). It is known that two hummingbirds occur on the islands and visit many of the native plants (Colwell, 1989), and these are likely to have had a profound effect on directional selection in floral features in many endemic taxa. Skottsberg (1928) surveyed the endemic flora for modes of pollination with the following results: anemophilous (34 species, 35.4% of total), entomophilous (57 species, 59.4%) and ornithophilous (five species, 5.2%). From our casual observations, these figures seem to exaggerate the levels of entomophily upward and ornithophily downward, but they do represent an attempt to survey pollination syndromes based on morphological and occasional field observations. Whatever the exact figures on distribution of pollinators on the islands, little evidence of differences in mechanical isolation exists. These two cases represent only 2.8% of the 72 total documented (Table 3.4).

Reproductive isolation by autogamy is uncommon in the archipelago. Only two instances recorded are in *Megalachne masafueranus* (Skottsb. et Pilg.) Matthei (Gramineae), which is cited as having cleistogamous flowers (Skottsberg, 1921), and *Dendroseris litoralis* (Compositae), which sets abundant seed in isolated cultivated individuals (personal observation).

One of the most important types of isolating mechanisms that keep species of continental angiosperms separated is internal genetic factors, both prezygotic and postzygotic. No detailed crossing studies have been carried out between endemic species on the Juan Fernandez Islands, in large measure due to logistical considerations and problems with cultivation. Cytogenetic studies on the Hawaiian Islands by Carr and associates (e.g., Carr & Kyhos, 1981, 1986; Carr *et al.*, 1989; see also Chapters 1 and 2) have revealed that in the silversword alliance (Compositae), the taxa are mostly quite cross-compatible and interfertile. That is, there are few incompatibility or sterility barriers between endemic species, both within and between some of the genera. Furthermore, genetic identities as evidenced from isozyme data are quite high among most of the species, particularly those endemic to the younger islands (Witter & Carr, 1988). The same situation prevails on the Juan Fernandez Islands, as far as data now show. Between species of *Dendroseris*, the genetic identities range from 0.80 to 0.99, the latter between very morphologically different species (*D. pinnata* and *D. berteroana*; Crawford, Stuessy & Silva O., 1987). There are exceptions, however, with genetic identities between species of *Robinsonia* much lower (Crawford *et al.*, 1992*b*). Chromosome numbers of closely related species are also the same (Sanders, Stuessy & Rodríguez R., 1983; Spooner *et al.*, 1987; Sun, Stuessy & Crawford, 1990). These facts suggest that few, if any, internal reproductive barriers exist between closely related species on the Juan Fernandez Islands. In the few cases in which hybridization is known, one is a natural intergeneric cross (as × *Margyracaena skottsbergii* Bitter) between an endemic species [*Margyricarpus digynus* (Bitter) Skottsb.] and the introduced and abundant *Acaena argentea* Ruiz et Pavón (confirmed by Crawford *et al.*, 1993). The second is between *Gunnera bracteata* Steud. ex Benn. and *G. peltata* Phil., but producing apparently robust and fertile interspecific hybrids especially in Valle Villagra (Pacheco *et al.*, 1991). The geographic and temporal isolation between most pairs of congeneric species prevent hybrids from being formed in most cases. In the two cases cited, human disturbance of the habitat has led to opportunities for crossing which ordinarily would not have occurred.

In summary, therefore, spatial isolating mechanisms are by far the most important for stimulating speciation among the endemic angiosperms of the Juan Fernandez flora (Table 3.4). Next comes environmental isolation (13 cases or 18.1% of occurrence), which is most important within individual islands. In general, major patterns of isolation between islands is not accompanied by significant ecological shifts. Reproductive barriers are basically poorly developed in the archipelago. Temporal isolation accounts for five instances and 6.9%, and mechanical isolation and autogamy together make up the remaining 5.6%. Little is known about actual genetic intercompatibilities between closely related congeners, but barriers are believed to be weak at best.

Modes of speciation in flowering plants

As with isolating mechanisms, many different classifications of modes of speciation exist (White, 1978; Grant, 1981; Templeton, 1981; Carson, 1985; King, 1993). Of these different approaches, we believe the outline of Grant (1981), directly oriented to vascular plants, most useful for our interests in the angiosperms of the Juan Fernandez Islands. We recognize and discuss here the following modes of speciation: (1) geographical (allopatric) speciation; (2) quantum speciation (or catastrophic speciation, including aneuploid speciation and speciation by peripheral isolates); (3) diploid (recombinational) hybrid speciation; and (4) allopolyploid speciation. Although this classification is more organismic than genetic in orientation (such as found in Templeton, 1981), it is more suited for relating observed patterns in isolating mechanisms to presumptive processes of speciation in the Juan Fernandez archipelago.

Modes of speciation in Juan Fernandez angiosperms

The previously listed modes of speciation for vascular plants can be further modified to parallel the different isolating mechanisms already discussed (Table 3.4) for the Juan Fernandez archipelago. That is, we will consider the following modes of speciation that emphasize: (1) geography (allopatry), including both major (i.e., between island) and minor (i.e., within island) isolation; (2) ecological differentiation; (3) chromosomal alterations (involving structural chromosomal rearrangements or involving aneuploidy); and (4) hybridization either at diploid (recombinational) or allopolyploid levels.

Obviously geographic speciation has been a major process in the production of endemic species in the Juan Fernandez archipelago. Although geographic isolation by itself does not stimulate speciation, it can subdivide a population and open the possibility for continued genetic divergence, especially that part of the original population that now occurs in newer and more ecologically open (and changing) environments. This would be particularly the case with species originally arriving on the older and closer island, Masatierra, and with propagules dispersing from Masatierra to the younger island, Masafuera, when it became available for colonization during the past one to two million years. Nearly 40% of all endemic species appear to have originated in this fashion.

Speciation via ecological differentiation and habitat isolation within islands has also occurred on the Juan Fernandez Islands, but not conspicuously. Comparisons of close congeners between islands (Table 3.3) reveal similar ecological tolerances, suggesting that distance alone is often sufficient when geographic distances are substantial. Within islands, ecological zonation is seen among species of the larger genera, such as *Dendroseris* on Masatierra and *Erigeron* on Masafuera. Strongly divergent adaptive zones within the older island, however, are not now present due in part to natural erosional changes characteristic of older oceanic islands and habitat destruction caused by human activities since discovery of the islands by Juan Fernandez in 1574 (Woodward, 1969). Masafuera, younger geologically and more isolated from human intervention, still retains more distinctive habitat zones. Only about 18% of closely related species pairs within an island can be inferred to have evolved in this fashion.

With regard to modes of speciation involving chromosomal alteration, a few facts can be mentioned based on previous cytological surveys on the Juan Fernandez Islands (Sanders, Stuessy & Rodríguez R., 1983; Spooner *et al.*, 1987; Sun, Stuessy & Crawford, 1990). No aneuploid or euploid relationship between closely related species is known within any genus of the archipelago. In fact, such events are unknown within any parts of genera on the islands. Further, no euploid origins seem evident for the origin of endemic taxa from continental relatives either, and only one case of possible aneuploid origins for taxa arriving and speciating within the archipelago is known (*Wahlenbergia*, Campanulaceae). All species of this genus on the islands are known as $n = 11$, but this number is not known elsewhere in the genus (Sanders, Stuessy & Rodríguez R., 1983). These facts argue against speciation in the archipelago involving catastrophic chromosomal rearrangements involving aneuploidy. Likewise, there is no evidence for reticulate evolution via autopolyploidy or allopolyploidy, so common in higher plants of continental areas (Grant, 1981; Soltis & Soltis, 1993).

Degrees of importance of modes of speciation in the context of isolating mechanisms in the endemic angiosperm flora of the Juan Fernandez Islands are summarized in Table 3.5. Once again, geographic isolation by itself has been the most important isolating mechanism that has stimulated geographic speciation. That closely related species on different islands are close genetically can be inferred by considering the species pair *Wahlenbergia berteroi* Hook. et Arn. and *W. masafuerae* (Phil.) Skottsb., on Masatierra and Masafuera, respectively (Table 3.6), which have an extremely high average genetic identity of 0.947. Ecological differentiation, or smaller-scale geographic differentiation, is the next most conspicuous mode of speciation within the archipelago. A good example of the close genetic correspondence of two very closely related species, yet ecologically distinct, are *Dendroseris pinnata* (high open ridges) and *D. berteroana* (dark moist fern forests), but with a genetic identity of 0.99 (Table 3.6).

Table 3.5. *Degrees of importance of different modes of speciation in angiosperms of the Juan Fernandez Islands in view of known (or suspected) isolating mechanisms*

Mode of speciation	Degree of importance	Reason(s)
GEOGRAPHIC		
Between islands	Very high	Many species have evolved inter-island
Within islands	Low (by itself); medium if combined with ecological and/or reproductive factors	Most intra-island speciation is accompanied by other factors, such as temporal or habitat shifts
ECOLOGICAL DIFFERENTIATION (Within islands)	Medium	Most species pairs within islands are so differentiated
REPRODUCTIVE DIFFERENTIATION (Within islands)	Low	Only a few species pairs are isolated this way
CHROMOSOMAL ALTERATION		
Structural rearrangements	Unknown	No data
Aneuploidy	Extremely low	No aneuploidy known in flora
HYBRIDIZATION		
Diploid (recombinational hybrid speciation)	Very low	Very few inter-taxon hybrids known in flora
Allopolyploidy	Extremely low	No euploidy known in flora

All facts point to rapid speciation within the Juan Fernandez Islands strongly facilitated by geographic isolation primarily between islands, and to a lesser extent within islands. Within the largest genus, *Dendroseris*, with 11 species, if the islands are no older than four million years, this gives an average rate of speciation of one species every 364 000 years. Molecular divergence values (Crawford *et al.*, 1992*a*) suggest actual values of 100 000 to 320 000 years between the close relatives *D. pinnata* and *D. berteroana*, depending upon the divergence rate used (0.05 or 0.16%; Wendel & Percival, 1990). These assessments for the Juan

Table 3.6. *Isozymic and cpDNA genetic divergences between closely related species of endemic angiosperms on the Juan Fernandez Islands*

Taxa	Average genetic identities	Average per cent sequence divergence in cpDNA
Wahlenbergia		
berteroi – masafuerae	0.947	–
berteroi – fernandeziana	0.725	–
Robinsonia		
sect. *Eleutherolepis*		
evenia – gracilis	0.604	0.12
sect. *Robinsonia*		
thurifera – gayana	0.706	0.02
Dendroseris		
subg. *Dendroseris*		
litoralis – macrantha	0.93	–
litoralis – marginata	0.88	0.065
marginata – macrantha	0.91	–
subg. *Rea*		
micrantha – neriifolia	0.80	0.049
micrantha – pruinata	0.944	0.016
pruinata – neriifolia	0.787	0.032
subg. *Phoenicoseris*		
pinnata – berteroana	0.99	0.016

Internal transcribed spacer regions (1 and 2) of ribosomal DNA show similar low sequence divergence values (Sang *et al.*, 1994, 1995), following down the table beginning with *R. evenia – gracilis*: 0.0212, 0; 0, 0, 0; 0.0191, 0, 0.0191; 0.0084
Source: From Crawford *et al.*, 1990, 1992*a,b*, 1993

Fernandez Islands are similar to those being revealed in other oceanic archipelagos (Crawford, Whitkus & Stuessy, 1987; see also Chapters 4 and 6).

Acknowledgements

It is a pleasure to acknowledge with appreciation: the National Science Foundation for continuous support under grants (to Tod F. Stuessy and Daniel J. Crawford) nos. INT-7721637, BSR-8306436, BSR-8906988, and DEB-9500499; FONDECYT of Chile for support to Mario Silva O. under grants 196-08-22 and 7-96-0015; CONAF (Corporación Nacional Forestal) of Chile for permission to collect and work in the Robinson Crusoe Islands National Park; the CONAF guides without whom it would have been impossible to carry out successful investigations on the islands, and whose friendship and positive spirit have been particularly meaningful through the years; Ohio State University Office of International Affairs and Department of Plant Biology which provided support for a trip to Tokyo which facilitated finishing this paper, and to Tokyo Metropolitan University for additional support for this same trip (arranged by Mikio Ono); and to the Departamento de Botánica of the Universidad de Concepción for the use of space and facilities (for Tod F. Stuessy and Daniel J. Crawford) during trips to Chile.

Literature cited

Carr, G. D. & Kyhos, D. W. (1981). Adaptive radiation in the Hawaiian silversword alliance (Compositae–Madiinae). I. Cytogenetics of spontaneous hybrids. *Evolution*, **35**, 543–56.

Carr, G. D. & Kyhos, D. W. (1986). Adaptive radiation in the Hawaiian silversword alliance (Compositae: Madiinae). II. Cytogenetics of artificial and natural hybrids. *Evolution*, **40**, 959–76.

Carr, G. D., Robichaux, R. H., Witter, M. S. & Kyhos, D. W. (1989). Adaptive radiation of the Hawaiian silversword alliance (Compositae–Madiinae): a comparison with Hawaiian picture-winged *Drosophila*. In *Genetics, Speciation and the Founder Principle*, ed. L. W. Giddings, K. Y. Kaneshiro & W. W. Anderson, pp. 79–97. New York: Oxford University Press.

Carson, H. L. (1985). Unification of speciation theory in plants and animals. *Systematic Botany*, **10**, 380–90.

Carlquist, S. (1980). *Hawaii: A Natural History. Geology, Climate, Native Flora and Fauna above the Shoreline*, 2nd edn. Honolulu: Pacific Tropical Botanical Garden.

Colwell, R. K. (1989). Hummingbirds of the Juan Fernandez Islands: natural history, evolution and population status. *Ibis*, **131**, 548–66.

Crawford, D. J. (1985). Electrophoretic data and plant speciation. *Systematic Botany*, **10**, 405–16.

Crawford, D. J., Brauner, S., Cosner, M. B. & Stuessy, T. F. (1993). Use of RAPD markers to document the origin of the intergeneric hybrid × *Margyracaena skottsbergii* (Rosaceae) on the Juan Fernandez Islands. *American Journal of Botany*, **80**, 89–92.

Crawford, D. J., Stuessy, T. F., Cosner, M. B., Haines, D. W., Silva O., M. & Baeza, M. (1992*a*). Evolution of the genus *Dendroseris* (Asteraceae: Lactuceae) on the Juan Fernandez Islands: evidence from chloroplast and ribosomal DNA. *Systematic Botany*, **17**, 676–82.

Crawford, D. J., Stuessy, T. F., Haines, D. W., Cosner, M. B., Silva O., M. & Lopez, P. (1992*b*). Allozyme diversity within and divergence among four species of *Robinsonia* (Asteraceae: Senecioneae), a genus endemic to the Juan Fernandez Islands, Chile. *American Journal of Botany*, **79**, 962–6.

Crawford, D. J., Stuessy, T. F., Lammers, T. G., Silva O., M. & Pacheco, P. (1990). Allozyme variation and evolutionary relationships among three species of *Wahlenbergia* (Campanulaceae) in the Juan Fernandez Islands. *Botanical Gazette*, **151**, 119–24.

Crawford, D. J., Stuessy, T. F. & Silva O., M. (1987). Allozyme divergence and the evolution of *Dendroseris* (Compositae: Lactuceae) on the Juan Fernandez Islands. *Systematic Botany*, **12**, 435–43.

Crawford, D. J., Whitkus, R. & Stuessy, T. F. (1987). Plant evolution and speciation on oceanic islands. In *Patterns of Differentiation in Higher Plants*, ed. K. Urbanska, pp. 183–99. London: Academic Press.

Grant, V. (1981). *Plant Speciation*, 2nd edn. New York: Columbia University Press.

Grant, V. (1985). *The Evolutionary Process: A Critical Review of Evolutionary Theory*. New York: Columbia University Press.

Hemsley, W. B. (1884). Report on the botany of Juan Fernandez, the south-eastern Moluccas and the Admiralty Islands. In *Report on the Scientific Results of the Voyage of HMS Challenger During the Years 1873–76*, vol. 1, part 3, *Botany*, ed. C. W. Thomson & J. Murray, pp. 1-96. London: HMSO.

Johow, F. (1896). *Estudios sobre la Flora de las Islas de Juan Fernandez*. Santiago: Gobierno de Chile.

King, M. (1993). *Species Evolution: The Role of Chromosome Change*. Cambridge: Cambridge University Press.

Levin, D. A. (1978). The origin of isolating mechanisms in flowering plants. *Evolutionary Biology*, **11**, 185–317.

Littlejohn, M. J. (1981). Reproductive isolation: a critical review. In *Evolution and Speciation: Essays in honor of M.J.D. White*, ed. W. R. Atchley & D. S. Woodruff, pp. 298–334. Cambridge: Cambridge University Press.

Lowrey, T. K. (1986). A biosystematic revision of Hawaiian *Tetramolopium* (Compositae: Astereae). *Allertonia*, **4**, 203–65.

Matthei, O., Marticorena, C. & Stuessy, T. F. (1993). La flora adventicia del archipélago de Juan Fernández. *Gayana Botánica*, **50**, 69–102.

Pacheco, P., Crawford, D. J., Stuessy, T. F. & Silva O., M. (1985). Flavonoid evolution in *Robinsonia* (Compositae) of the Juan Fernandez Islands. *American Journal of Botany*, **72**, 989–98.

Pacheco, P., Crawford, D. J., Stuessy, T. F. & Silva O., M. (1991). Flavonoid evolution in *Dendroseris* (Compositae, Lactuceae) from the Juan Fernandez Islands, Chile. *American Journal of Botany*, **78**, 534–43.

Pacheco, P., Stuessy, T. F. & Crawford, D. J. (1991). Natural interspecific hybridization in *Gunnera* (Gunneraceae) of the Juan Fernandez Islands, Chile. *Pacific Science*, **45**, 389–99.

Sanders, R. W., Stuessy, T. F. & Marticorena, C. (1982). Recent changes in the flora of the Juan Fernandez Islands, Chile. *Taxon*, **31**, 284–9.

Sanders, R. W., Stuessy, T. F., Marticorena, C. & Silva O., M. (1987). Phytogeography and evolution of *Dendroseris* and *Robinsonia*, tree-Compositae of the Juan Fernandez Islands, Chile. *Opera Botanica*, **92**, 195–215.

Sanders, R. W., Stuessy, T. F. & Rodríguez R., R. (1983). Chromosome numbers from the flora of the Juan Fernandez Islands. *American Journal of Botany*, **70**, 799–810.

Sang, T., Crawford, D. J., Kim, S. C. & Stuessy, T. F. (1994). Radiation of the endemic genus *Dendroseris* (Asteraceae) on the Juan Fernandez Islands: evidence from sequences of the ITS regions of nuclear ribosomal DNA. *American Journal of Botany*, **81**, 1494–1501.

Sang, T., Crawford, D. J., Stuessy, T. F. & Silva O., M. (1995). ITS sequences and the phylogeny of the genus *Robinsonia* (Asteraceae). *Systematic Botany, 20*, 55–64.

Skottsberg, C. (1921). The phanerogams of the Juan Fernandez Islands. In *The Natural History of Juan Fernandez and Easter Island*, vol. 2, *Botany*, ed. C. Skottsberg, pp. 95–240. Uppsala: Almqvist & Wiksells.

Skottsberg, C. (1928). Pollinationsbiologie und Samenverbreitung auf den Juan Fernandez-Inseln. In *The Natural History of Juan Fernandez and Easter Island*, vol. 2, *Botany*, ed. C. Skottsberg, pp. 503–47. Uppsala: Almqvist & Wiksells.

Skottsberg, C. (1951). A supplement to the pteridophytes and phanerogams of the Juan Fernandez Islands. In *The Natural History of Juan Fernandez and Easter Island*, vol. 2, *Botany*, ed. C. Skottsberg, pp. 763–92. Uppsala: Almqvist & Wiksells.

Skottsberg, C. (1957). Une seconde espèce de *Centaurodendron* Johow. *Bulletin du Jardin Botanique de l'Etat Bruxelles, 27*, 585–9.

Soltis, D. E. & Soltis, P. S. (1993). Molecular data and the dynamic nature of polyploidy. *Critical Reviews in Plant Sciences, 12*, 243–73.

Spooner, D., Stuessy, T. F., Crawford, D. J. & Silva O., M. (1987). Chromosome numbers from the flora of the Juan Fernandez Islands. II. *Rhodora, 89*, 351–6.

Stuessy, T. F., Crawford, D. J. & Marticorena, C. (1990). Patterns of phylogeny in the endemic vascular flora of the Juan Fernandez Islands, Chile. *Systematic Botany, 15*, 338–46.

Stuessy, T. F., Foland, K. A., Sutter, J. F., Sanders, R. W. & Silva O., M. (1984). Botanical and geological significance of potassium-argon dates from the Juan Fernandez Islands. *Science, 255*, 49–51.

Stuessy, T. F., Marticorena, C., Rodríguez R., R., Crawford, D. J. & Silva O., M. (1992). Endemism in the vascular flora of the Juan Fernandez Islands. *Aliso, 13*, 297–307.

Sun, B. Y., Stuessy, T. F. & Crawford, D. J. (1990). Chromosome counts from the flora of the Juan Fernandez Islands, Chile. III. *Pacific Science, 44*, 258–64.

Sun, B. Y., Stuessy, T. F., Humaña, A. M., Riveros G., M. & Crawford, D. J. (1996). Evolution of *Rhaphithamnus venustus* (Verbenaceae), a gynodieocious hummingbird-pollinated endemic of the Juan Fernandez Islands, Chile. *Pacific Science, 50*, 55–65.

Suzuki, K. (1992). Bumblebee pollinators and pollination ecotypes of *Isodon umbrosus* and *I. shikokianus* (Lamiaceae). *Plant Species Biology, 7*, 37–48.

Suzuki, K. (1993). Disruptive selection in flowering time of *Wyethia amplexicaulis* (Asteraceae). *Plant Species Biology, 8*, 51–9.

Suzuki, K. (1994). Pollinator restriction in the narrow-tube flower type of *Mertensia ciliata* (James) G. Don (Boraginaceae). *Plant Species Biology, 9*, 69–73.

Templeton, A. R. (1981). Mechanisms of speciation – A population genetic approach. *Annual Review of Ecology and Systematics, 12*, 23–48.

Valdebenito, H., Stuessy, T. F., Crawford, D. J. & Silva O., M. (1992a). Evolution of *Peperomia* (Piperaceae) in the Juan Fernandez Islands, Chile. *Plant Systematics and Evolution, 182*, 107–19.

Valdebenito, H. A., Stuessy, T. F., Crawford, D. J. & Silva O., M. (1992b). Evolution of *Erigeron* (Compositae) in the Juan Fernandez Islands, Chile. *Systematic Botany, 17*, 470–80.

Wendel, J. F. & Percival, A. E. (1990). Molecular divergence in the Galapagos Islands – Baja California species pair *Gossypium klotzschianum* and *G. davidsonii* (Malvaceae). *Plant Systematics and Evolution*, **171**, 99–115.

White, M. J. D. (1978). *Modes of Speciation*. San Francisco: W. H. Freeman.

Witter, M. S. & Carr, G. D. (1988). Adaptive radiation and genetic differentiation in the Hawaiian silversword alliance (Compositae: Madiinae). *Evolution*, **42**, 1278–87.

Woodward, R. L. (1969). *Robinson Crusoe's Island: A History of the Juan Fernandez Islands*. Chapel Hill: University of North Carolina Press.

4

Dendroseris (Asteraceae: Lactuceae) and *Robinsonia* (Asteraceae: Senecioneae) on the Juan Fernandez Islands: similarities and differences in biology and phylogeny

DANIEL J. CRAWFORD, TAO SANG,
TOD F. STUESSY, SEUNG-CHUL KIM
AND MARIO SILVA O.

Abstract

The biology and phylogeny of *Dendroseris* and *Robinsonia* (Asteraceae), two genera endemic to the Juan Fernandez Islands, are discussed and contrasted. Morphological and molecular data were used to test the monophyly and to generate phylogenetic hypotheses for the two genera. Restriction site mutations in the chloroplast DNA (cpDNA) and inter-genic spacer (IGS) region of the nuclear ribosomal DNA (rDNA) were used to produce a phylogeny of each genus, and sequences from the internal transcribed spacer (ITS) region of the nuclear rDNA were also employed to generate phylogenies. All molecular data sets strongly support the monophyly of each genus despite the morphological and ecological diversity found in them. For *Dendroseris*, all molecular information, in concordance with morphology, shows that the two subgenera *Dendroseris* and *Phoenicoseris* are monophyletic, but this is not true for subg. *Rea*. The restriction site data for *Robinsonia* do not provide any phylogenetic resolution whereas the ITS sequences produce a completely resolved tree that is concordant with relationships inferred from morphology. Allozyme diversity is higher in species of *Robinsonia* than in those of *Dendroseris* and divergence between species is likewise higher in *Robinsonia*. Differences in diversity may reflect factors such as amount of genetic variation carried to the island by the original colonizers, population sizes and breeding systems. Higher allozyme divergence between species of *Robinsonia* could result from sorting of alleles from polymorphic ancestors during speciation. Also, *Robinsonia* may have been on the Juan Fernandez Islands longer than *Dendroseris*, thereby allowing time for the accumulation of unique alleles via mutation subsequent to speciation. Comparison of ITS sequence divergence between species in each genus supports the hypothesis of *Robinsonia* having been on the Juan Fernandez Islands longer than *Dendroseris*, assuming that the rate of divergence has been comparable in each. The two genera are different in many respects despite the fact that both are endemic to the same oceanic archipelago, members of the same family and share the same insular syndrome of morphological features.

When thinking of endemic plants on oceanic islands, several points often come to mind. One envisions plants so different morphologically from continental relatives that it is difficult to ascertain the progenitors of the insular endemics. Perhaps the most spectacular and well-known example of this is the silversword alliance of Hawaii (Carr, 1985). In addition to morphological divergence from their continental relatives, endemic congeners often show unusual morphological differences among themselves. Again, the silverswords are a good example, as are Hawaiian *Bidens* (Ganders & Nagata, 1984), where there is more diversity in the *c.* 20 species than in the rest of this large and widespread genus (over 200 species on five continents). *Tetramolopium* (Lowrey, 1986, 1995), *Cyanea* (Givnish, Sytsma & Hahn, 1995), and *Schiedea* and *Alsinidendron* (Wagner, Weller & Sakai, 1995) from Hawaii are other outstanding examples. Another salient aspect of insular endemics is small population sizes, although there are notable exceptions. The species may be rare not only because of small population sizes but also because of few populations. Different endemic congeners often occur in a diverse array of habitats; again the silverswords provide excellent examples with species occurring in some of the wettest terrestrial habitats on earth (in terms of annual rainfall) to extremely dry areas (Carr, 1985). The occurrence of congeneric species in different habitats is a result of the process of adaptive radiation and has been discussed in detail for islands by Carlquist (1974, his chapter 3).

Allozyme studies on insular endemics in a variety of angiosperm families have demonstrated that, in general, species show low diversity, both within populations as well as total diversity, as compared to the mean for continental species (summary in De Joode & Wendel, 1992; see also Hamrick & Godt, 1989). Another generalization that has emerged from electrophoretic studies is that there is low divergence between species as compared to most congeneric species from continental areas (Crawford, 1989). That is, there may be little divergence between congeners at allozyme loci despite pronounced morphological differences and occurrence in a diverse array of habitats. This situation has been interpreted as the result of rapid morphological and ecological divergence via adaptive radiation with insufficient time for divergence at allozyme loci (Lowrey & Crawford, 1985; Helenurm & Ganders, 1985). This decoupling of morphological and molecular evolution will be discussed later.

The Juan Fernandez Islands are 667 km west of mainland Chile at 33° S latitude. The archipelago is comprised of the three main islands of Masatierra, Santa Clara and Masafuera (Fig. 4.1). The estimated ages of the islands from radiometric dating are about four million years for Masatierra (including Santa Clara which is just offshore from Masatierra) and one to two million years for Masafuera (Stuessy *et al.*, 1984*a*; Stuessy, Sanders & Silva, 1984*b*). The islands are strictly of volcanic origin, and there is no evidence that they were ever connected to a continental land mass or that they were in contact with South America through a series of sea mounts (Stuessy, Sanders & Silva O., 1984*b*). Sanders *et al.* (1987) review data indicating that both islands are now only about 5% of their original size. Thus, the Juan Fernandez Islands represent a rather simple system for studying endemic plant species because the species presumably evolved *in situ* on the islands where they now occur.

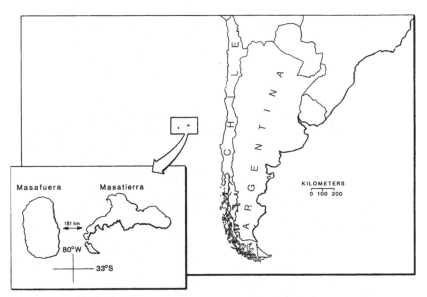

Fig. 4.1. Location of the Juan Fernandez archipelago. The small island off the tip of Masatierra is Santa Clara. [From Pacheco *et al.* (1985). Used with permission of the Botanical Society of America.]

Fig. 4.2. Representative species of *Dendroseris* from each of the three subgenera. A, *D. litoralis* subg. *Dendroseris*; B, *D. pruinata* subg. *Rea*; C, *D. pinnata* subg. *Phoenicoseris*.

In this chapter, we will compare and contrast the two largest endemic genera on the Juan Fernandez Islands for a variety of features, and consider the possible causes of similarities and differences between them. This includes a comparison of phylogenies generated from morphology and two different molecular data sets. The process of adaptive radiation might confound phylogenies produced from morphological characters because the environment could select similar features in different lineages or different features could be selected in the same lineage. Because mutations at the molecular level are probably under lower selection than some morphological features, they may provide a more accurate phylogeny. We, therefore, wished to construct and compare phylogenies from morphological and molecular data. Another focus of this chapter is to compare the level and apportionment of allozyme diversity within species of the two genera, and to examine allozyme divergence between species in the two genera. The data will be used to present contrasting scenarios for the origin and evolution of the two genera on the Juan Fernandez Islands. Insular endemics are good systems for comparative studies of molecular evolution, and another purpose of the present study, therefore, was to compare the tempo of molecular evolution in the two endemic genera.

Adaptive radiation and phylogeny of *Dendroseris* and *Robinsonia*

Carlquist (1974) discussed adaptive radiation in the flora of the Juan Fernandez Islands, and indicated that because of the small size of the archipelago (Masatierra and Masafuera are known now to each have an area of about 50 km^2; Stuessy, 1995) there are few clear examples of the process. He considered the two largest genera, both endemics, to be the best examples; these are *Dendroseris* (Asteraceae: Lactuceae) with 11 species and *Robinsonia* (Asteraceae: Senecioneae) with seven species. There is extensive morphological diversity within each genus. In the past, four genera have been recognized to accommodate the variation found in what is now recognized as the single genus *Dendroseris* (discussion in Sanders *et al.*, 1987). Plants of *Dendroseris* are rosette trees or shrubs, and variation occurs in both floral and vegetative features (Fig. 4.2). In a similar manner, three genera were once recognized for what is now treated as *Robinsonia* (Sanders *et al.*, 1987), and the plants are tree-like rosettes (Fig. 4.3).

Sanders *et al.* (1987) provided the most recent taxonomic treatments for the two genera and generated the first explicit phylogenetic

Fig. 4.3. Representative species of *Robinsonia*. A, *R. berteroi* subg. *Rhetinodendron*; B, *R. gayana* subg. *Robinsonia* sect. *Robinsonia*; C, *R. gracilis* subg. *Robinsonia* sect. *Eleutherolepis*.

hypotheses for them. These workers concluded, primarily on the basis of morphology, that only one genus should be recognized in each group and that each resulted from a single introduction, that is, each genus was viewed as monophyletic. Relationships within each genus were portrayed by undirected networks, and then each network was rooted using the genus *Hieracium* for *Dendroseris* and *Senecio* (the 'Senecioid' complex) for *Robinsonia* (Fig. 4.4). In *Dendroseris*, two of the three recognized subgenera, *Dendroseris* and *Phoenicoseris*, are holophyletic whereas subg. *Rea* is paraphyletic, with *D. micrantha* and *D. pruinata* representing one clade, and *D. gigantea* and *D. neriifolia* another clade (Fig. 4.4). One species from each subgenus (enclosed in boxes in Fig. 4.4A) occurs on Masafuera. In *Robinsonia*, the phylogeny indicates that the two subgenera are monophyletic, and within subg. *Robinsonia* each of the three sections is monophyletic (Fig. 4.4B).

Crawford *et al.* (1992*a*) used restriction site mutations in the chloroplast DNA (cpDNA) and the intergenic spacer (IGS) region of the nuclear ribosomal DNA (rDNA) to construct a phylogeny for *Dendroseris*. This cladogram, with only the cpDNA mutations shown, is given in Fig. 4.5A, and a comparison with the phylogenetic hypothesis based on morphology (Fig. 4.4A) shows that the two are essentially identical topologically. Subgenera *Dendroseris* and *Phoenicoseris* appear

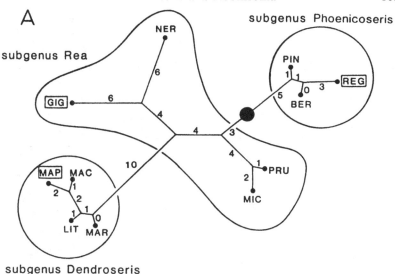

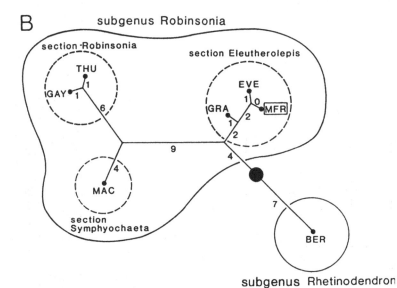

Fig. 4.4. Network of relationships in *Dendroseris* and *Robinsonia* based on morphological characters, as described in Sanders *et al.* (1987). Numbers along branches indicate distances. The species from Masafuera are indicated by boxes. A, *Dendroseris*. BER = *D. berteroana*; GIG = *D. gigantea*; LIT = *D. litoralis*; MAC = *D. macrantha*; MAP = *D. macrophylla*; MAR = *D. marginata*; MIC = *D. micrantha*; NER = *D. neriifolia*; PIN = *D. pinnata*; PRU = *D. pruinata*; REG = *D. regia*; large dot designates rooting of the network of *Hieracium* if assumed to be ancestral. B, *Robinsonia*. BER = *R. berteroi*; EVE = *R. evenia*; GAY = *R. gayana*; GRA = *R. gracilis*; MAC = *R. macrocephala*; MFR = *R. masafuerae*; rooting assumes *Senecio* as ancestor.

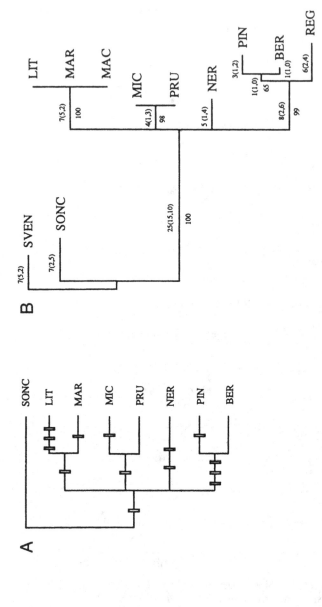

Fig. 4.5. Phylogenetic relationships in *Dendroseris* based on molecular data. (From Sang *et al.*, 1994. Used with permission of the Botanical Society of America.) A, Cladogram based on cpDNA restriction site mutations. There is no homoplasy; bars represent mutations. Species designations same as Fig. 4.4, with SONC indicating outgroup *Sonchus canariensis*. (Redrawn from Crawford *et al.*, 1992.) B, Strict consensus tree from four equally most parsimonious trees from ITS sequences. The two outgroups are *Sonchus fruticosus* (SONC) and *Sventenia bupleuroides* (SVEN). Numbers above lines indicate substitutions for a branch, followed by transitions and transversions, respectively, in parentheses; numbers below branches designate percentages of group occurrence in 1000 bootstrap replications.

to be holophyletic whereas subg. *Rea* is not. Furthermore, *D. micrantha* and *D. pruinata* are sister taxa (as with morphology) and *D. neriifolia* is on a branch by itself. Four of the species were not available for the restriction site studies, so the cpDNA and morphological phylogenies cannot be compared for all taxa.

Sang *et al.* (1994) used sequences from the internal transcribed spacer (ITS) region of the nuclear rDNA to assess evolutionary relationships in *Dendroseris*, and because PCR was used to amplify DNAs, it was possible to include two species not examined in the cpDNA study due to lack of plant material (Fig. 4.5B). The ITS phylogeny is concordant with the other two phylogenies, but provides some additional insights relative to the cpDNA phylogeny because of the additional species included. Subgenera *Dendroseris* and *Phoenicoseris* are strongly supported clades, and subg. *Rea* consists of one well supported clade with *D. micrantha* and *D. pruinata* and another branch containing *C. neriifolia* (Fig. 4.5B). Within subg. *Phoenicoseris*, *D. regia* from the younger island of Masafuera is sister to the two other species from Masatierra. This indicates that *D. regia* split first from an ancestral lineage that subsequently diverged to produce *D. berteroana* and *D. pinnata*. The ITS data provide much stronger support than the other data sets for the monophyly of *Dendroseris*. None of the phylogenies clearly resolve relationships among the subgenera and this is particularly true with the two molecular data sets where there is a polytomy at the base (Fig. 4.5).

Crawford *et al.* (1993) used restriction site mutations in cpDNA and the IGS of nuclear rDNA to produce a phylogeny for *Robinsonia*. The result was a 'star' phylogeny with mutations restricted to single species and thus there was no useful phylogenetic information (Fig. 4.6A). By contrast, ITS sequences produced a well-resolved cladogram that is essentially identical to the one generated from morphological data (cf. Fig. 4.4B, 4.6B)(Sang *et al.*, 1995). *Robinsonia berteroi*, the only member of subg. *Rhetinodendron*, is sister to the rest of the genus (*R. macrocephala*, which comprises sect. *Symphyochaeta* is presumed extinct and no DNA could be amplified from herbarium material) which consists of subg. *Robinsonia*. Within the latter subgenus, sections *Eleutherolepis* and *Robinsonia* form well supported monophyletic groups (Fig. 4.6B). The only species of *Robinsonia* from the younger island, *R. masafuerae*, is sister to *R. evenia* which suggests that these two taxa evolved from a common ancestor after the lineage split that produced *R. gracilis* and their common ancestor (Fig. 4.6B). Also, it is possible that *R. masafuerae* evolved from *R. evenia*, but the cladogram cannot show that possibility.

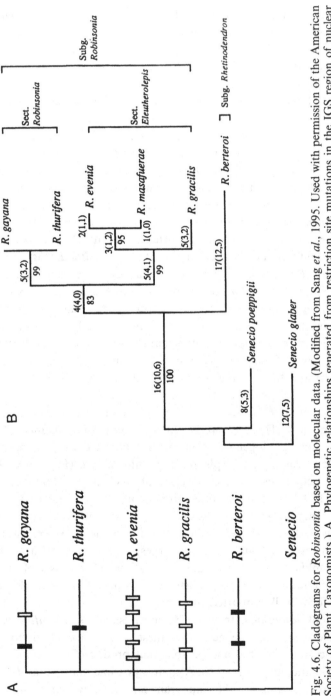

Fig. 4.6. Cladograms for *Robinsonia* based on molecular data. (Modified from Sang *et al.*, 1995. Used with permission of the American Society of Plant Taxonomists.) A, Phylogenetic relationships generated from restriction site mutations in the IGS region of nuclear rDNA (solid bars) and cpDNA restriction site mutations (open bars). *Senecio glaber and S. poeppigii* were both used as outgroups, with the same results. (Redrawn from Crawford *et al.*, 1993). B, Single most parsimonious tree for *Robinsonia* produced from ITS sequences. Number of nucleotide substitutions given above branches with transitions and transversions, respectively, shown in parentheses. Percentage of 1000 bootstrap replications in which a group occurred is given below branches.

The ITS sequences provide strong support for monophyly of *Robinsonia;* one could recognize subg. *Rhetinodendron* as a genus, as was done prior to Sanders *et al.* (1987), and all taxa would be monophyletic. ITS sequences have produced a more highly resolved phylogeny for *Robinsonia* than for *Dendroseris*; possible reasons for this will be discussed later. The phylogenies from ITS will be used as frameworks for the discussion of other data.

Because the phylogenies produced from sequence data and morphology in each genus are fully concordant, the morphological features used in constructing the phylogenies represent primarily characters shared by common descent rather than parallelisms or convergences resulting from selection in similar habitats. Carlquist (1974, p. 206) suggested that *Dendroseris* and *Robinsonia* 'show what could be called adaptive radiation' but clearly did not consider the two genera outstanding examples of radiation in oceanic archipelagos, ostensibly because of the small sizes of the islands. It may be that there is no conflict between the molecular (ITS) and morphological phylogenies simply because there has not been strong adaptive radiation. Another possible explanation, however, exists for *Robinsonia*. Carlquist (1974) examined anatomical features of leaves of *Robinsonia* species and showed a correlation between leaf anatomy and habitats in which the species occur. For example, species from more xeromorphic habitats have leaves with adaptive features such as more highly developed hypodermis, greater presence of secretory canals, greater development of fibres near vascular bundles, etc. It so happens, however, that species from the same sections occur in habitats that appear to be similarly 'mesomorphic' or 'xeromorphic'. This seems to suggest that the ancestors of the different sections radiated into particular kinds of habitats with subsequent speciation. Some anatomical-morphological features characteristic of a given section are adaptive because they were selected early in the radiation of *Robinsonia*, but perhaps subsequent speciation within each section has not been accompanied by radiation into radically different habitats.

Carlquist (1967, 1974) carried out a detailed anatomical study of *Dendroseris* and commented on its adaptive radiation because, as a genus, it occupies a greater range of habitats than *Robinsonia*. Carlquist (1967) found that less mesomorphic species, such as *D. litoralis* (subg. *Dendroseris*) and *D. pruinata* (subg. *Rea*) share the features of thicker leaves with a better developed hypodermis. Both species occur in open sunny habitats at very low elevations. By contrast, the leaves of *D. macrophylla* (subg. *Dendroseris*) are quite thin with a poorly developed

Table 4.1. *Total gene diversity* (H_T), *mean number of alleles per locus* (A), *percentage of polymorphic loci* (Ap) *and proportion of polymorphic loci* (P) *for species of* Dendroseris *and* Robinsonia

	Species	Measures of genetic variation			
		H_T	A	Ap	P
Dendroseris					
	D. berteroana	0.000	1.00	—	0.00
	D. litoralis	0.071	1.14	2.00	0.14
	D. micrantha	0.065	1.14	2.00	0.14
	D. neriifolia	0.053	1.05	2.00	0.05
	D. pinnata	0.029	1.05	2.00	0.05
	D. pruinata	0.000	1.00	—	0.00
Robinsonia					
	R. evenia	0.071	1.40	2.20	0.33
	R. gayana	0.145	1.53	2.67	0.33
	R. gracilis	0.177	1.79	2.38	0.57
	R. thurifera	0.033	1.14	2.00	0.14

hypodermis and large air spaces. This species occurs in shady, more mesic habitats at higher elevations. In these instances, aspects of the leaf anatomy suggest a parallel development of similar features in different lineages and different features within a lineage. There is some evidence of adaptive radiation in both *Robinsonia* and *Dendroseris*, but it is not spectacular by any means. Sanders *et al.* (1987) attempted to characterize habitat differences (primarily edaphic factors and the associated taxa) between species in the two genera but were not able to demonstrate clear-cut differences in most instances. One possible explanation is that evidence of original habitat differentiation has been lost through erosion, with concomitant compression of the flora into the present small size of the islands.

Allozyme diversity and divergence for *Dendroseris* and *Robinsonia*: patterns and possible processes

Allozyme diversity within and divergence among species in the two genera has been examined. The total gene diversity (Nei, 1973), mean number of alleles per locus, percentage of polymorphic loci and proportion of polymorphic loci are given in Table 4.1 for each species of *Dendroseris* (Crawford, Stuessy & Silva O., 1987; Crawford *et al.*, unpublished).

When values for *Dendroseris* are compared with mean levels of diversity in other island endemics (De Joode & Wendel, 1992), it is apparent that species of *Dendroseris* are depauperate at allozyme loci. For example, three species have total diversities similar to the mean of 0.064 while the remaining three species have much lower values (Table 4.1). All species of *Dendroseris* are below the mean value for insular endemics for both mean number of alleles per locus (1.32) and proportion of polymorphic loci (0.25) (Table 4.1).

The results for species of *Robinsonia* (Crawford *et al.*, 1992*b*) contrast with *Dendroseris* in having much higher levels of allozyme diversity (Table 4.1). The mean diversity within species of *Robinsonia* is 0.107 whereas it is 0.036 for *Dendroseris*. All other measures of allozyme variation are likewise higher in *Robinsonia* and variation exceeds the mean for island endemics. Also, in both genera there is a wide range of values among the species; this is particularly true for *Robinsonia* (Table 4.1).

There are rather striking differences between the interspecific genetic identities in the two genera (Table 4.2). The mean value for all species comparisons in *Dendroseris* is 0.84 compared to 0.63 in *Robinsonia*. Other notable aspects of the allozyme identities include high values (0.99) between *D. berteroana* and *D. pinnata* of subg. *Phoenicoseris* as well as the high similarity (0.94) between *D. micrantha* and D. *pruinata* of subg. *Rea* (Table 4.2). Also, identities between species of subg. *Dendroseris* and those in subg. *Phoenicoseris* are quite high (average of 0.908). It is also notable that all three species now assigned to subg. *Rea* show low identities with members of the other two subgenera, and *D. neriifolia* has a lower identity with *D. micrantha* and *D. pruinata* (0.84 and 0.78, respectively) than the latter two species have with each other. The low allozyme divergence between members of subg. *Dendroseris*, and between *D. micrantha* and *D. pruinata*, is concordant with the strong clades each group forms in the ITS tree and the identical sequences for members of the respective clades (Fig. 4.5B). Also, the low identity between *D. neriifolia* and the other two members of subg. *Rea* corresponds with their highly divergent ITS sequences (Fig. 4.5B). The high isozyme similarity between *D. berteroana* and *D. pinnata* contrasts somewhat with the four mutations by which they differ in ITS sequences; the high identities between subg. *Dendroseris* and subg. *Phoenicoseris* also contrast with the high ITS divergence between them. Although the ITS sequences show clearly that *Robinsonia gayana* and *R. thurifera* are sister taxa and that *R. evenia* and *R. gracilis* are in the same strongly supported

Table 4.2. *Genetic identities between species of* Dendroseris *(A) and* Robinsonia *(B)*

Acronyms same as in Fig. 4.4.

A. *Dendroseris*

	subg. *Phoenicoseris*		subg. *Dendroseris*			subg. *Rea*		
	BER	PIN	LIT	MAC	MAR	MIC	NER	PRU
BER	X	0.99	0.84	0.93	0.91	0.81	0.82	0.76
PIN		X	0.85	0.94	0.90	0.81	0.80	0.77
LIT			X	0.93	0.88	0.81	0.78	0.80
MAC				X	0.91	0.81	0.80	0.80
MAR					X	0.77	0.82	0.73
MIC						X	0.84	0.94
NER							X	0.78
PRU								X

B. *Robinsonia*

	sect. *Eleutherolepis*		sect. *Robinsonia*	
	R. evenia	R. gracilis	R. gayana	R. thurifera
R. evenia	X	0.60	0.56	0.61
R. gracilis		X	0.63	0.64
R. gayana			X	0.71
R. thurifera				X

From Crawford, Stuessy & Silva O., 1987; Crawford *et al*., 1992*b*; Crawford *et al*., unpublished

Table 4.3. *Distribution of alleles (a–d) between sections of* Robinsonia

Gpi = glucose-6-phosphate isomerase; *Idh* = isocitrate dehydrogenase; *Mdh* = malate dehydrogenase; *Tpi* = triose-phosphate isomerase.

Taxon	Allozyme loci				
	Gpi-1	Gpi-2	Idh-1	Mdh-1	Tpi-1
Sect. *Robinsonia*					
R. *gayana*	b,c	a,b,c	a	a	a
R. *thurifera*	d	b	a	b	a,b
Sect. *Eleutherolepis*					
R. *evenia*	b,c	a,b	a	c	a
R. *gracilis*	b,c	a,c,d	b	b,d	a,b

clade (Fig. 4.6B), all four taxa are about equally divergent at isozyme loci (Table 4.1); this is discussed below.

Possible reasons for lower isozyme divergence between species of *Dendroseris* as compared to *Robinsonia* must consider the two processes, certainly not mutually exclusive, by which divergence can occur at allozyme loci. One is by the sorting of alleles from a polymorphic ancestor during radiation and speciation on the islands, and another is by the accumulation of mutations subsequent to speciation (Witter & Carr, 1988; Crawford *et al.*, 1992*b*). The former is no doubt the more rapid process and the distribution of alleles among species may offer clues as to which process has occurred. The taxonomic distribution of several alleles in *Robinsonia* shows that some of the same alleles occur in taxa belonging to different strongly supported clades (Table 4.3). This distribution could be the result of allele sorting during radiation when population sizes may have been small, and random loss of alleles could have occurred rather easily. Species of *Robinsonia* also contain unique alleles, with seven in *R. gracilis*, six each in *R. evenia* and *R. gayana*, and three in *R. thurifera* (Crawford *et al.*, 1992*b*). Unique alleles could result either from sorting in a variable ancestor with loss in all but one species or the accumulation of mutations subsequent to speciation. It is possible, therefore, that both processes have been involved in generating allozyme divergence between species of *Robinsonia* and in producing the pattern of allele distribution seen among the species. In *Dendroseris*, by contrast, the same alleles are not distributed between species in different subgenera. Also, there are unique alleles in only two species, with one having one allele and the other three. It appears that in *Dendroseris* there has been little, if any, divergence via allelic sorting, and other than in *D. neriifolia*, there has been little accumulation of mutations subsequent to speciation.

There are several historical and biological factors that could generate both higher diversities in species of *Robinsonia* than in *Dendroseris* and higher divergences between species of *Robinsonia*. An important initial factor is the amount of genetic variation brought to the islands by the ancestor in the colonization event. If the progenitor of *Robinsonia* were more variable than the ancestor of *Dendroseris*, then this could help explain the differences between the two genera. Baker (1955, 1967) suggested that the ancestors of many island endemics were self-compatible and self-pollinating plants. The basic rationale for this is that only one propagule is needed for reproduction, and with the small islands in oceans acting like sieves, this is a more likely event than having several propagules reach them. Also, since pollinators are no doubt initially limited on islands, a self-pollinating plant would have a distinct advantage. Carlquist (1974) does not feel that most colonizing events include a single propagule, and if this were the case, then the compatibility of the ancestral taxon would not be such a critical factor for establishment and reproduction on the island. Although we do not know with certainty the ancestors of *Dendroseris* and *Robinsonia*, the former genus belongs to subtribe Sonchinae (Kim, Crawford & Jansen, 1996) and its closest relatives are self-incompatible. The ancestor of *Robinsonia* was very likely to have been a member of *Senecio* and self-incompatibility is prevalent in this large genus (T. Barkley, personal communication). Also, the virtual absence of dioecy in *Senecio* suggests that the condition existing in *Robinsonia* is likely to have evolved on the islands. There is no evidence, therefore, from the breeding systems of probable progenitors that the ancestor of *Robinsonia* brought more diversity to the islands than the progenitor of *Dendroseris*, although this can never be known with certainty.

Another factor that may have had a large effect on genetic diversity is population size. The two genera differ greatly in this regard, with the most abundant species of *Dendroseris* being rarer than all species of *Robinsonia* except *R. berteroi*, which is known from only one extant male plant. The loss of alleles via drift within small populations as well as in the founding of new populations during speciation may well account for the lower gene diversity within as well as divergence among species of *Dendroseris* as compared to *Robinsonia*. It is unknown whether rare species in either genus have always been rare or have become that way through time. With regard to *Dendroseris*, the answer may be that certain species have become rare whereas others have always been so. For example, certain species from lower elevations such as *D. litoralis* and *D. pruinata* were no doubt at one time more abundant, with historical accounts of *D. litoralis* being very common at low elevations where it is now known in nature from a few plants on Santa Clara and the nearby rock called Morro Spartan. Other species such as *D. berteroana* occur at upper elevations where there is no indication of disturbance and perhaps this species has always been rare. Given, however, that Masatierra is much smaller than its original size (Stuessy *et al.*, 1984*b*), it is possible that even the areas of pristine, undisturbed habitat have shrunken a great deal through time.

With regard to *Robinsonia*, *R. macrocephala* was once known from several locations on Masatierra (Skottsberg, 1921), but is now thought to be extinct. It has not been located nor found elsewhere despite repeated searches on several expeditions. *Robinsonia berteroi* was reported from eight localities, with the observation of many specimens at one locality (Skottsberg, 1921), but the species is now known from one male plant. It is not at all apparent that the recent demise of the species can be attributed to destruction of habitats by humans or other animals. A comparison of rarity in closely related species of *Robinsonia* is also instructive. *Robinsonia gayana* and *R. thurifera* are sister species with identical ITS sequences (Fig. 4.6); the latter is a rare species known from several very small populations whereas the former is quite abundant. Note also that *R. gayana* has over four times the total allozyme diversity found in *R. thurifera*. *Robinsonia gracilis* and *R. evenia* are not sister species but they belong to the same strongly supported clade (Fig. 4.6B). Although the split between *R. gracilis* and the ancestor of *R. evenia* (*R. masafuerae*) is unlikely to be as recent as between *R. gayana* and *R. thurifera*, there are still similarities in the two situations. For example, *R. gracilis* is much more common than *R. evenia* and the former species has over 2.5 times as much total allozyme diversity as the latter (Table 4.1). It appears, therefore, that with each lineage-splitting on Masatierra one species tends to be much more common than the other, with one having gone through or presently going through a bottleneck. By contrast, *R. masafuerae* is rather common on Masafuera despite its recent origin. This may be because it is free of competition from its congeners, particularly its ancestor, on the other island.

Breeding systems may also be a factor accounting for differences in genetic diversities in the two genera. Species of *Robinsonia*, being dioecious, are outcrossers. In *Dendroseris*, *D. litoralis* sets fruit when capitula are bagged whereas heads of *D. neriifolia* and *D. pruinata* do not set fruit even when self-pollinated and bagged (Anderson, unpublished). As mentioned above, the probable ancestor of *Dendroseris* was self-incompatible; if this is true, then within *Dendroseris* there has been the evolution of self-compatibility in at least one species while two species are self-incompatible.

Comparative molecular evolution in *Dendroseris* and *Robinsonia*

Endemic genera on oceanic islands represent good systems for examining the tempo of molecular evolution. One advantage is that the maximum ages of the islands are known (assuming no sea mounts) and it may be assumed that divergence has occurred within this time period. *Dendroseris* and *Robinsonia* offer the opportunity to compare cpDNA and ITS sequences, both within the same genus and between genera.

In *Dendroseris*, the rate of ITS sequence divergence is not significantly different among the four lineages when a relative rate test is applied (Sang *et al.*, 1994). The average pair-wise sequence divergence between the species is about 2.7%, which is some 38 times faster than the 0.07% estimated for cpDNA (Crawford *et al.*, 1992a). The average sequence divergence between species in the four different lineages is 3.15% and because there are no informative substitutions for resolving relationships among lineages, divergence between species in different lineages may be taken as sequence divergence since the ancestor of *Dendroseris* colonized Masatierra. If one assumes that the colonization occurred shortly after the formation of Masatierra some four million years ago (Stuessy *et al.*, 1984a), then with an average of 3.15% between the four lineages, the average rate of nucleotide substitutions/site per year would be a little less than 4.0×10^{-9} (Sang *et al.*, 1994). Based on cpDNA divergence and an assumption of a divergence rate of 0.07% per million years, Crawford *et al.* (1992a) estimated a divergence time of 2.6 million years for the most divergent lineages of *Dendroseris*. If this age is used for the origin of *Dendroseris* on Masatierra, then the average rate of ITS sequence divergence is about 6.0×10^{-9}/site per year.

Consider next molecular divergence between lineages in *Robinsonia*. The estimated average sequence divergence between species in cpDNA is 0.056% whereas the average divergence between species in ITS is

3.65% (Crawford *et al.*, 1993; Sang *et al.*, 1995), suggesting that the latter has evolved about 65 times faster than the former. Relative rate tests showed that the molecular clock hypothesis cannot be rejected for ITS between any of the lineages of *Robinsonia* (Sang *et al.*, 1995). Species in the two earliest splitting ITS lineages in *Robinsonia* subg. *Rhetinodendron* and subg. *Robinsonia* (Fig. 4.6B), differ by an average of 6.26%. If we assume that the ancestor of *Robinsonia* colonized Masatierra shortly after its formation about four million years ago, then the rate of sequence divergence would be a little less than 8.0×10^{-9}/site per year (Sang *et al.*, 1995) and this represents the slowest possible rate of ITS sequence evolution in *Robinsonia*. If the ancestor of *Robinsonia* arrived later, then the average rate of ITS sequence divergence could be considerably higher.

Dispersal and radiation of *Dendroseris* and *Robinsonia*: speculations from available data

As indicated earlier, the Juan Fernandez Islands are a good system for studying evolution and speciation because they apparently have never been connected to a continental area by a series of sea mounts. Also, there is no evidence that they were ever available for the early origin of endemic taxa followed by subsequent dispersal to the present islands (Stuessy *et al.*, 1984a). Thus, unlike Hawaii (Carson & Clague, 1995; Givnish, Sytsma & Hahn, 1995), the historical significance of sea mounts does not have to be considered in the evolution of Juan Fernandez endemics. Given this situation, it is possible to put the maximum time on the ages of *Dendroseris* and *Robinsonia*. As discussed above, when considering rates of sequence evolution, it is not possible to ascertain when within the last four million years the ancestor of each genus arrived in the islands. Calculations of average sequence divergence between the four basal unresolved lineages in *Dendroseris* and the average divergences between *Robinsonia* subgenera *Rhetinodendron* and *Robinsonia*, which is the basal most split in *Robinsonia* (Fig. 4.6B), reveals a value for *Robinsonia* (6.26%) that is about twice that of *Dendroseris* (3.15%). If an equal rate of substitution is assumed in the ITS regions of the two genera, then it suggests that *Robinsonia* has been on Masatierra twice as long as *Dendroseris*.

There are several contrasting aspects of *Robinsonia* and *Dendroseris* that are at least consistent with the former genus having evolved on the islands before the latter. If the ancestor of *Robinsonia* colonized Masatierra earlier than the ancestor of *Dendroseris* then there may have been more open habitats into which it could radiate during evolution and speciation. The reduced competition, relative lack of pathogens, etc., presumably would have allowed relatively large population sizes to build up quickly. As discussed earlier, there are species of *Robinsonia* (e.g., *R. gayana* and *R. gracilis*) with large population sizes and these are in two different lineages within subg. *Robinsonia* (Fig. 4.6). If, by contrast, *Dendroseris* radiated later on Masatierra, then the size and number of open habitats may have been much smaller than were available for *Robinsonia*. This may be a factor in the rarity of most species of *Dendroseris*, but it clearly does not apply to all. As mentioned earlier, *D. litoralis* was once much more abundant but human influences have made it quite rare other than in cultivation. Other species such as *D. berteroana* and *D. micrantha* appear to be in undisturbed habitats, but perhaps these habitats have undergone contraction as the size of Masatierra decreased from its original size (Sanders *et al.*, 1987).

Additional evidence congruent with an older age for *Robinsonia* is the lower genetic identities between species at allozyme loci compared to *Dendroseris* (Table 4.2). Particularly significant are the number of unique alleles found in species of *Robinsonia* as contrasted with the almost total lack of species-specific alleles in *Dendroseris* (see earlier discussion). Presumably, if the unique alleles result from mutations subsequent to speciation (rather than by lineage sorting), then it suggests an older age for species of *Robinsonia*.

Within *Robinsonia*, ITS sequence data suggest the following scenario for its initial and subsequent radiations. If the colonizer arrived when there were many large open habitats, then there may have been large population sizes built up following the initial bottleneck presumably associated with dispersal to and initial establishment on the island. The initial cladogenic event produced one lineage now represented by subgenera *Rhetinodendron* and *Robinsonia*. Available evidence indicates that the former radiated very little, if at all, because only one species, *R. berteroi*, is known from this lineage (Fig. 4.4B). It is possible, however, that extinctions in this lineage have left *R. berteroi* as the only remaining species. As mentioned earlier, *R. berteroi* is now on the brink of extinction and there is no evidence that it was ever a common species. By contrast, subg. *Robinsonia* presumably underwent two lineage splittings (although ITS data are not available for *R. macrocephala* sect. *Symphyochaeta*) to produce these very distinct sections. The morphological data (Fig. 4.4B) suggest that the initial cladogenic event involved one lineage now represented by sect. *Eleutherolepis* and another ancestral to sections *Symphyochaeta* and *Robinsonia*. If this is the case, then both lineages were equally species rich, but again, the second cladogenic event produced one lineage, sect. *Symphyochaeta* with only the one species, *R. macrocephala*, which was apparently never common as far as is known from collections (Skottsberg, 1921), and is now thought to be extinct. As mentioned earlier, within sect. *Eleutherolepis* and *Robinsonia* the sister taxa consist of rare and quite common species.

The data suggest that in the genus following the initial establishment of the 'original' *Robinsonia*, subsequent cladogenic events produced one lineage that was not successful at radiating and speciating whereas the other lineage in each case was much more successful. This scenario must be considered somewhat speculative because it is supported only by data from past collections and plants now extinct on the islands. That is, the role of extinctions in producing the pattern now seen in *Robinsonia* cannot be assessed. Given that two species representing very distinct lineages have either gone extinct or are on the brink of extinction, it may well be that a number of species have suffered extinction during the evolutionary history of *Robinsonia*.

Acknowledgements

Work discussed in this chapter has been supported by National Science Foundation grants INT-7721637, INT-8317088, BSR-8306436, BSR-8906988 and DEB-9500499 to Daniel Crawford and Tod Stuessy and by FONDECYT (Fomento Nacional para el Desarrollo de Ciencia y Tecnología) of Chile under grants 1093, 1087, 1509, 0652, 196-08-22 and 7-96-0015 to Mario Silva O. Tim Lowrey provided valuable comments on an earlier draft of the manuscript, and Bette Hellinger did the word processing with her usual accuracy and care.

Literature cited

Baker, H. G. (1955). Self-compatibility and establishment after 'long distance' dispersal. *Evolution*, **9**, 347–9.
Baker, H. G. (1967). Support for Baker's Law – as a rule. *Evolution*, **21**, 853–6.
Carlquist, S. (1967). Anatomy and systematics of *Dendroseris* (*sensu lato*). *Brittonia*, **19**, 99–121.
Carlquist, S. (1974). *Island Biology*. New York: Columbia University Press.
Carr, G. D. (1985). Monograph of the Hawaiian Madiinae (Asteraceae): *Argyroxiphium*, *Dubautia*. and *Wilkesia*. *Allertonia*, **4**, 1–123.

Carr, G. D. (1987). Beggar's ticks and tarweeds: masters of adaptive radiation. *Trends in Ecology and Evolution*, **2**, 192–5.

Carson, H. L. & Clague, D. A. (1995). Geology and biogeography of the Hawaiian Islands. In *Hawaiian Biogeography: Evolution on a Hot Spot Archipelago*, ed. W. L. Wagner & V. A. Funk, pp. 14–29. Washington DC: Smithsonian Institution Press.

Crawford, D. J. (1989). Enzyme electrophoresis and plant systematics. In *Isozymes in PlantBiology*, ed. D. E. Soltis & P. S. Soltis, pp. 146–64. Portland: Dioscorides.

Crawford, D. J., Stuessy, T. F., Cosner, M. B., Haines, D. W., Silva O., M. & Baeza, M. (1992*a*). Evolution of the genus *Dendroseris* (Asteraceae: Lactuceae) on the Juan Fernandez Islands: evidence from chloroplast and ribosomal DNA. *Systematic Botany*, **17**, 676–82.

Crawford, D. J., Stuessy, T. F., Haines, D. W., Cosner, M. B., Silva O., M. & Lopez, P. (1992*b*). Allozyme diversity within and divergence among four species of *Robinsonia* (Asteraceae: Senecioneae), a genus endemic to the Juan Fernandez Islands, Chile. *American Journal of Botany*, **79**, 962–6.

Crawford, D. J., Stuessy, T. F., Cosner, M. B., Haines, D. W. & Silva O., M. (1993). Ribosomal and chloroplast DNA restriction site mutations and the radiation of *Robinsonia* (Asteraceae: Senecioneae) on the Juan Fernandez Islands. *Plant Systematics and Evolution*, **184**, 233–9.

Crawford, D. J., Stuessy, T. F. & Silva O., M. (1987). Allozyme divergence and the evolution of *Dendroseris* (Compositae: Lactuceae) on the Juan Fernandez Islands. *Systematic Botany*, **12**, 435–43.

De Joode, D. R. & Wendel, J. F. (1992). Genetic diversity and origin of the Hawaiian Islands cotton, *Gossypium tomentosum*. *American Journal of Botany*, **79**, 1311–19.

Ganders, F. R. & Nagata, K. M. (1984). The role of hybridization in the evolution of *Bidens* on the Hawaiian Islands. In *Plant Biosystematics*, ed. W. F. Grant, pp. 179–94. Toronto: Academic Press.

Givnish, T. J., Sytsma, K. J. & Hahn, W. J. (1995). Molecular evolution, adaptive radiation, and geographic speciation in *Cyanea* (Campanulaceae: Lobelioideae). In *Hawaiian Biogeography: Evolution on a Hot Spot Archipelago*, ed. W. L. Wagner & V. A. Funk, pp. 288-337. Washington DC: Smithsonian Institution Press.

Hamrick, J. L. & Godt, M. J. W. (1989). Allozyme diversity in plant species. In *Plant Population Genetics, Breeding, and Genetic Resources*, ed. A. H. D. Brown *et al.*, pp. 43–63. Sunderland: Sinauer.

Helenurm, K. & Ganders, F. R. (1985). Adaptive radiation and genetic differentiation in Hawaiian *Bidens*. *Evolution*, **39**, 753–65.

Kim, S-C., Crawford, D. J. & Jansen, R. K. (1996). Phylogenetic relationships among the genera of the subtribe Sonchinae (Asteraceae): evidence from ITS sequences. *Systematic Botany*, **21**, 417–32.

Lowrey, T. K. (1986). A biosystematic revision of Hawaiian *Tetramolopium* (Compositae: Astereae). *Allertonia*, **4**, 203–65.

Lowrey, T. K. (1995). Phylogeny, adaptive radiation and biogeography of Hawaiian *Tetramolopium* (Asteraceae: Astereae). In *Hawaiian Biogeography: Evolution on a Hot Spot Archipelago*, ed. W. L. Wagner & V. A. Funk, pp. 195–220. Washington DC: Smithsonian Institution Press.

Lowrey, T. K. & Crawford, D. J. (1985). Allozyme divergence and evolution in *Tetramolopium* (Compositae: Astereae) on the Hawaiian Islands. *Systematic Botany*, **10**, 64–72.

Nei, M. (1973). Analysis of gene diversity in subdivided populations. *Proceedings of the National Academy of Sciences USA*, **70**, 3321–23.

Pacheco, P., Crawford, D. J., Stuessy, T. F. & Silva O., M. (1985). Flavonoid evolution in *Robinsonia* (Compositae) of the Juan Fernandez Islands. *American Journal of Botany*, **72**, 989–98.

Sanders, R. W., Stuessy, T. F., Marticorena, C. & Silva O., M. (1987). Phytogeography and evolution of *Dendroseris* and *Robinsonia*, tree-Compositae of the Juan Fernandez Islands, Chile. *Opera Botanica*, **92**, 195–215.

Sang, T., Crawford, D. J., Kim S-C. & Stuessy, T. F. (1994). Radiation of the endemic genus *Dendroseris* (Asteraceae) on the Juan Fernandez Islands: evidence from sequences of the ITS regions of the nuclear ribosomal DNA. *American Journal of Botany*, **81**, 1494–1501.

Sang, T., Crawford, D. J., Stuessy, T. F. & Silva O., M. (1995). ITS sequences and the phylogeny of the genus *Robinsonia* (Asteraceae). *Systematic Botany*, **20**, 55–64.

Skottsberg, C. (1921). The phanerogams of the Juan Fernandez Islands. In *The Natural History of Juan Fernandez and Easter Island*, vol. 2, *Botany*, ed. C. Skottsberg, pp. 95–240. Uppsala: Almqvist & Wiksells.

Stuessy, T. F. (1995). Juan Fernandez Islands, Chile. In *Centres of Plant Diversity: A Guide and Strategy For Their Conservation*, vol. 2, *Asia, Australia and the Pacific*, ed. S. D. Davis, V. H. Heywood & A. C. Hamilton, pp. 565–8. Cambridge: WWF; IUCN.

Stuessy, T. F., Foland, K. A., Sutter, J. F., Sanders, R. W. & Silva O., M (1984a). Botanical and geological significance of potassium-argon dates from the Juan Fernandez Islands. *Science*, **225**, 49–51.

Stuessy, T. F., Sanders, R. W. & Silva O., M. (1984b). Phytogeography and evolution of the flora of Juan Fernandez Islands: a progress report. In *Biogeography of the Tropical Pacific*, ed. F. J. Radovsky, S. H. Sohmer & P. H. Raven, pp. 55–69. Honolulu: Association of Systematics Collections and Bishop Museum.

Wagner, W. L., Weller, S. G. & Sakai, A. (1995). Phylogeny and biogeography in *Schiedea and Alsinidendron*. In *Hawaiian Biogeography: Evolution on a Hot Spot Archipelago*, ed. W. L. Wagner & V. A. Funk, pp. 221–58. Washington DC: Smithsonian Institution Press.

Witter, M. S. & Carr, G. D. (1988). Adaptive radiation and genetic differentiation in the Hawaiian silversword alliance (Compositae: Madiinae). *Evolution*, **42**, 1278–87.

5

Island biogeography of angiosperms of the Juan Fernandez archipelago

TOD F. STUESSY, DANIEL J. CRAWFORD,
CLODOMIRO MARTICORENA AND ROBERTO
RODRÍGUEZ

Abstract

MacArthur & Wilson's (1967) equilibrium theory of island biogeography predicts that numbers of species on oceanic islands are dependent upon the size of islands, distance from major source area, and rates of immigration and extinction. Despite healthy criticisms of this theory, it has been enormously stimulating for understanding species diversity on islands as well as other island-like habitats. More recently, other workers have attempted more complex models to find better ways of predicting species diversity in oceanic archipelagoes. The Juan Fernandez Islands are well suited for specific model assessment due to the small size of the endemic and native flora (156 angiosperm species), the few major islands (Masatierra and Masafuera), their small size ($50\,km^2$ each) and proximity to the major source area (southern South America). Factors taken into consideration include size (both ancient and modern), distance from source area, extinction, speciation within and between islands, and differing modes of dispersal. A model for predicting species diversity is first developed to explain numbers of species on Masatierra, the island closest to the continent ($667\,km$ in the Pacific Ocean). This approach is then applied to explain species diversity on Masafuera, with encouraging results. The most important point is that numerous geological, ecological and historical factors impinge on determining the total number of species that can be supported on a particular oceanic island. Changing island size through geological time is especially important to consider.

An important contribution to understanding species diversity on oceanic islands has been the equilibrium theory of island biogeography of MacArthur & Wilson (1967). This theory predicts levels of species diversity on islands in response to size and distance from source areas, and interacting dynamics of rates of immigration and extinction. It builds, in part, on the long-known general ecological relationship of increased species diversity with increasing geographic area (He & Legendre, 1996). The importance of the theory of island biogeography is not its ability to predict with great precision the *exact* number of species that a particular island system should have, but rather in its simplicity in attempting to focus on the most significant factors in determining species diversity. Some experimental data [e.g., Simberloff & Wilson, 1969, 1970; Simberloff, 1974 (review), 1976] and additional theoretical considerations (e.g., Gilpin & Diamond, 1976) have helped confirm basic aspects of the model.

Numerous other studies, however, have shown that although size and distance from source areas are certainly important considerations, these factors alone are insufficient to explain species diversity on islands (e.g., Brown & Kodric-Brown, 1977; Gilbert, 1980; Brown, 1986; Case & Cody, 1987; Williamson, 1989). Workers have specifically emphasized the need also to consider habitat diversity (Buckley, 1982; van der Werff, 1983; Boomsma *et al.*, 1987; Quinn, Wilson & Mark, 1987), including island elevation (Buckley, 1985) and niche shifts (Diamond, 1970), 'assembly rules', which suggest that the presence, and hence biology, of certain existing species can help determine successful establishment of new colonizers (this is still controversial, however; e.g., Diamond, 1975; Connor & Simberloff, 1979; Diamond & Gilpin, 1982; Gilpin & Diamond, 1982; Roberts & Stone, 1990), and historical human-introduced factors (Connor & Simberloff, 1978; Crowell, 1986; Kadmon & Pulliam, 1995). A recent attempt has also been made to develop a computer model for simultaneously dealing with several of these factors in hypothetical island situations (Haydon, Radtkey & Pianka, 1993).

Despite acknowledged limitations, the theory of island biogeography has been applied efficaceously to help understand species diversity in several archipelagoes, such as on the western Pacific Islands (Adler & Dudley, 1994), Hawaii (Juvik & Austring, 1979) and the Galapagos Islands (van der Werff, 1983). An archipelago that has not yet been examined within the context of equilibrium theory biogeography, however, is the Juan Fernandez Islands. These islands are located at 33°40′ S latitude off the coast of continental Chile (Figs. 4.1, 5.1). There are two major

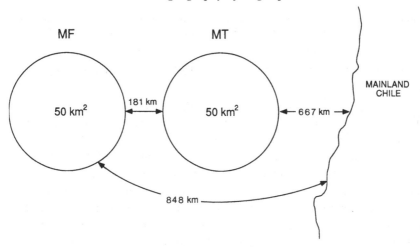

Fig. 5.1. Phytogeographic model of the Juan Fernandez archipelago showing the two major islands (MF = Masafuera; MT = Masatierra, including Santa Clara), their relative sizes, and distances (not to scale) between each other and from the major source area (continental Chile).

islands, Masatierra (48 km²) and Masafuera (50 km²). There is also a third smaller island, Santa Clara (2 km²), lying 1.8 km off the western tip of Masatierra. Radiometric dates for these islands give approximately four million years for Masatierra (and Santa Clara) and one to two million years for Masafuera (Stuessy *et al.*, 1984). In many ways the archipelago is well suited for island biogeographic analysis. The islands are few in number, close to the principal source area in southern South America (Skottsberg, 1956), lie in an almost direct east–west line to the continent and have few vascular plants (383 species). (Recent studies are adding even more introduced species, found especially in and around the only village, San Juan Bautista; Swenson *et al.*, 1997.) Furthermore, previous evolutionary studies have provided considerable information regarding speciation within and between islands (see, for example, Chapter 4).

Over the past 17 years, investigations on patterns and processes of evolution in the endemic vascular plants have been carried out collaboratively by workers at the Ohio State University and the Universidad de Concepción, Chile (for an earlier review, see Stuessy, Sanders & Silva O., 1984). Studies have focused on many dimensions including patterns of phylogeny (Stuessy, Crawford & Marticorena, 1990), chromosomal evolution (Sanders, Stuessy & Rodríguez R., 1983; Spooner *et al.*, 1987; Sun, Stuessy & Crawford, 1990), evolution of secondary products (e.g.,

Pacheco *et al.*, 1985, 1991), phylogeny of selected groups, especially Compositae (Sanders *et al.*, 1987; Crawford, Stuessy & Silva O., 1987; Crawford *et al.*, 1992*a, b*, 1993; Sang *et al.*, 1994, 1995) and biogeography of particular genera (e.g., *Peperomia*; Valdebenito, Stuessy & Crawford, 1990).

The focus of this chapter, therefore, is to examine levels of specific plant diversity within the Juan Fernandez Islands and attempt to understand these observed levels in response to size and distance from source areas plus additional biological, geological, ecological and historical factors derived from previous evolutionary studies. We deal only with angiosperms in this paper because of our greater understanding of them from previous work, and because they disperse with greater difficulty than do ferns and hence show more specific geographic patterns within the archipelago (Stuessy, Crawford & Marticorena, 1990). Specifically, this chapter will: (1) present statistics on the number and distribution of angiosperms in Juan Fernandez; (2) examine the factors that most probably influence species diversity in Juan Fernandez; (3) develop a model that adequately explains species diversity on Masatierra; (4) use that same model to predict species diversity on Masafuera; (5) compare predicted versus observed results in the context of general approaches to island biogeographic models.

Statistics of angiosperms in Juan Fernandez

The basic inventory of the angiosperms of the Juan Fernandez archipelago has been completed through a series of excellent floristic studies by Hemsley (1884), Johow (1896) and Skottsberg (1921, 1951). Our own eight expeditions during 1980–97 have revealed only one new angiosperm species (*Erigeron* sp. nov., H. Valdebenito, personal communication); a new fern, *Gleichenia lepidota*, has also been described (Rodríguez R., 1990). More conspicuous is the absence of taxa cited previously due to disturbance of the environment largely by human intervention (e.g., Sanders, Stuessy & Marticorena, 1982). In fact, recent collecting has revealed an increase in the number of introduced weeds, especially in and around San Juan Bautista (Matthei, Marticorena & Stuessy, 1993; Swenson *et al.*, 1997).

Table 5.1 shows the number and percentage of endemic, native and introduced species of angiosperms on the Juan Fernandez Islands. A total of 104 endemic, 52 native and 227 introduced species yields 383 angiosperms. Most of this diversity is dicotyledonous (82%). Table 5.2

Table 5.1. *Number and percentage of endemic, native and introduced*
species of angiosperms on the Juan Fernandez Islands

	Endemic	Native	Introduced	Total
DICOTS	90 (29%/71%)	29 (9%/36%)	195 (62%/86%)	314 (82%)
MONOCOTS	14 (20%/11%)	23 (33%/30%)	32 (46%/14%)	69 (18%)
TOTALS	104 (27%)	52 (14%)	227 (59%)	383
	156 (41%)			

Table 5.2. *Relationship of different distributional categories of angiosperms*
to occurrence on major islands of the Juan Fernandez archipelago

	MF only	MT only	Both MF and MT
ENDEMIC	35 (50%)	55 (29%)	14 (11%)
NATIVE	15 (21%)	26 (14%)	11 (9%)
INTRODUCED	20 (29%)	110 (58%)	97 (80%)
TOTALS	70	191	122

MF = Masafuera
MT = Masatierra

presents these data with respect to occurrence on the two major islands, Masatierra (including Santa Clara) and Masafuera. The most significant datum is that the highest level of introduced species occurs on Masatierra (110 species on Masatierra vs. 20 on Masafuera; 97 species are found on both islands) which would be expected due to the constant traffic from the mainland and elsewhere. Table 5.3 gives the specific number of endemic families (1), genera (10) and species (104) in the archipelago, yielding levels of endemism for angiosperms of 14% at the generic level and 67% for species.

Obviously, for biogeographic studies dealing with equilibrium concepts, species introduced in historical time must be discounted. This leaves a total of 104 endemic and 52 native angiosperms (156) in the Juan Fernandez archipelago. The endemic taxa are those which have arrived on the islands and speciated whereas the native taxa are those that arrived via natural dispersal means and have yet to change appreciably. In general, species arriving early in the archipelago have a higher

Table 5.3. *Number (and percentage of native flora) of endemic angiosperm taxa on the Juan Fernandez Islands*

	Families	Genera		Species	
DICOTS	1 (8%)	7 (12%) }	14%	90 (58%) }	67%
MONOCOTS	0 (0%)	3 (14%)		14 (9%)	
TOTALS	1	10		104	

probability of speciating while those arriving relatively late will maintain continental features (and genotype).

Factors likely to influence species diversity on Juan Fernandez

Of all the factors instrumental in determining angiosperm species diversity on the Juan Fernandez Islands, clearly size of island and distance from continental source area are important.

These two factors, however, are just part of the more complex set of determinants of species diversity within these two oceanic islands: (1) distance from source area(s); (2) distance between islands; (3) size (area and elevation) of island; (4) rates of extinction; (5) levels of speciation, both within each island and between islands; (6) habitat diversity (relating to niche variation); and (7) age of islands. Each of these factors will be discussed briefly below.

Distance from the source area is relatively straightforward with regard to the main source for propagules for the Juan Fernandez flora, i.e., continental southern South America. The difficulty is that not all of the flora of the islands came from this source. Based on geographical floristic and phylogenetic affinities, an estimate of the sources of the entire flora can be made (see Skottsberg, 1956). The major sources (Table 5.4) for the angiospermous flora can be viewed as coming from Chile (79.6%), the Pacific region (10.2%), the Neotropics (6.5%) and other areas (3.7%). The distance between islands is 181 km.

Size of islands might seem a very simple matter, but two dimensions merit comment. The first is that we are treating the islands as if they were planar, without vertical relief. This is not the case, obviously, and hence the model will be in error to the extent that the islands differ in relief features (relating to habitat diversity) that affect species diversity. The second point is that the present sizes of the islands are most assuredly not

Table 5.4. *Numbers of species (and percentage) of endemic and native angiosperms of the Juan Fernandez Islands that have originated from four general source areas (excluding species which have originated within the archipelago)*

	Endemic	Native	Total	Per cent
PACIFIC	10	1	11	10.2
NEOTROPICS	7	0	7	6.5
CHILE	35	51	86	79.6
OTHER	4	0	4	3.7
TOTALS	56	52	108	100.0

the same as when they were first formed, in the case of Masatierra some four million years ago. It will be important, therefore, to attempt to reconstruct the original size of the islands for proper area considerations at the time of original immigration. Such an attempt has already been provided (Sanders *et al.*, 1987) based on submarine erosional surfaces at 200 m (by analogy to the Hawaiian Islands; Macdonald, Abbott & Peterson, 1983). These estimates lead to calculations of 1092.5 km^2 for the original size of Masatierra when it first arose from the sea floor by volcanic action from a hot spot on the Nazca Plate. Masafuera, on the other hand, due to its relative youth, is estimated to have been approximately 69.2 km^2 at time of origin. Likewise, based on calculated erosional patterns of over four million years for Masatierra, it may have been approximately 90.1 km^2 one million years ago at the time of the appearance of Masafuera.

Differential rates of extinction could well have an impact on the present species diversity on the Juan Fernandez Islands. Because Masatierra is an older island, one might expect greater levels of habitat disappearance than on the younger island, Masafuera, and this might have caused greater levels of extinction. Certainly the former island is now slightly lower than the latter, with 950 vs. 1300 m, respectively, and with much greater estimated loss of area (95% vs. 28%, respectively). It is likely that both were considerably higher at time of their origin, with Masatierra possibly up to 3000 m tall (Sanders *et al.*, 1987). We have estimated previously that perhaps 25–50% of the original number of species on Masatierra has become extinct (Stuessy *et al.*, 1997). Although recognizing the casual nature of these estimates, for purposes of this paper we assume a 30% loss of species diversity on Masatierra over the four million years of its existence.

Due to numerous phyletic studies on genera of the archipelago by us and our collaborators, we can with some certainty provide figures on numbers of speciation events occurring within and between each island in the Juan Fernandez archipelago. The best estimates are that 25 such events occurred within Masatierra and six took place within Masafuera. These calculations assume that within any genus of very closely related species, if more than one endemic species occurs, then the others must have been derived within the island. The number of anagenetic speciation events (Stuessy, Crawford & Marticorena, 1990) between Masatierra and Masafuera is judged to be 17.

Habitat diversity is related directly to the size of island both in area and elevation. At the present time, the existing vegetation zones are quite similar on both islands, with the exception of an 'alpine' zone on Masafuera but not on Masatierra. This is a high altitude (over 1000 m) zone with plants close to the surface of the ground, perhaps as an adaptation to surviving the strong winds and occasional frost (and rarely snow, fide CONAF guides). It is possible that this zone earlier existed on Masatierra but has now become eroded away through inexorable landscape changes due to wind and rain damage, and island subsidence.

Another factor that must be taken into account in assessment of species diversity is the age difference between the two major islands. With Masatierra being approximately four million years old and Masafuera only one to two, they have simply not been available for colonization for the same amount of time. Masatierra has been above the surface of the sea for up to four times as long as Masafuera. Hence, the probability of a propagule arriving on the older island is up to four times greater than the younger island, all other factors being equal.

We also must assume, for purposes of simplicity, that there have been no complicating back migrations from Masafuera to Masatierra, an assumption that in at least one case, *Wahlenbergia berteroi* Hook. & Arn. (Campanulaceae) appears to be incorrect (Lammers, 1996). This taxon has already been excluded from the calculations, however, because it has evolved autochthonously within the archipelago. We also assume that there has been no loss of dispersal ability in propagules of taxa becoming established on Masatierra, which would have limited capability for dispersal to Masafuera. This is no doubt dubious, based on studies showing clear loss of dispersal structures in diaspores of Compositae and other angiosperm families in other archipelagoes (Carlquist, 1966*a, b*).

Model to explain species diversity on Masatierra

The basic approach to understanding species diversity in the Juan Fernandez archipelago in this paper is to develop a model that satisfactorily explains species diversity on one of the islands, Masatierra, and to use that same model to predict the species diversity on the other island, Masafuera. If the factors chosen result in a reasonable approximation of the number of species on Masafuera (which is already known), then these factors can be judged to realistically reveal species diversity in this archipelago, and perhaps in other oceanic archipelagos as well.

There are 156 native and endemic species of angiosperms on both islands of the archipelago, with 81 on Masatierra, 50 on Masafuera and 25 on both islands (i.e., 106 total for Masatierra and 75 for Masafuera). But these figures by themselves do not represent appropriate figures for the purpose of modelling species diversity for either island; modified figures must be developed by adding species to the totals for Masatierra to compensate for presumed loss due to extinction, subtracting differential speciation events within each of the islands and considering the geographic source for immigrants. First, assuming that 30% of species diversity has been lost on Masatierra over the four million years of its existence (especially during the first three million years before Masafuera was born), we must add 35 more species to the 81 ($0.70x = 81$; $x = 116$), plus 25 found on both islands, for an original total of 141 species for Masatierra or 191 for the archipelago. Second, we subtract known numbers of speciation events within each of the islands, as this has no bearing on the success of dispersal and establishment of propagules. Some 25 speciation events are calculated to have occurred within Masatierra and six within Masafuera, which reduces the total species diversity to 160 ($191 - 31 = 160$; 116 for Masatierra and 69 for Masafuera). Finally, we adjust these values in view of the major source of the flora of the islands. For the purpose of calculating probabilities of arrival from a source area to the islands, it is necessary to establish the size of the source flora. Because four major sources for the Juan Fernandez flora exist (Table 5.4) and because some of these areas have no precise numbers of extant species (e.g., Neotropics or Pacific region), it was decided, therefore, to work only with that part of the island flora that came originally from southern South America (i.e., continental Chile, the closest source area) for which the total number of angiosperms is reasonably well known (Marticorena & Quezada, 1985). Some 80% of the flora has come from this region, which gives a further adjusted value of 153 species total (80% of 191 = 153) that need to be considered in the calculations. Of these, 73 are known only from Masatierra (80% of 81 + 35 − 25) and 20 (80% of 25) are found on both islands, for a total of 93 species found on Masatierra that are pertinent for our phytogeographic purposes in this paper. Some 44 species are found only on Masafuera, plus the 20 species found on both islands, yields a total of 64 species.

With these adjusted numbers of species on Masatierra (and for later use on Masafuera), we can attempt a model that will help explain the diversity of species first on this older island. For island equilibrium biogeographic considerations, we assume all diaspores came from the adjacent region of the Chilean continent. Hence, the effective dispersal distance is 667 km. We also know that the total size of the angiosperm flora in Chile has recently been documented at 5684 species (5788 − 104 endemic in Juan Fernandez; Marticorena, 1990). What is lacking is a probability of colonization (Cp) that is based on distance from the source area. Since we already know the adjusted species diversity on Masatierra, we use this to develop Cp values for different distances. We estimate that for each 100 km out into the ocean, the Cp becomes lowered by 0.539 (or 53.9%). This was determined by taking the total species diversity in the source area (5684) and calculating what level of Cp multiplied 6.67 times for 667 km would give the known 93 species. The Cp values of interest, therefore, are for 667 km (distance of Masatierra from the continent; 0.0176), 181 km (the distance between the two islands; 0.345) and 848 km (the distance of Masafuera from the continent; 0.0058).

Prediction of species diversity on Masafuera

Having established a model for calculating species diversity on Masatierra, therefore, we now can attempt to calculate species diversity on Masafuera. These calculations become more involved because several additional factors must be considered.

First, we must consider the original sizes of the two islands and not just their areas as seen today. It has been estimated that Masatierra has lost 95% of its land area during the four million years of its existence (Sanders *et al.*, 1987; Fig. 5.2), whereas Masafuera as a younger island much less (although similar initial percentage loss in the first million years). These calculations can be used to establish probability of the islands as target areas for arriving propagules (Tp). The Tp between the islands now is obviously 1.0, but one million years ago it can be

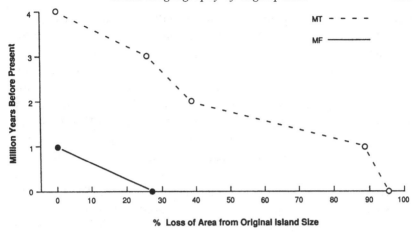

Fig. 5.2. Hypothetical loss of island surface area (in per cent) on both islands of the Juan Fernandez archipelago. Based on data in Sanders *et al.* (1987). MT = Masatierra; MT = Masafuera.

estimated at 0.768 (with Masatierra at 90.1 km^2 and Masafuera 69.2 km^2) in comparison of the former with the latter. Another needed Tp is that between the estimated original size of Masatierra (1092.5 km^2) and that of Masafuera (69.2 km^2) when it formed one million years ago, or Tp = 0.063.

Second, we need to consider different types of dispersal to the archipelago and the influence each might have had on using island size as a significant factor in our calculations. For diaspores arriving to an oceanic island via wind or water, the size of the island is the limiting factor in arrival (e.g., Buckley & Knedlhans, 1986). The larger the island, the higher the probability that the propagule will arrive there. With bird dispersed diaspores (either externally or internally; McAtee, 1947), however, this is not quite the case. Surely a bird finding itself high above the Pacific Ocean and seeking land would deliberately head toward any island, however large or small, and stop to rest. The important point is obviously that the bird initially would see at least some part of the island and the larger the island the greater the probability that it would be seen. With the considerable mobility of birds in flight, however, this seems a much less important factor than land area would be for passively dispersed fruits or seeds by wind or water.

Several studies have attempted to determine the types of dispersal for the flora of the Juan Fernandez Islands. Based on morphological attributes of diaspores in the flora, Johow (1896) estimated that 60% of the species arrived by bird dispersal. Skottsberg (1928) calculated 36% (both ingested and on feathers) but with 43% of the flora listed as 'uncertain'. Carlquist (1974) suggested that 93% of the flora was bird transported (40% ingested; 29% barbs on feathers; 21% mud on feet; 3% through viscid fruits/seeds). Ono (1992) compared these same categories of bird dispersal for the Bonin Islands, Samoa, Hawaii, the Galapagos Islands and Easter Island, and obtained an average 71% for all bird types over all archipelagoes. Perhaps more pertinent would be the per cent bird dispersal in the continental source area rather than on the islands themselves, recognizing that change of dispersability (especially loss) is characteristic of oceanic island floras (Carlquist, 1974). The mainland Chilean flora has not been thoroughly examined in this regard, but Carlquist (1983) has estimated the modes of dispersal of temperate herbs in Mediterranean areas of Chile and California; bird dispersal is calculated at 97%. As for woody plants, Macedo & Prance (1978) determined that bird dispersal accounted for 59% of species in the Caatinga vegetation of Brazil. A correlative study was done by Mori & Brown (1994) who assessed the degree of wind dispersed diaspores in a lowland moist forest in French Guiana (16.2%), a low level compatible with those postulated by others. Hence, although no absolute values exist for source areas in continental Chile, nor for the islands themselves, we suggest that at least 70% of the native and endemic species of the Juan Fernandez Islands might have arrived in the archipelago through bird dispersal.

Calculation of the species diversity on Masafuera, therefore, is as follows. Species diversity = [number of bird dispersed species (70% of the total source area angiosperms, or 3979 species) × the Cp for 848 km (0.0058)] + [number of wind and water dispersed species (30% of source area angiosperms, or 1705 species) × the Cp for 848 km × the Tp relative to Masatierra at four million years before present (0.063)] + [number of species originating via anagenetic speciation from Masatierra (17)] + [the number of bird dispersed species on Masatierra (70% of 93 or 65) × the Cp for 181 km (0.345] + [wind and sea dispersed species on Masatierra (30% of 93 or 28) × Cp for 181 km × Tp relative to Masatierra at 1 mybp (0.768)]. This calculates to 70.5 species expected on Masafuera. The actual adjusted number is 64.

Discussion

Despite the attractive simplicity and predictive quality of MacArthur & Wilson's (1967) equilibrium theory of island biogeography, it has been realized by many workers (e.g., Gilbert, 1980; Williamson, 1989; Wu & Vankat, 1991) that this is a beginning, rather than an end, to understanding species diversity in oceanic island archipelagoes. Difficulties with the MacArthur & Wilson model have led others to attempt more complex models that might be even more explanatory. The most ambitious recent effort has been developed by Haydon, Radtkey & Pianka (1993) involving computer simulations with factors of area and distance from source, and risk of extinction, niche widths, phylogenetic constraints, dispersal abilities and competitive prowess. In particular, they stress the importance of historical events in the development of species diversity within archipelagoes, which are not simply random equilibrium concepts.

On the Juan Fernandez Islands, simple concepts of present-day size of the two islands and distance from the mainland can be used to provide a rough estimate of levels of species diversity now seen. This is, however, a superficial and misleading view of explaining this diversity. The more detailed model presented here attempts to take into account as many factors as realistically can be assessed, while realizing that this attempt is also incomplete. The model estimates 70.5 species on Masafuera whereas there really are 64 (adjusted value), i.e., the model estimates 10% more species than actually are found on the younger island. That the values are close suggests some degree of confidence in the incorporation of factors used in the analysis. Our model suffers from the use of several estimated values (such as levels of extinction on Masatierra) and from not dealing effectively with the obvious importance of habitat differentiation. Although one sees similar habitats on the two islands at the present time, it is suspected that ecological change has occurred differentially on the two islands. It is likely that Masatierra has suffered much more ecological alteration due to the impact of greater surface area loss. Also, in modern historical time, especially during the 19th century, Masatierra has continued to be lumbered (Woodward, 1969), which no doubt has also contributed to the reduction of species diversity.

The most instructive aspect of the model is to allow us to examine the phytogeographic contributions of its different components. Several points are noteworthy. First, despite the very large numbers of species in continental Chile, few propagules have the likelihood of arriving directly on Masafuera. This is especially true of taxa dispersing by water or wind (only 0.33 species). Bird dispersed species are essentially the only taxa (12 species) that would realistically have a probability of arriving on the younger and further island. This is concordant with observed high morphological adaptations for bird dispersal of taxa now on the islands (see earlier discussion). Second, the most important colonizers (33 species; more than one-half of the adjusted total number) arrive as propagules from the older island, Masatierra, either through animal (i.e., bird; 25 species) or abiotic (wind and ocean currents; eight species) vectors. Third, the second most influential colonization comes via dispersal from Masatierra and subsequent anagenetic speciation (17 species). The overall pattern, therefore, is that nearly 80% of successful colonization and establishment comes from Masatierra. This correlates very well with our own observations through previous evolutionary studies, in which we know of only one genus, *Erigeron* (Compositae) with six endemic species (Valdebenito *et al.*, 1992), that appears to have successfully dispersed to, colonized upon and speciated within Masafuera.

An important message of this chapter is to stress the need to consider other physical and biological factors in attempting to determine species diversity in oceanic archipelagoes. Because oceanic islands have a short life span of not more than six million years, they erode quickly and change shape due to subsidence, and erosion by wind, rain and wave action. Renewed volcanic activity can also play a role in modifying landscape in a very young island (such as seen now on the big island of Hawaii). Speciation is also very important in adding to species diversity, as discussed more fully for the Juan Fernandez archipelago in Chapters 3 and 4 of this book.

Acknowledgements

We acknowledge with gratitude: the National Science Foundation for support from grant nos. INT-7721637, BSR-8306436, BSR-8906988 and DEB-9500499 which have made our investigations on the Juan Fernandez Islands possible; CONAF of Chile for permission to carry out field work in the Robinson Crusoe Islands National Park; and the many CONAF guides without whom it would have been impossible to complete our field studies, e.g., Alfonso Andauer, Oscar Chamorro, Miguel García, Bernardo López and Ramón Schiller.

Literature cited

Adler, G. H. & Dudley, R. (1994). Butterfly biogeography and endemism on tropical Pacific islands. *Biological Journal of the Linnean Society*, **51**, 151–62.

Boomsma, J. J., Mabelis, A. A., Verbeek, M. G. M. & Los, E. C. (1987). Insular biogeography and distribution ecology of ants on the Frisian islands. *Journal of Biogeography*, **14**, 21–37.

Brown, J. H. (1986). Two decades of interaction between the MacArthur–Wilson model and the complexities of mammalian distributions. *Biological Journal of the Linnean Society*, **28**, 231–51.

Brown, J. H. & Kodric-Brown, A. (1977). Turnover rates in insular biogeography: effect of immigration on extinction. *Ecology*, **58**, 445–9.

Buckley, R. C. (1982). The habitat-unit model of island biogeography. *Journal of Biogeography*, **9**, 339–44.

Buckley, R. C. (1985). Distinguishing the effects of area and habitat type on island plant species richness by separating floristic elements and substrate types and controlling for island isolation. *Journal of Biogeography*, **12**, 527–35.

Buckley, R. C. & Knedlhans, S. B. (1986). Beachcomber biogeography: interception of dispersing propagules by islands. *Journal of Biogeography*, **13**, 69–70.

Carlquist, S. (1966a). The biota of long-distance dispersal. II. Loss of dispersability in Pacific Compositae. *Evolution*, **20**, 30–48.

Carlquist, S. (1966b). The biota of long-distance dispersal. III. Loss of dispersability in the Hawaiian flora. *Brittonia*, **18**, 310–35.

Carlquist, S. (1974). *Island Biology*. New York: Columbia University Press.

Carlquist, S. (1983). Intercontinental dispersal. *Sonderbände des Naturwissenschaftlichen Vereins in Hamburg*, **7**, 37–47.

Case, T. J. & Cody, M. L. (1987). Testing theories of island biogeography. *American Scientist*, **75**, 402–11.

Conner, E. F. & Simberloff, D. (1978). Species number and compositional similarity of the Galápagos flora and avifauna. *Ecological Monographs*, **48**, 219–48.

Conner, E. F. & Simberloff, D. (1979). The assembly of species communities: chance or competition? *Ecology*, **60**, 1132–40.

Crawford, D. J., Stuessy, T. F., Cosner, M. B., Haines, D. W., Silva O., M. & Baeza, M. (1992a). Evolution of the genus *Dendroseris* (Asteraceae: Lactuceae) on the Juan Fernandez Islands: evidence from chloroplast and ribosomal DNA. *Systematic Botany*, **17**, 676–82.

Crawford, D. J., Stuessy, T. F., Haines, D. W., Cosner, M. B., Silva O., M. & Lopez, P. (1992b). Allozyme diversity within and divergence among four species of *Robinsonia* (Asteraceae: Senecioneae), a genus endemic to the Juan Fernandez Islands, Chile. *American Journal of Botany*, **79**, 962–6.

Crawford, D. J., Stuessy, T. F., Cosner, M. B., Haines, D. W. & Silva O., M. (1993). Ribosomal and chloroplast DNA restriction site mutations and the radiation of *Robinsonia* (Asteraceae: Senecioneae) on the Juan Fernandez Islands. *Plant Systematics and Evolution*, **184**, 233–9.

Crawford, D. J., Stuessy, T. F. & Silva O., M. (1987). Allozyme divergence and the evolution of *Dendroseris* (Compositae: Lactuceae) on the Juan Fernandez Islands. *Systematic Botany*, **12**, 435–43.

Crowell, K. L. (1986). A comparison of relict versus equilibrium models for insular mammals of the Gulf of Maine. *Biological Journal of the Linnean Society*, **28**, 37–64.

Diamond, J. M. (1970). Ecological consequences of island colonization by southwest Pacific birds. I. Types of niche shifts. *Proceedings of the National Academy of Sciences USA*, **67**, 529–36.

Diamond, J. M. (1975). Assembly of species communities. In *Ecology and Evolution of Communities*, ed. M. L. Cody & J. M. Diamond, pp. 342–444. Cambridge: Cambridge University Press.

Diamond, J. M. & Gilpin, M. E. (1982). Examination of the 'null' model of Connor and Simberloff for species co-occurrences on islands. *Oecologia*, **52**, 64–74.

Gilbert, F. S. (1980). The equilibrium theory of island biogeography: fact or fiction? *Journal of Biogeography*, **7**, 209–35.

Gilpin, M. E. & Diamond, J. M. (1976). Calculation of immigration and extinction curves from the species-area-distance relation. *Proceedings of the National Academy of Sciences USA*, **73**, 4130–4.

Gilpin, M. E. & Diamond, J. M. (1982). Factors contributing to non-randomness in species co-occurrences on islands. *Oecologia*, **52**, 75–84.

Haydon, D., Radtkey, R. R. & Pianka, E. R. (1993). Experimental biogeography: interactions between stochastic, historical and ecological processes in a model archipelago. In *Species Diversity in Ecological Communities: Historical and Geographical Perspectives*, ed. R. R. Ricklefs & D. Schluter, pp. 117–30. Chicago: University of Chicago Press.

He, F. & Legendre, P. (1996). On species-area relations. *American Naturalist*, **148**, 719–37.

Hemsley, W. B. (1884). Report on the botany of Juan Fernandez, the south-eastern Moluccas and the Admiralty Islands. In *Report on the Scientific Results of the Voyage of HMS Challenger During the Years 1873-76*, vol. 1, part 3, *Botany*, ed. C. W. Thomson & J. Murray, pp. 1-96. London: HMSO.

Johow, F. (1896). *Estudios sobre la Flora de las Islas de Juan Fernandez*. Santiago: Gobierno de Chile.

Juvik, J. O. & Austring, A. P. (1979). The Hawaiian avifauna: biogeographic theory in evolutionary time. *Journal of Biogeography*, **6**, 205–24.

Kadmon, R. & Pulliam, H. R. (1995). Effects of isolation, logging and dispersal on woody-species richness of islands. *Vegetatio*, **116**, 63–8.

Lammers, T. (1996). Phylogeny, biogeography and systematics of the *Wahlenbergia fernandeziana* complex (Campanulaceae: Campanuloideae). *Systematic Botany*, **21**, 397–415.

MacArthur, R. H. & Wilson, E. O. (1967). *The Theory of Island Biogeography*. Princeton: Princeton University Press.

Macedo, M. & Prance, G. T. (1978). Notes on the vegetation of Amazonia II. The dispersal of plants in Amazonian white sand campinas: the campinas as functional islands. *Brittonia*, **30**, 203–15.

Macdonald, G. A., Abbott, A. T. & Peterson, F. L. (1983). *Volcanoes in the Sea: The Geology of Hawaii*, 2nd edn. Honolulu: University of Hawaii.

Marticorena, C. (1990). Contribución a la estadística de la flora vascular de Chile. *Gayana, Botánica*, **47**, 85–113.

Marticorena, C. & Quezada, M. (1985). Catálogo de la flora vascular de Chile. *Gayana Botánica*, **42**, 1–155.

Matthei, O., Marticorena, C. & Stuessy, T. F. (1993). La flora adventicia del archipiélago de Juan Fernández. *Gayana Botánica*, **50**, 69–102.

McAtee, W. L. (1947). Distribution of seeds by birds. *American Midland Naturalist*, **38**, 214–23.

Mori, S. A. & Brown, J. L. (1994). Report on wind dispersal in a lowland moist forest in central French Guiana. *Brittonia*, **46**, 105–25.

Ono, M. (1991). Flora of the Bonin Islands: endemism and dispersal mode. *Aliso*, **13**, 95–105.

Pacheco, P., Crawford, D. J., Stuessy, T. F. & Silva O., M. (1985). Flavonoid evolution in *Robinsonia* (Compositae) of the Juan Fernandez Islands. *American Journal of Botany*, **72**, 989–98.

Pacheco, P., Crawford, D. J., Stuessy, T. F. & Silva O., M. (1991). Flavonoid evolution in *Dendroseris* (Compositae, Lactuceae) from the Juan Fernandez Islands, Chile. *American Journal of Botany*, **78**, 534–43.

Quinn, S. L., Wilson, J. B. & Mark, A. F. (1987). The island biogeography of Lake Manapouri, New Zealand. *Journal of Biogeography*, **14**, 569–81.

Roberts, A. & Stone, L. (1990). Island-sharing by archipelago species. *Oecologia*, **83**, 560–7.

Rodríguez R., R. (1990). *Gleichenia lepidota* n. sp. y la familia Gleicheniaceae del archipiélago de Juan Fernández, Chile. *Gayana Botánica*, **47**, 37–45.

Sanders, R. W., Stuessy, T. F. & Marticorena, C. (1982). Recent changes in the flora of the Juan Fernandez Islands, Chile. *Taxon*, **31**, 284–9.

Sanders, R. W., Stuessy, T. F., Marticorena, C. & Silva O., M. (1987). Phytogeography and evolution of *Dendroseris* and *Robinsonia*, tree-Compositae of the Juan Fernandez Islands, Chile. *Opera Botanica*, **92**, 195–215.

Sanders, R. W., Stuessy, T. F. & Rodríguez R., R. (1983). Chromosome numbers from the flora of the Juan Fernandez Islands. *American Journal of Botany*, **70**, 799–810.

Sang, T., Crawford, D. J., Kim, S. C. & Stuessy, T. F. (1994). Radiation of the endemic genera *Dendroseris* (Asteraceae) on the Juan Fernandez Islands: evidence from sequences of the ITS regions of nuclear ribosomal DNA. *American Journal of Botany*, **81**, 1494–1501.

Sang, T., Crawford, D. J., Stuessy, T. F. & Silva O., M. (1995). ITS sequences and the phylogeny of the genus *Robinsonia* (Asteraceae). *Systematic Botany*, **20**, 55–64.

Simberloff, D. S. (1974). Equilibrium theory of island biogeography and ecology. *Annual Review of Ecology and Systematics*, **5**, 161–82.

Simberloff, D. S. (1976). Experimental zoogeography of islands: effects of island size. *Ecology*, **57**, 629–48.

Simberloff, D. S. & Wilson, E. O. (1969). Experimental zoogeography of islands: the colonization of empty islands. *Ecology*, **50**, 278–89.

Simberloff, D. S. & Wilson, E. O. (1970). Experimental zoogeography of islands. A two-year record of colonization. *Ecology*, **51**, 934–7.

Skottsberg, C. (1921). The phanerogams of the Juan Fernandez Islands. In *The Natural History of Juan Fernandez and Easter Island*, vol. 2, *Botany*, ed. C. Skottsberg, pp. 95–240. Uppsala: Almqvist & Wiksells.

Skottsberg, C. (1928). Pollinationsbiologie und Samenverbreitung auf den Juan Fernandez-Inseln. In *The Natural History of Juan Fernandez and Easter Island*, vol. 2, *Botany*, ed. C. Skottsberg, pp. 503–47. Uppsala: Almqvist & Wiksells.

Skottsberg, C. (1951). A supplement to the pteridophytes and phanerogams of the Juan Fernandez Islands. In *The Natural History of Juan Fernandez and Easter Island*, vol. 2, *Botany*, ed. C. Skottsberg, pp. 763–92. Uppsala: Almqvist & Wiksells.

Skottsberg, C. (1956). Derivation of the flora and fauna of Juan Fernandez and Easter Islands. In *The Natural History of Juan Fernandez and Easter Island*, vol. 2, *Botany*, ed. C. Skottsberg, pp. 193–405. Uppsala: Almqvist & Wiksells.

Spooner, D., Stuessy, T. F., Crawford, D. J. & Silva O., M. (1987). Chromosome numbers from the flora of the Juan Fernandez Islands. II. *Rhodora*, **89**, 351–6.

Stuessy, T. F., Crawford, D. J. & Marticorena, C. (1990). Patterns of phylogeny in the endemic vascular flora of the Juan Fernandez Islands, Chile. *Systematic Botany*, **15**, 338–46.

Stuessy, T. F., Foland, K. A., Sutter, J. F., Sanders, R. W. & Silva O., M. (1984). Botanical and geological significance of potassium–argon dates from the Juan Fernandez Islands. *Science*, **255**, 49–51.

Stuessy, T. F., Sanders, R. W. & Silva O., M. (1984). Phytogeography and evolution of the flora of the Juan Fernandez Islands: a progress report. In *Biogeography of the Tropical Pacific*, ed. F. J. Radovsky, S. H. Sohmer & P. H. Raven, pp. 55–69. Honolulu: Association of Systematics Collections and Bishop Museum.

Stuessy, T. F., Swenson, U., Crawford, D. J., Anderson, G. & Silva O., M. (1997). Plant conservation in the Juan Fernandez archipelago, Chile. *Aliso*, **16**, 89–102.

Sun, B. Y., Stuessy, T. F. & Crawford, D. J. (1990). Chromosome counts from the flora of the Juan Fernandez Islands, Chile. III. *Pacific Science*, **44**, 258–64.

Swenson, U., Stuessy, T. F., Baeza, M. & Crawford, D. J. (1997). New and historical plant introductions, and potential pests, in the Juan Fernandez Islands, Chile. *Pacific Science*, **51**, 233–53.

Valdebenito, H. A., Stuessy, T. F. & Crawford, D. J. (1990). A new biogeographic connection between islands in the Atlantic and Pacific Oceans. *Nature*, **347**, 549–50.

Valdebenito, H. A., Stuessy, T. F., Crawford, D. J. & Silva O., M. (1992). Evolution of *Erigeron* (Compositae) in the Juan Fernandez Islands, Chile. *Systematic Botany*, **17**, 470–80.

van der Werff, H. (1983). Species number, area and habitat diversity in the Galapagos Islands. *Vegetatio*, **54**, 167–75.

Williamson, M. (1989). The MacArthur and Wilson theory today: true but trivial. *Journal of Biogeography*, **16**, 3–4.

Woodward, R. L. (1969). *Robinson Crusoe's Island: A History of the Juan Fernandez Islands*. Chapel Hill: University of North Carolina Press.

Wu, J. & Vankat, J. L. (1991). A system dynamics model of island biogeography. *Bulletin of Mathematical Biology*, **53**, 911–40.

Part three

Southern and western Pacific Islands

Introduction

There are many oceanic islands scattered in the huge area of the western and southern Pacific Ocean. Their origin is diverse; most of these islands are volcanic, but some of them are coral reefs and/or even of continental origin, having previously been connected to other land masses. The flora of these islands is also diverse, with floristic elements of tropical South East Asia, including Taiwan and the Philippines, of East Asia such as China, the Korean Peninsula and the Japanese Archipelago, and of Polynesia or Australia. Among these areas, remarkable plant speciation and evolution has occurred on the Bonin Islands (Japan), Ullung Island (Republic of Korea) and on some South Pacific Islands such as Fiji, New Caledonia, Samoa, the Society Islands, the Solomon Islands (including Vanuatu) and the Marquesas Islands. These archipelagos provide the focus for this part of the book.

Chapters 6 to 8 discuss aspects of the Bonin (Ogasawara) Islands, located 26°30' N–27°40' N and 142°00' E–142°15' E. The area occupied by the islands is very small; even the largest island is only c. 24 km^2. They are isolated from any other continental or large land mass such as Taiwan, the Philippines or the Japanese mainland. The flora of the Bonin Islands consists of floristic elements of tropical South East Asia including Taiwan and the Ryukyu (Okinawa) Archipelago, of continental East Asia such as China, the Korean Peninsula and the Japanese mainland, and even of the Polynesian Islands and Hawaii (e.g., *Santalum* and *Metrosideros*). Genetic diversity within some endemic genera of flowering plants and their origins are discussed in Chapter 6. Cryptic sex expression in the endemic species of *Callicarpa* (Verbenaceae) is described and discussed in Chapter 7. In Chapter 8 are reported the present status of endangered plant taxa on the Bonin Islands and factors impinging upon them.

In Chapter 9 are discussed examples of morphological and chromosomal variation on Ullung Island, which is a volcanic island located in the Japan (East) Sea, about 150 km east of mainland South Korea. It is an isolated oceanic island having never been connected to the continent and it is geologically young, less than two million years old (Kim, 1985). The flora of Ullung Island contains about 500 species of vascular plants considered as native (Lee & Yang, 1981), with close floristic relationships to Korea and Central Japan (Nakai, 1919). The flora contains, however, 37 endemic taxa showing nearly 10% specific endemism, which is high considering its location only 150 or 300 km from Korea and Japan, respectively, and its youthful geological age.

A remarkable example is described and discussed in Chapter 10 of differentiation and evolution both in morphological and molecular phylogenetic features in *Crossostylis* (Rhizophoraceae) of the South Pacific Islands. The distribution of this characteristic inland mangrove genus extends over New Caledonia, the Society Islands, the Solomon Islands (including Vanuatu), the Samoa Islands, Fiji and the Marquesas Islands.

Literature cited

Kim, K. Y. (1985). Petrology of Ulreung volcanic island, Korea. Geology. *Journal Japanese Association Mines, Petrology, and Economic Geology*, **80**, 128–35.

Lee, W. T. & Yang, I. S. (1981). The flora of Ulreung and Dogdo Island. In *A Report on the Scientific Survey of the Ulreung and Dogdo Islands*, ed. Anonymous, pp. 61–95. Seoul: The Korean Association for Conservation of Nature.

Nakai, T. (1919). *Report on the Vegetation of the Island Ooryongto or Dagelet Island, Corea*. Seoul: The Government of Chosen. [In Japanese.]

6

Genetic diversity of the endemic plants of the Bonin (Ogasawara) Islands

MOTOMI ITO, AKIKO SOEJIMA AND MIKIO ONO

Abstract

In this chapter we report the results of studies of genetic diversity of plants on the Bonin Islands, which are oceanic islands in the western Pacific, and compare these results with previous studies in other oceanic archipelagoes.

Oceanic island endemics have lowered genetic diversity in comparison with continental taxa. In populations of endemic plants on the Bonin Islands, low genetic diversity is also observed, and population size and heterozygosity are highly correlated.

In plants of the Bonin Islands, some genera show adaptive radiation. We have studied electrophoretically genetic divergence among these taxa. High genetic identities were observed in *Pittosporum* and in some species pairs of both *Crepidiastrum* and *Symplocos*. This result indicates low genetic differentiation among species in spite of their morphological divergence.

All the endemic plants now found on the Bonin Islands are descendants of progenitors that were immigrants in the past and these lineages have been isolated from original ancestral populations. Average genetic identities between the endemics and mainland species are comparable to those between congeneric species (0.63), as estimated by Crawford (1983). We have estimated divergence time of these endemic lineages from ancestral source populations to be 2.25 million years for *Pittosporum*, 3.15 million years for *Crepidiastrum* and 2.10 million years for *Symplocos*. On the other hand, preliminary data of widespread plants show relatively high genetic identity between insular and mainland populations.

141

Several studies have measured the genetic diversity of endemic plants on oceanic islands, such as on the Hawaiian Islands (Lowrey & Crawford, 1985; Helenurm & Ganders, 1985; Witter & Carr, 1988), the Juan Fernandez Islands (Crawford *et al.*, 1987a, 1992) and the Galapagos Islands (Wendel & Percival, 1990). In this chapter we report the results of studies of genetic diversity (via enzyme electrophoresis) of plants on the Bonin Islands, which are oceanic islands in the western Pacific, and compare these results with previous studies in other oceanic archipelagoes.

The Bonin Islands are located in the Pacific Ocean about 1000 km south of Tokyo (Fig. 6.1). This archipelago consists of more than 20 small islands aggregated into two major groups: Chichijima and Hahajima (north to south, respectively.) All have volcanic origins (Asami, 1970).

Every island of the Bonin archipelago is very small and has very low relief, unlike other oceanic islands. Compared to the Hawaiian Islands which are 16 000 km² and reach a height of 4100 m, or the Galapagos Islands which are 7700 km² and 1500 m high, the largest of the Bonin

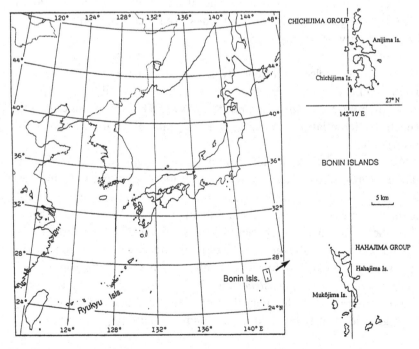

Fig. 6.1. Location of the Bonin (Ogasawara) Islands in the western Pacific.

Islands, Chichijima, is only approximately 24 km^2 and 498 m above sea level. The diversity of the available habitats is therefore much lower on the Bonin Islands, in comparison to these larger oceanic islands in the Pacific.

The level of endemism in the Bonin archipelago is high. In our recent survey (Ono, Kobayashi & Kawakubo, 1986), about 43% of the vascular flora is considered endemic to the archipelago. This high level of endemism is comparable to that of the Galapagos Islands (45%; Porter, 1979), but it is lower than that of the Hawaiian Islands (91%; Wagner, Herbst & Sohmer, 1990). There are only two endemic genera on the Bonin Islands: the monotypic genus *Dendrocacalia* Nakai (Asteraceae) and *Boninia* Planchon (Rutaceae), which includes two species.

The floristic origin of approximately 500 indigenous taxa on the Bonin Islands is complex (Ono, Kobayashi & Kawakubo, 1986); a large component of the flora has affinities with southeastern tropical and subtropical Asia (*c.* 50%), such as the Philippines, Taiwan and the Ryukyu Islands (Fig. 6.1). Another major group is related to plants of Japan itself (*c.* 40%), especially its southwestern portions, such as Kyushu, Shikoku or the Izu Islands. Finally, other taxa have affinities with the Mariana Islands' flora, including several widespread pantropical species (*c.* 10%).

In this chapter, we discuss electrophoretic studies of endemic species of the following three genera *Crepidiastrum* Nakai (Asteraceae), *Pittosporum* Banks ex Gaertner (Pittosporaceae) and *Symplocos* Jacq. (Symplocaceae). *Crepidiastrum* is a genus with seven species distributed in eastern Asia (Koyama, 1995). They are rosette perennials or shrubs and all have the chromosome number $2n = 10$ (Nishioka, 1956; Ono & Nagai, 1958; Ono, 1975). Three species of *Crepidiastrum* occur on the Bonin Islands and all are endemic (Table 6.1). Two species, *C. ameristo-*

Table 6.1. *Genera that contain more than two species on the Bonin Islands*

Genus	Number of Species
Pittosporum	4
Callicarpa	3
Crepidiastrum	3
Ficus	3
Ilex	3
Machilus	3
Symplocos	3

phyllum (Nakai) Nakai and *C. linguaefolium* (A. Gray) Nakai, are arborescent and grow to 2 m. The third species, *C. grandicollum* Nakai is a small perennial herb of less than 50 cm. The two arborescent species have white petals, whereas *C. grandicollum* has yellow petals, similar to the remaining four species of this genus.

Pittosporum includes about 160 species (Ohba, 1989), four of which are endemic to the Bonin Islands (Table 6.1). They are distinguished mainly by fruit and leaf morphology, but each also has a characteristic habitat: *P. beecheyi* Tuyama grows in coastal regions of the Hahajima Islands; *P. boninense* Koidzumi grows in mountain forests; *P. chichijimense* Nakai et Tuyama lives in sparse inland forests; and *P. parvifolium* Hayata occurs on the hill tops of dry dwarf forests. This latter species is critically endangered, with only two individuals known from the wild. There is no variation in chromosome number; all species are $2n = 24$ (Ono & Masuda, 1981).

Symplocos includes approximately 400 species (Soejima *et al.*, 1994). They are evergreen or deciduous trees that are distributed mainly in tropical and subtropical regions. Three species of *Symplocos* occur on the Bonin Islands (Table 6.1). All of them are endemic to the islands and are restricted to very small areas. Two of them, *S. kawakamii* Hayata and *S. pergracilis* (Nakai) Yamazaki, are endemic to the Chichijima and Anijima Islands (Chichijima Group; Fig. 6.1). *Symplocos boninensis* Rehd. et Wils. is distributed only on the small island of Mukōjima, located near Hahajima (Fig. 6.1). All of these three endemics belong to sect. *Palaeosymplocos*, which is distributed in Indo-Malaysia and eastern Asia.

Genetic diversity within populations of endemic plants

It is now well known that in general oceanic island endemics have lowered genetic diversity in comparison with continental taxa (Crawford, Whitkus & Stuessy, 1987*b*; Crawford *et al.*, 1992). Low genetic diversity is observed in populations of endemic plants of the Bonin Islands (Table 6.2).

In *Crepidiastrum*, *Pittosporum* and *Symplocos*, mean expected heterozygosity is 0.028, 0.053 and 0.040 in the Bonin populations, and 0.074, 0.084 and 0.179 in mainland populations, respectively. The average value of the outcrossed flowering plants is estimated as 0.086 (Gottlieb, 1981). The present values observed in mainland populations are comparable. On the other hand, those in the Bonin populations are very low.

Table 6.2. *Parameters of genetic diversity of the populations from oceanic islands and the mainland*

Species names followed by localities in parentheses are continental

Species	% polymorphic loci	Number of alleles per polymorphic locus	Number of alleles per locus	Expected Heterozygosity
Crepidiastrum[a]				
C. grandicollum	11.8	2.00	1.12	0.020
C. linguaefolium-1	17.6	2.00	1.18	0.014
C. linguaefolium-2	23.5	2.00	1.24	0.051
C. platyphyllum (Kanagawa)	5.9	2.00	1.06	0.030
C. lanceolatum (Okinawa)	35.3	2.09	1.38	0.105
C. lanceolatum (Amami)	35.3	2.17	1.41	0.092
C. keiskeanum (Wakayama)	35.3	2.17	1.41	0.091
C. keiskeanum (Kochi-1)	17.6	2.00	1.18	0.059
C. keiskeanum (Kochi-2)	23.5	2.25	1.29	0.065
Pittosporum[b]				
P. parvifolium	10.0	2.00	1.10	0.046
P. chichijimense	0.0	–	1.00	0.000
P. beecheyi	30.0	2.17	1.35	0.049
P. boninense-1	37.5	2.20	1.45	0.024
P. boninense-2	35.0	2.14	1.40	0.089
P. tobira (Chiba)	32.0	2.15	1.37	0.084
P. tobira (Kagoshima)	30.0	2.00	1.30	0.090
P. tobira (Okinawa)	25.0	2.40	1.35	0.068
P. tobira (Amami)	50.0	2.20	1.60	0.113
P. tobira (Ishigaki)	30.0	2.17	1.35	0.086
Symplocos[c]				
S. boninensis	12.5	2.00	1.13	0.038
S. kawakamii	0.0	–	1.00	0.000
S. pergracilis	18.8	2.67	1.31	0.081
S. kuroki (Fukuoka)	30.8	2.50	1.46	0.124
S. nakaharae (Okinawa-1)	53.8	2.57	1.85	0.240
S. nakaharae (Okinawa-2)	46.2	2.67	1.77	0.174
Robinsonia[d]				
R. evenia	33.3	2.20	1.40	0.071
R. gayana	33.3	2.67	1.53	0.145
R. gracilis	57.1	2.38	1.79	0.177
R. thurifera	14.3	2.00	1.14	0.033

Table 6.2. *(cont.)*

Species	% polymorphic loci	Number of alleles per polymorphic locus	Number of alleles per locus	Expected Heterozygosity
Hawaiian Madiinae[e]	24.0	2.20	1.29	0.075
Tetramolopium[f]	7.0	2.13	1.09	< 0.01
Bidens[g]	12.0	2.47	1.16	0.045

Source: Data from: [a]Ito & Ono, 1990; [b]Ito, Soejima & Ono, 1997; [c]Soejima *et al.*, 1994; [d]Crawford *et al.*, 1992; [e]Witter & Carr, 1988; [f]Lowrey & Crawford, 1985; [g]Helenurm & Ganders, 1985

There are many factors that affect genetic diversity (Nei, 1987). Among them, population size, bottleneck effect and breeding system are considered to be important in oceanic island endemics, excluding enzyme factors such as molecular weight, subunit structure or selection pressure. All species examined are estimated as outbreeders.

Because oceanic islands are separated from other land masses by the ocean, dispersal to islands is difficult. Long-distance dispersal to oceanic islands, therefore, should be a rare event and founder populations on these islands should be small in size. This implies that only a small portion of the total genetic diversity of the ancestral populations will arrive and become established on the islands; hence a bottleneck effect will have occurred. According to computer simulations, such a decrease of genetic diversity in founder populations is expected to last for hundreds of thousands of years even after the recovery of population size (Nei, Maruyama & Roychoudhury, 1975). Population sizes in oceanic island endemics, however, often are small. The expected heterozygosity (He) or gene diversity is given by $He = 4Nv/(1 + 4Nv)$ where N and v are effective population size and mutation rate, respectively. If the mutation rate is constant, it is expected that a large population will have a high heterozygosity. In oceanic island endemics, small population sizes should result in lowered heterozygosity. On the Bonin islands, populations of some endemic plants are very small, even to the extent of the species being endangered (e.g., *Pittosporum parvifolium*). To examine population size and heterozygosity in endemic plants of the Bonin Islands, population size was estimated for each population analysed electrophoretically and plotted against expected heterozygosity (Fig. 6.2). The result shows a high

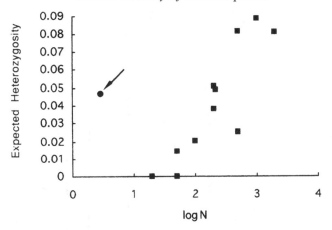

Fig. 6.2. Relationships between population size (logarithm; horizontal axis) and expected heterozygosity (vertical axis) for each population of Bonin Island endemics. Solid circle (arrow): population of *Pittosporum parvifolium.*

correlation between these two factors, except in *Pittosporum parvifolium.* Only a few individuals of *P. parvifolium* are known on the islands at this time, but another population with more than ten individuals did exist ten years ago before being destroyed by construction (Ono, Kobayashi & Kawakubo, 1986).

Genetic diversity among congeneric island species

Adaptive radiation is the evolution of many related species into different ecological zones within a relatively short time (Futuyma, 1986). Since the early studies of Charles Darwin (1859), it has been shown that adaptive radiation is a conspicuous component of evolution on oceanic islands. In plants of the Bonin Islands, some genera do indeed show adaptive radiation. However, as mentioned above, the Bonin Islands have a very small area and low environmental diversity; only a few species have evolved from an ancestor in the archipelago (Table 6.1). We have studied electrophoretically the genetic divergences among these taxa.

In *Crepidiastrum*, which has three species endemic to the Bonin Islands, *C. ameristophyllum* and *C. linguaefolium* are genetically very similar, showing 0.992 identity. This value is very high and is nearly the same as those observed between conspecific populations (Gottlieb, 1977, 1981; Crawford, 1983). In contrast, *C. grandicollum* has relatively low genetic identity with the former two species; 0.766 with *C. ameristophyllum* and 0.750 with *C. linguaefolium* (Fig. 6.3, Table 6.3). In *Pittosporum*, genetic

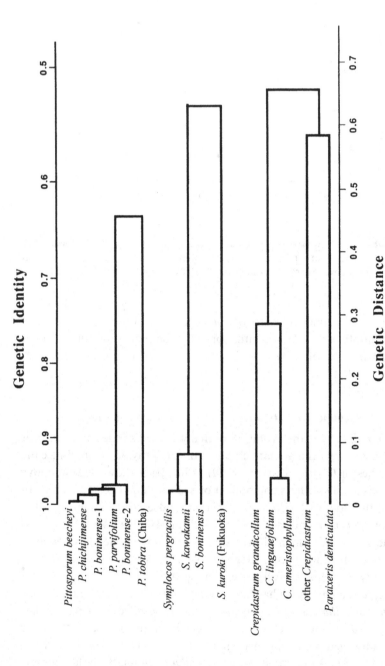

Fig. 6.3. UPGMA (unweighted pair-group method with arithmetic mean) dendrogram of electrophoretic data showing genetic identities and distances among species of *Crepidiastrum*, *Pittosporum* and *Symplocos*.

Table 6.3. *Average genetic identities among congeneric species on oceanic islands*

n = number of populations examined; s = number of species. Numbers in parentheses are ranges of identities.

Island/Genus	Average genetic identity	Reference
Bonin Islands		
Crepidiastrum (n = 3, s = 3)	0.836 (0.750–0.992)	Ito & Ono, 1990
Ficus (n = 2, s = 2)	0.968	Ito *et al.*, unpublished
Machilus (n = 6, s = 3)	0.994 (0.988–0.997)	Ito *et al.*, unpublished
Pittosporum (n = 5, s = 4)	0.982 (0.970–0.999)	Ito, Soejima & Ono, 1997
Symplocos (n = 3, s = 3)	0.95 (0.93–0.97)	Soejima *et al.*, 1994
Hawaiian Islands		
Silversword alliance:		Witter & Carr, 1988
Wilkesia (n = 2, s = 2)	0.968	
Dubautia (n = 21, s = 18)	0.769 (0.426–0.997)	
chromosome number *n* = 14		
(n = 10, s = 7)	0.713 (0.426–0.973)	
chromosome number *n* = 13		
(n = 9, s = 9)	0.902 (0.724–0.997)	
Bidens (n = 22, s = 11)	0.957 (0.886–0.996)	Helenurm & Ganders, 1985
Tetramolopium (n = 11, s = 6)	0.95 (0.87–1.00)	Lowrey & Crawford, 1985
Juan Fernandez Islands		
Robinsonia (n = 4, s = 4)	0.625 (0.560–0.706)	Crawford *et al.*, 1992

identities among endemic species are also very high; they range from 0.970 to 0.999 with an average of 0.982 (Fig. 6.3, Table 6.3). In *Symplocos*, the genetic identity of 0.97 which exists between *S. pergracilis* and both *S. kawakamii* and *S. boninensis*, is slightly higher than the value of 0.93 between *S. kawakamii* and *S. boninensis* (Fig. 6.3; Table 6.3). The high genetic identities observed in *Pittosporum* and in some species pairs of both *Crepidiastrum* and *Symplocos* indicate low genetic differentiation among species in spite of their morphological divergence. This result is concordant with a hypothesis of recent evolutionary differentiation from Nei's (1972) calculations based on genetic distance.

The relatively low value of genetic identity between *C. grandicollum* and the other two species of this genus on the Bonin Islands may suggest relatively old diversification and speciation, possibly in an early stage of establishment of the Bonin Islands' flora. In *Crepidiastrum*, the genetic distances correspond to the morphological features of these endemic species; *C. ameristophyllum* and *C. linguaefolium* are small shrubs with white flowers, whereas *C. grandicollum* is a perennial herb with yellow flowers (like the rest of the species of the genus). Low genetic diversity among endemic congeneric species also has been reported on other oceanic islands such as Hawaii and the Juan Fernandez Islands (Table 6.3).

Because increasing genetic distance can be correlated with increasing time of populational divergence, we can estimate when the adaptive radiations occurred. The formula is $t = D/2a$; where t is the time since divergence of two populations, D is the genetic distance between them and a is an average rate of codon substitution detectable by electrophoresis (Nei, 1987). In practice, we must decide the parameter a. Here we employ the value $a = 10^{-7}$ which has been suggested by Nei (1987). Applying this formula, divergence times for pairs of congeneric species can be estimated (Table 6.4).

High genetic identities have also been found in endemic plants evolving via adaptive radiation in other oceanic islands (Table 6.3). These results suggest that rapid speciation accompanied by dramatic morphological and ecological differentiation occurs on oceanic islands. However, it is not true that all speciation events are recent. For example, in the Hawaiian silverswords, mean genetic identity is 0.713 among species of *Dubautia* with a chromosome number of $n = 14$. On the other hand, among species of the genus with $n = 13$, mean genetic identity is 0.902 (Witter & Carr 1988). This disparity might suggest different times of

Table 6.4. *Estimation of divergence time among congeneric species, and between them and their putative ancestors*

Genus	Among endemic species		Between endemics and their putative ancestors	
	Genetic identity	Divergence time	Genetic identity	Divergence time
Crepidiastrum	–	–	0.533	3.15 my
C. ameristophyllum – *C. linguaefolium*	0.992	0.05 my	–	–
C. grandicollum – the other endemics	0.750	1.44 my	–	–
Pittosporum	0.970	0.16 my	0.637	2.25 my
Symplocos	0.950	0.51 my	0.675	2.10 my

adaptive radiation, species with $n = 14$ radiating earlier than those species with $n = 13$. Witter & Carr (1988) considered that the $n = 13$ species originated on the younger islands, Hawaii and Maui, and that they had a biogeographic history similar to *Tetramolopium* in which genetic identity is also high (Lowrey & Crawford, 1985). On the Bonin islands, most of the species pairs show high genetic identities and suggest a youthful origin. Differentiation may relate to availability of new habitats, such as dry dwarf forests.

Genetic differentiation between endemics and their putative ancestors

All the endemic plants now found on the Bonin Islands are descendants of progenitors which were immigrants in the past, and these lineages have been isolated from original ancestral populations. Genetic identities and distances have also been calculated between Bonin endemics and their putative ancestors on the Japanese mainland or on the Ryukyu Islands (Table 6.4). In *Pittosporum*, average genetic identity between Bonin endemics and *P. tobira* is 0.637. Both in *Crepidiastrum* and *Symplocos*, the Bonin endemics are considered to be monophyletic assemblages based on electrophoretic studies. Average genetic identities between the endemics and mainland species are 0.533 and 0.675, respectively (Ito & Ono, 1990; Soejima *et al.*, 1994). These values are comparable to those between congeneric species (0.63), as estimated by Crawford (1983). We estimate divergence time of these endemic lineages from ancestral source populations using the formula given above. The result is 2.25 million years for *Pittosporum*, 3.15 million years for *Crepidiastrum* and 2.10 million years for *Symplocos* (Table 6.4).

Genetic diversity of widespread species

In addition to endemic taxa, on the Bonin Islands there are also many plant species that have wide distributions, found far beyond the archipelago and representing successful long-distance colonizers. They are geographically isolated on the islands from source populations and no gene flow has occurred between them. As they are included taxonomically in the same species, they do not differ morphologically from their conspecific relatives. It is of interest to learn whether these populations differ genetically and to what degree. Scanty available data concerning widespread plants on oceanic islands do not allow a clear answer to this question. Preliminary data, however, are shown in Table 6.5. Between

Table 6.5. *Mean genetic identities between conspecific populations on the Bonin Islands and mainland Japan*

Species	Mean genetic identity
Piper kazura (Chois.) Ohwi	0.78
Trachelospermum asiaticum Nakai	0.80

the Bonin Island and mainland Japanese populations of *Piper kazura* (Chois.) Ohwi (Piperaceae) and *Trachelospermum asiaticum* Nakai (Apocynaceae), the results show lowered genetic identities, compared to those of typical congeneric populations in continental areas (Crawford, 1983). Whether this is the general situation is unknown; the species pairs examined were too few. Clearly what is needed is an extensive study of genetic divergence between island and mainland populations at various taxonomic levels, i.e., species, subspecies and variety.

Acknowledgements

The authors thank T. Yahara, J. Yokoyama and D. J. Crawford for giving valuable advice and encouragement. This research was supported in part by a Grant-in-Aid for Scientific Research from the Ministry of Education, Science and Culture, Japan, no. 04640643 to Mikio Ono and Motomi Ito, and a Grant-in-Aid for Natural History from the Fujiwara Natural History Foundation, Botany in 1995 and 1996 to Motomi Ito.

Literature cited

Asami, S. (1970). Topography and geology in the Bonin Islands. In *The Nature of the Bonin Islands*, ed. T. Tuyama & S. Asami, pp. 91-108. Tokyo: Hirokawa Shoten.

Crawford, D. J. (1983). Phylogenetic and systematic inferences from electrophoretic studies. In *Isozymes in Plant Genetics and Breeding*, part A, ed. S. D. Tanksley & T. J. Orton, pp. 257–87. New York: Elsevier.

Crawford, D. J., Stuessy, T. F., Haines, D. W., Cosner, M. B., Silva O., M. & Lopez, P. (1992). Allozyme diversity within and divergence among four species of *Robinsonia* (Asteraceae: Senecioneae), a genus endemic to the Juan Fernandez Islands, Chile. *American Journal of Botany*, **79**, 962–6.

Crawford, D. J., Stuessy, T. F., & Silva O., M. (1987a). Allozyme divergence and the evolution of *Dendroseris* (Compositae: Lactuceae) on the Juan Fernandez Islands. *Systematic Botany*, **12**, 435–43.

Crawford, D. J., Whitkus, R. & Stuessy, T. F. (1987*b*). Plant evolution and speciation on oceanic islands. In *Patterns of Differentiation in Higher Plants*, ed. K. Urbanska, pp. 183–99. London: Academic Press.

Darwin, C. (1859). *On the Origin of Species by Means of Natural Selection*. London: Murray.

Futuyma, D. J. (1986). *Evolutionary Biology*, 2nd edn. Massachusetts: Sinauer.

Gottlieb, L. D. (1977). Electrophoretic evidence and plant systematics. *Annals of the Missouri Botanical Garden*, **64**, 161–80.

Gottlieb, L. D. (1981). Electrophoretic evidence and plant populations. *Progress in Phytochemistry*, **7**, 1–45.

Helenurm, K. & Ganders, F. R. (1985). Adaptative radiation and genetic differentiation in Hawaiian *Bidens*. *Evolution*, **39**, 753–65.

Ito, M. & Ono, M. (1990). Allozyme diversity and the evolution of *Crepidiastrum* (Compositae) on the Bonin (Ogasawara) Islands. *Botanical Magazine Tokyo*, **103**, 449–59.

Ito, M., Soejima, A. & Ono, M. (1997). Allozyme diversity of *Pittosporum* (Pittosporaceae) on the Bonin (Ogasawara) Islands. *Journal of Plant Research*, **111**, 455–62.

Koyama, H. (1995). *Crepidiastrum*. In *Flora of Japan, IIIb*, ed. K. Iwatsuki, T. Yamazaki, D. E. Boufford, & H. Ohba, pp. 13–15. Tokyo: Kodansha.

Lowrey, T. K. & Crawford, D. J. (1985). Allozyme divergence and evolution in *Tetramolopium* (Compositae: Astereae) on the Hawaiian Islands. *Systematic Botany*, **10**, 64–72.

Nei, M. (1972). Genetic distance between populations. *American Naturalist*, **106**, 283–92.

Nei, M. (1987). *Molecular Evolutionary Biology*. New York: Columbia University Press.

Nei, M., Maruyama, T. & Roychoudhury, A. K. (1975). The bottleneck effect and genetic variability in populations. *Evolution*, **29**, 1–10.

Nishioka, T. (1956). Karyotype analysis in Japanese Cichorieae. *Botanical Magazine Tokyo*, **69**, 586–92.

Ohba, H. (1989). Pittosporaceae. In *Wild Flora of Japan: Woody Plant, I.*, ed. Y. Satake, H. Hara, S. Watari, & T. Tominari, pp. 177–8. Tokyo: Heibonsha.

Ono, M. (1975). Chromosome numbers of some endemic species of the Bonin Islands. I. *Botanical Magazine Tokyo*, **88**, 323–8.

Ono, M., Kobayashi, S. & Kawakubo, N. (1986). Present situation of endangered plant species in the Bonin (Ogasawara) Islands. *Ogasawara Research*, **12**, 1–32.

Ono, M. & Masuda, Y. (1981). Chromosome numbers of some endemic species of the Bonin Islands. *Ogasawara Research*, **4**, 1–24.

Ono, F. & Nagai, S. (1958). The hybrid of *Crepidiastrum platyphyllum* and *C. keiskeanum*. *Japanese Journal of Genetics*, **33**, 12–55.

Porter, D. M. 1979. Endemism and evolution of Galapagos Islands vascular plants. In *Plants and Islands*, ed. D. Bramwell, pp. 225–56. London: Academic Press.

Soejima, A., Nagamasu, H., Ito, M. & Ono, M. (1994). Allozyme diversity and the evolution of *Symplocos* (Symplocaceae) on the Bonin (Ogasawara) Islands. *Journal of Plant Reserach*, **107**, 221–7.

Wagner, W. L., Herbst, D. R. & Sohmer, S. H. (1990). *Manual of the Flowering Plants of Hawaii*. Honolulu: University of Hawaii Press; Bishop Museum Press.

Wendel, J. F. & Percival, A. E. (1990). Molecular divergence in the Galapagos Islands – Baja California species pair, *Gossypium klotzchianum* and *G. davidsonii* (Malvaceae). *Plant Systematics and Evolution*, **171**, 99–115.

Witter, M. S. & Carr, G. D. (1988). Adaptive radiation and genetic differentiation in the Hawaiian silversword alliance (Compositae: Madiinae). *Evolution*, **42**, 1278–87.

7

Evolution of cryptic dioecy in *Callicarpa* (Verbenaceae) on the Bonin Islands

NOBUMITSU KAWAKUBO

Abstract

Examination of flower morphology and function of three species of genus *Callicarpa* (Verbenaceae) endemic to the Bonin Islands, *C. glabra*, *C. nishimurae* and *C. subpubescens*, reveals all three species to be dioecious. Male plants have short-styled flowers with sterile ovaries, whereas female plants have long-styled flowers with non-germinating pollen grains. Although the non-germinating pollen grains of female flowers are inaperturate, their size and content are almost the same as the normal 3-colpate pollen grains of male flowers. On the Bonin Islands, the insect fauna for the pollination of *Callicarpa* plants is sparse and the flowers of the three species produce little nectar. The non-germinating pollen grains of the female flowers may have been maintained as a reward for insect pollinators. Such an unusual dioecism has never been reported from Verbenaceae. The sexual system of *Callicarpa* on the Bonin Islands might have evolved under biological conditions unique to small oceanic islands.

The evolution of dioecy in flowering plants has been discussed as one of the interesting phenomena in evolutionary biology (e.g., Darwin, 1877; Lewis, 1942; Ross & Weir, 1976; Charlesworth & Charlesworth, 1978; Bawa, 1980; Ross, 1980; Lloyd, 1980; Charnov, 1982; Willson, 1983). Furthermore, the existence of dioecy in island plants has received considerable attention by several workers (Baker, 1955, 1967; Carlquist, 1974; Bawa & Opler, 1975; Ehrendorfer, 1979; Bawa, 1980, 1982; Baker & Cox, 1984; Cox, 1985).

Concerning the colonization of dioecious plants on islands, Bawa (1980, 1982) examined various ecological attributes of dioecy and suggested that dioecious taxa may have been disproportionately more successful in colonizing islands than hermaphroditic taxa. Carlquist (1974) opined that dioecy appears to have evolved *in situ* in some genera on Hawaii, although Bawa (1980) claimed that many species of the same genera were also dioecious elsewhere. Autochthonously evolved dioecy on islands has not been examined sufficiently in the context of various environments peculiar to islands.

Although the advantage of outcrossing has been suggested as a selective pressure in the evolution of dioecy in island plants (e.g., Carlquist, 1974), no stringent evidence has been brought forward. Other factors that appear to be connected with the evolution of dioecy on islands, such as the influence of pollinator and seed dispersers, must be examined (Bawa, 1982). The relationship between dioecious plants and animals that have an impact on them on islands has not been investigated adequately.

In this context I have attempted to examine the dioecious conditions of *Callicarpa* on the Bonin Islands. Unfortunately, not all aspects of the relationship between plants of *Callicarpa* and their pollinators or seed dispersers on the islands is known at present and therefore selective pressures for the evolution of dioecy in the Bonin *Callicarpa* can not be fully addressed. The curious nature of dioecy in Bonin *Callicarpa*, however, can be examined as an evolutionary product of insular environments.

In this paper, I show the unusual morphological features of dioecy of Bonin *Callicarpa* and then attempt to clarify their functional role. Comments are also added regarding the evolution of dioecy on small islands, in which the pollinator fauna is viewed as a restriction of the evolutionary process toward dioecy rather than as a promoter of it.

Dioecy in *Callicarpa* on the Bonin Islands

The genus *Callicarpa* of the Bonin Islands is composed of three species, *C. glabra* Koidz., *C. nishimurae* Koidz. and *C. subpubescens* Hook. et Arn., which are endemic to the archipelago (Kawakubo, 1986). The dioecious conditions of these three species have been revealed and the non-germinating pollen grains of female flowers noted previously (Kawakubo, 1990). This sex expression was called 'cryptic dioecy' by Mayer & Charlesworth (1991).

Dioecious features of *C. subpubescens* are summarized in Fig. 7.1. Female plants of the species have long pistils and long stamens in their flowers (Fig. 7.1A). The morphology of these flowers is superficially similar to that of hermaphrodites. However, their pollen grains cannot

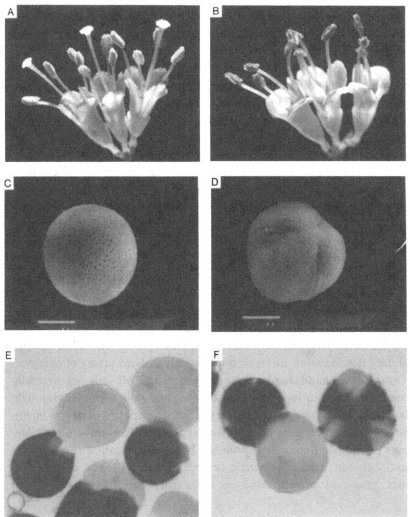

Fig. 7.1. Dioecy in *Callicarpa subpubescens*. Features of female flowers (A, C, E) and male flowers (B, D, F). The female flowers have long pistils and apparently long stamens (A). The anthers of female flowers contain only pseudo-pollen grains (C, bar 10 μm). The pollen grains (E) include generative cells but do not germinate. The male flowers have short pistils (B) but their ovaries do not develop fruits. The short styles are often broken in the bud stage and consequently many male flowers lack styles. The anthers of male flowers contain only normal 3-colpate pollen grains (D) with generative cells (F).

germinate under any conditions and are entirely sterile (Fig. 7.1C), although the ovaries of these flowers develop fruits very easily. On the other hand, male flowers have short pistils and long stamens (Fig. 7.1B). Although the ovaries do not develop fruits, the anthers of these flowers contain normal 3-colpate pollen grains (Fig. 7.1D), typical for the genus. The other two species, *C. glabra* and *C. nishimurae*, show almost the same floral morphology (Kawakubo, 1990).

Notable morphological features of dioecy in Bonin *Callicarpa* are the occurrence of seemingly useless organs in male and female flowers. Male flowers have short pistils, whereas female flowers have long stamens and sterile pollen grains. If these characteristics are actually non-functional, their maintenance would be unexpected in the context of resource allocation theory (Charnov, Smith & Bull, 1976; Charnov, 1982). Male plants do not have the energetic cost of fruit and seed production, and the short pistils of their flowers are probably lower in energetic cost compared with the long pistils of female flowers. Further, styles of the short pistils are often broken off from ovaries before flowering and the remains of a brown, broken style are often observed inside the corolla tube of the male flowers of *C. subpubescens* (see Kawakubo, 1990). The pistils of male flowers, therefore, may be regarded as apparently useless organs and their degeneration seems to result in only a slight saving of energy.

On the other hand, female flowers do not show obvious degeneration of male organs, still retaining long stamens and pollen grains (although sterile). In fact, there is no difference in morphology of the stamens between male and female flowers (Kawakubo, 1990). Numbers of pollen grains per anther were strictly counted, but no significant difference was found between female and male flowers (Table 7.1). The contents of pollen grains at present have not fully been investigated. However,

Table 7.1. *Number of pollen (or pseudo-pollen) grains of* Callicarpa
subpubescens *in male and female flowers*

No significant difference between sexes was found by Mann-Whitney U-test.

Sex expression	No. of pollen or pseudo-pollen grains per stamen (mean: min.~max.)		No. of observations
	Fertile or substantial	Sterile or empty	
male	2558.5 : 1927~2959	81.1 : 7~88	7
female	2414.1 : 1796~3366	36 : 4~254	8

observations of crushed pollen grains stained by acetocarmine reveal that the sterile pollen grains of female flowers always include generative cells (Fig. 7.1E). Morphologically, therefore, the stamens and sterile pollen grains of female flowers barely show degeneration except for the disappearance of germ slits on the pollen wall (Fig. 7.1C).

Sterile pollen grains of cryptic female flowers have been reported for few species (see Mayer & Charlesworth, 1991). A dioecious condition similar to that of Bonin *Callicarpa* has been documented in Australian *Solanum*. Female plants have morphologically hermaphroditic flowers with anthers containing inaperturate non-germinating pollen grains (Anderson & Symon, 1989). Differences between Bonin *Callicarpa* and the Australian S*olanum* can be seen in the characteristics of non-germinating pollen grains of the female flowers. In Australian *Solanum*, the non-germinating pollen grains are smaller than the normal pollen grains of male flowers and many of the former are not stained by aniline blue in lactophenol (Anderson & Symon, 1989).

To avoid confusion in terminology, the stamens and pollen grains of female flowers are described in the following as 'pseudo-stamens' and 'pseudo-pollen grains', respectively.

Pseudo-pollen grains as rewards for pollinators

It is generally known that many flowering plants attract pollinators by the shape, colour and/or smell of flowers, and the flowers receive pollen grains (or disperse them) via the pollinators. At the same time, these pollinators obtain nectar and/or pollen grains as a reward for pollination (e.g., Barth, 1985). As Anderson & Symon (1989) pointed out in the case of Australian *Solanum*, the pseudo-stamens and pseudo-pollen grains of female flowers can be regarded as attractants and rewards, respectively, for pollinators, which also applies to the Bonin *Callicarpa*. Since nectar is found only in low quantity inside the corollas of Bonin *Callicarpa*, the pseudo-pollen grains may be the primary rewards for pollinators that transport normal pollen grains from male to female flowers.

To examine the effectiveness of the pseudo-stamens and pseudo-pollen grains of the female flowers on pollination, pollinator behaviour on 'emasculated' female inflorescences of *C. subpubescens* were observed in the field ('emasculation' in this context means the removal of pseudo-stamens from female flowers). For the observations, two inflorescences on opposite sides of the stem were used (Fig. 7.2). In one inflorescence, the projecting parts of the pseudo-stamens from the corolla tubes

Nobumitsu Kawakubo

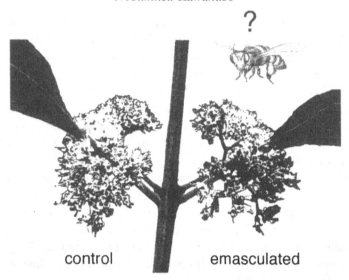

control   emasculated

Fig. 7.2. Method of 'emasculation' test on female flowers of *Callicarpa subpubescens*. All anthers of open flowers were removed (emasculated) on one of two opposite inflorescences after confirmation of normal pollinator visits. Number of pollinator visits and length of stay per visit were observed in each test.

(anthers and upper parts of the filaments) of the open flowers were removed with small scissors ('emasculated') after the confirmation of normal visits of pollinators to the two inflorescences. Pollinator behaviour (the number of pollinator visits and duration of each visit) was recorded for 30 minutes on ten emasculated inflorescences.

Table 7.2. *Comparison in pollinator visits between normal and emasculated inflorescences of* Callicarpa subpubescens.

Sample No.	NORMAL		EMASCULATED		
	No. of visits	Stay length (mean sec.)	No. of visits	Stay length (mean sec.)	Time (a.m.) & date
1	52	9.0	10	5.1	10:50–11:20 3 Jul 87
2	60	11.6	27	4.4	10:50–11:20 3 Jul 87
3	33	12.3	12	3.7	11:38–12:10 3 Jul 87
4	14	19.6	1	1.0	11:38–12:10 6 Jul 87
5	38	22.0	12	3.8	09:45–10:15 6 Jul 87
6	39	17.8	2	1.0	09:45–10:15 6 Jul 87
7	38	15.2	3	6.0	09:23–09:53 7 Jul 87
8	34	15.1	3	5.0	09:23–09:53 7 Jul 87
9	41	14.7	15	3.9	09:05–09:36 8 Jul 87
10	13	22.3	0	–	09:05–09:36 8 Jul 87

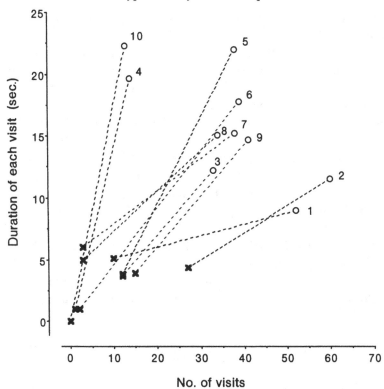

Fig. 7.3. Variation in pollinator visit obtained from emasculations of *Callicarpa subpubescens*. ○ = visits on normal inflorescences. X = visits on emasculated inflorescences. Lines connect paired inflorescences on opposite sides of the same stem. Numbers correspond to sample numbers in Table 7.2.

Artificial effects of emasculation, such as damage to flower parts, were minimized by experience in preliminary experiments. Observations were carried out in a sunny area at the forest edge on the way to Kopepehama beach, the southwest part of Chichijima, in the mornings of 3, 6, 7 and 8 July 1987. According to climatic data from the Chichijima Meteorological Station, on those days, there was almost no wind, and the minimum and maximum temperatures were 21.1 °C and 26.5 °C, respectively.

The total number of pollinators recorded during these observations was over 700, which consisted mostly of honey bees (*Apis mellifera*), although a few carpenter bees (*Xylocopa ogasawaraensis*) were also observed. The results of observations (after emasculations) are shown in Table 7.2 and Fig. 7.3. The number of visits and/or the length of each visit on the emasculated inflorescences were remarkably low as compared with the control (normal). Figure 7.4 shows the changing pattern of pollinator visits through time (Sample No. 7 in Table 7.2). Before emasculation of the two inflorescences, the visits were not significantly different between the two inflorescences. The number of visits and the duration of each visit on one inflorescence, respectively, were 76 and 16.8 seconds (mean), and those on the other inflorescence were 69 and 15.4 seconds (mean) (Fig. 7.4A). After emasculation (Fig. 7.4B), however, pollinator visits were dramatically different between the two inflorescences. Only three bees visited the emasculated inflorescence and quickly flew away (Fig. 7.4C, lower graph), while the control inflorescence was visited by 38 bees (Fig. 7.4C, upper graph). Patterns similar to the above results were also obtained from observations on the other plants of *C. subpubescens* (Table 7.2 and Fig. 7.3). Pseudo-stamens of female flowers seem important for attracting pollinators.

Attractiveness of flowers with pseudo-stamens may be promoted by mass-flowering in the cymose inflorescence. As shown in the upper graph

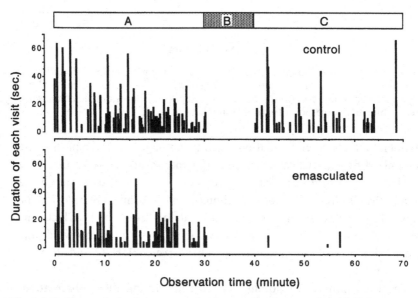

Fig. 7.4. Pollinator behaviour on a pair of normal and emasculated inflorescences of *Callicarpa subpubescens*. Upper: control inflorescence. Lower: emasculated inflorescence. A, Confirmation of normal pollinator visits (first 30 minutes of an observation); B, Emasculation (next 10 minutes); C, Comparison between normal and emasculated inflorescences (last 30 minutes).

of Fig. 7.4, visits by bees to the control inflorescence are reduced (from 76 to 38) after emasculation on the opposite one, although the length of the visits is not significantly different. Hence, pseudo-stamens of female flowers may have to be similar to the normal ones of male flowers, not only in morphology, but also in quantity, to attract pollinators.

On the other hand, pseudo-pollen grains must act as rewards for pollinators. Observations of the pollen basket on the hind legs of honey bees showed that the bees did not distinguish between normal pollen grains of male flowers and pseudo-pollen grains of female flowers. Both types of pollen grains are found equally in pollen baskets of bees (Fig. 7.5).

Cellular contents of pseudo-pollen grains may be important for pollinators. Pollen grains of many flowering plants are recognized as nutritionally rich foods for pollinators (Barth, 1985). It is well known that honey bees, as social insects, learn the most effective means of pollen-collecting (Real, 1981; Barth, 1985). Pseudo-pollen grains without contents are likely to be much less attractive.

The evolution of dioecy in Bonin *Callicarpa*

An hypothesis of the evolution of cryptic dioecy in Bonin *Callicarpa* is presented in Fig. 7.6. In this scenario, the development of pseudo-pollen grains of female plants plays an important role in the context of the restricted insect fauna (Kato, 1992*a,b*) on the Bonin Islands.

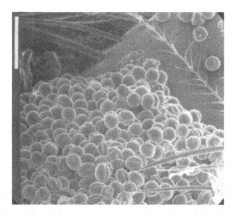

Fig. 7.5. Pollen grains in the pollen basket on the hind leg of a honey bee (*Apis mellifera*). Both normal and pseudo-pollen grains are collected equally by bees.

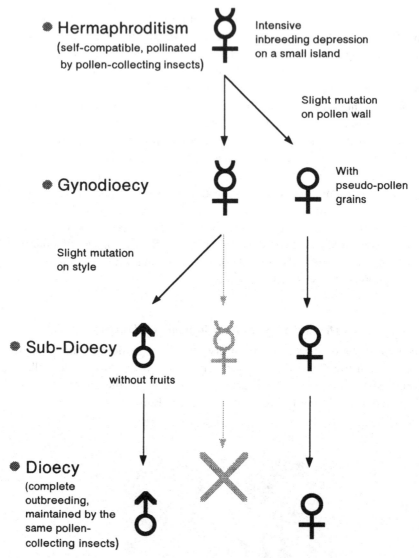

Fig. 7.6. Hypothetical evolution of dioecious *Callicarpa* on the Bonin Islands. In this scheme, populations of *Callicarpa* achieve complete outbreeding without change of pollinators.

I speculate that flowers of Bonin *Callicarpa* were originally hermaphroditic because this is the condition found in almost all other species of the genus (with more than 100 species). Ancestral hermaphroditic flowers of Bonin *Callicarpa* may have been self-compatible and flowers may have produced little nectar. These assumptions are not unreasonable because very little nectar is found inside the corolla of present-day Japanese mainland *Callicarpa* and several species of the genus (i.e., *C. dichotoma* K. Koch, *C. formosana* Rolfe and *C. shikokiana* Makino) are self-compatible (Kawakubo, unpublished). Under these circumstances, if inbreeding depression occurred in ancestral hermaphroditic populations, any mutation avoiding selfing would be favoured by natural selection.

One way to avoid selfing would be mutations favouring self-incompatibility. However, the genetic structure of self-incompatibility is complex and it may be difficult in some cases to evolve by simple mutations (Heslop-Harrison, 1957; Richards, 1986). Other mechanisms for outcrossing, however, such as dioecy and gynodioecy, may be achievable through simpler mutations (e.g., Heslop-Harrison, 1957; Baker, 1959; Richards, 1986). Female plants in species with altered sex expression can avoid inbreeding depression. Another sexual dimorphism, androdioecy (with hermaphrodite and male plants), can also avoid selfing on male plants, but this is a rare sexual system in flowering plants (Lloyd, 1975; Willson, 1983; Richards, 1986). A slight mutation affecting the pollen wall and eliminating germ slits, therefore, could be the first step in this evolutionary process.

Such a slight mutation might have been more favoured than those causing loss of entire pollen grains and anthers, because female flowers without pseudo-stamens generally require other pollinators, such as nectar-collecting insects. Further, on small oceanic islands such as the Bonin Islands, female flowers without both nectar and pollen grains would have difficulty attracting the very restricted number of pollinators. Female flowers with both pseudo-stamens and pseudo-pollen grains, however, could be maintained by the same pollinators as before.

Bawa (1982) has suggested that ecological correlates of dioecism, such as pollination by limited insects, may explain the evolution of dioecism on islands. Sun *et al.* (1996) hypothesize that the evolution of floral features (gynodioecism) in *Rhaphithamnus venustus* (Verbenaceae) is the result of a change from primarily insect to hummingbird pollination on the Juan Fernandez Islands. However, the pollinator fauna may have acted as a retardant in the evolution of dioecy in Bonin *Callicarpa* rather than a promoter. The pseudo-stamens and pseudo-pollen grains of female flowers may be maintained by the obviously limited insect fauna. In plants with nectarless hermaphroditic flowers, the limited insect fauna (even if it is composed mainly of small generalist insects) is not likely to be the selective agent in the evolution of dioecy.

Past insect fauna on the Bonin Islands is not the same as that found today. The present predominant pollinator of Bonin *Callicarpa* is the honey bee (*Apis mellifera*), an introduced and naturalized insect on the islands (Kato, 1992a). Several endemic bee species (e.g., four species of *Hylaeus* and *Xylocopa ogasawarensis*) do live now on the Bonin Islands (Kato, 1992a, b) and are potential additional pollinators of *Callicarpa*. In the past, Bonin *Callicarpa* might have been pollinated predominantly by these endemic bees.

In the gynodioecious hypothetical evolutionary stage, female plants succeeded in outbreeding, but hermaphroditic plants would still have suffered serious inbreeding depression. In such a gynodioecious condition, a mutation leading to female sterility of ancestral hermaphrodites would have been favoured. This mutation may have caused an insufficient development of the styles, because male flowers now sometimes have short styles or often lack them entirely. In such populations, female sterility (male plants) may have been selected as one possible way of avoiding inbreeding depression. At this evolutionary step, therefore, hermaphroditic, female and male plants may have co-existed. Subsequently, genes of ancestral hermaphroditic plants may have decreased in the population, eventually disappearing and leaving the dioecious condition now seen in Bonin *Callicarpa*.

Acknowledgements

I thank Professor M. Ono of Tokyo Metropolitan University and Professor T. F. Stuessy of the Institut für Botanik und Botanischer Garten, University of Vienna for reading the manuscript and for helpful comments. I also appreciate the kind help of Dr M. Mishio of Gifu University.

Literature cited

Anderson, G. J. & Symon, D. E. (1989). Functional dioecy and andromonoecy in *Solanum*. *Evolution*, **43**, 204–19.

Baker, H. G. (1955). Self-compatibility and establishment after 'long distance' dispersal. *Evolution* **9**, 347–9.

Baker, H. G. (1959). Reproductive methods as factors in speciation in flowering plants. *Cold Spring Harbor Symposium on Quantitative Biology* **24**, 177–91.

Baker, H. G. (1967). Support for Baker's Law – as a rule. *Evolution*, **21**, 853–6.

Baker, H. G. & Cox, P. A. (1984). Further thoughts on dioecism and islands. *Annals of the Missouri Botanical Garden*, **71**, 244–53.

Barth, F. G. (1985). *Insects and Flowers, the Biology of a Partnership.* Princeton: Princeton University Press.

Bawa, K. S. (1980). Evolution of dioecy in flowering plants. *Annual Review of Ecology and Systematics,* **11**, 15–39.

Bawa, K. S. (1982). Outcrossing and incidence of dioecism in island floras. *American Naturalist,* **119**, 866–71.

Bawa, K. S. & Opler, P. A. (1975). Dioecism in tropical forest trees. *Evolution,* **29**, 167–79.

Carlquist, S. (1974). *Island Biology.* New York: Columbia University Press.

Charlesworth, B. & Charlesworth, D. (1978). A model for the evolution of dioecy and gynodioecy. *American Naturalist,* **112**, 975–97.

Charnov, E. L. (1982). *The Theory of Sex Allocation.* Princeton: Princeton University Press.

Charnov E. L., Smith, J. M. & Bull, J. J. (1976). Why be an hermaphrodite? *Nature,* **263**, 125–6.

Cox, P. A. (1985). Islands and dioecism: insights from the reproductive ecology of *Pandanus tectorius* in Polynesia. In *Studies on Plant Demography,* ed. J. White, pp. 359–72. London: Academic Press.

Darwin, C. (1877). *The Different Forms of Flowers on Plants of the Same Species,* 1986 edn. Chicago: University of Chicago Press.

Ehrendorfer, F. (1979). Reproductive biology in island plants. In *Plants and Islands,* ed. D. Bramwell, pp. 293–306. London: Academic Press.

Heslop-Harrison, J. (1957). The experimental modification of sex expression in flowering plants. *Biological Review,* **32**, 38–90.

Kato, M. (1992*a*). Partnership between plants and insect pollinator in Ogasawara Islands. *World Wide Fund for Nature Japan Science Report,* **1** (part 1), 51–62.

Kato, M. (1992*b*). List of insects in Ogasawara. *World Wide Fund for Nature Japan Science Report,* **1** (part 1), 73–106.

Kawakubo, N. (1986). Morphological variation of three endemic species of *Callicarpa* (Verbenaceae) in the Bonin (Ogasawara) Islands. *Plant Species Biology* ,**1**, 59–68.

Kawakubo, N. (1990). Dioecism of the genus *Callicarpa* (Verbenaceae) in the Bonin (Ogasawara) Islands. *Botanical Magazine Tokyo,* **103**, 57–66.

Lewis, D. (1942). The evolution of sex in flowering plants. *Biological Review,* **17**, 46–67.

Lloyd, D. G. (1975). The maintenance of gynodioecy and androdioecy in angiosperms. *Genetica,* **45**, 325–39.

Lloyd, D. G. (1980). The distribution of gender in four angiosperm species illustrating two evolutionary pathways to dioecy. *Evolution,* **34**, 123–34.

Mayer, S. S. & Charlesworth, D. (1991). Cryptic dioecy in flowering plants. *Trends in Ecology and Evolution,* **6**, 320–5.

Real, L. (1981). Uncertainty and pollinator–plant interactions; the foraging behaviour of bees and wasps on artificial flowers. *Ecology,* **62**, 20–6.

Richards, A. J. (1986). *Plant Breeding Systems.* London: George Allen & Unwin.

Ross, M. D. (1980). The evolution and decay of overdominance during the evolution of gynodioecy, subdioecy, and dioecy. *American Naturalist,* **116**, 607–20.

Ross, M. D. & Weir, B. S. (1976). Maintenance of males and females in hermaphrodite populations and the evolution of dioecy. *Evolution,* **30**, 425–41.

Sun, B. Y., Stuessy T. F., Humaña, A. M., Riveros G., M. & Crawford, D. J. (1996). Evolution of *Rhaphithamnus venustus* (Verbenaceae), a gynodioecious hummingbird-pollinated endemic of the Juan Fernandez Islands, Chile. *Pacific Science,* **50**, 55–65.

Willson, M. F. (1983). *Plant Reproductive Ecology.* New York: John Wiley & Sons.

8

Conservation of the endemic vascular plant species of the Bonin (Ogasawara) Islands

MIKIO ONO

Abstract

The Bonin Islands, consisting of about 20 small islands, are located in the western Pacific Ocean about 1000 km south of the Japanese mainland. The flora of the Bonin Islands is characterized by a high proportion of endemism: 30.4% of indigenous angiosperms, or c. 43% of trees and shrubs. Although humans have inhabited the Bonin Islands for less than 200 years, their impact on the flora has been substantial. Because each island is so small and the diversity of available habitats low, plant species have few options for survival. In addition to the negative effects of human impact, climatic disasters such as droughts or typhoons have exerted considerable damage on certain species. After World War II, the Bonin Islands were left almost uninhabited during more than 20 years of military occupation. During this period many crops and cultivated species escaped and outcompeted original vegetation. Feral goats also contributed to the damage. Of the approximately 460 taxa of flowering plants now found native to the Bonin Islands (including c. 140 endemic taxa), about 80 taxa are considered endangered. Recent efforts toward conservation of selected endangered endemic taxa are discussed.

The Bonin Islands are located in the western Pacific Ocean about 1000 km south of the Japanese mainland. They consist of about 20 small islands scattered in the area of 26°30' to 27°40' N, and 142°00' to 142°15' E. These islands are aggregated into three groups from north to south; Mukojima, Chichijima and Hahajima.

The climate of the islands is subtropical with an annual mean temperature of 23.0 °C and annual precipitation of nearly 1300 mm (mean of 20 years from 1971 to 1990). Typhoons strike the islands every year bringing heavy rainfall that supplies a large part of the annual precipitation level. Typhoons also often cause serious damage to the vegetation through strong wind and sea water spray.

When the Bonin Islands were discovered in the 16th century, they were recorded as covered by dense forest dominated by subtropical evergreen trees and shrubs such as *Morus* (Moraceae), *Machilus* (Lauraceae), *Boninia* (Rutaceae, one of two endemic genera of flowering plants on the islands), *Hibiscus* (Malvaceae), *Pouteria* (Sapotaceae) and *Ochrosia* (Apocynaceae).

The Bonin Islands are completely isolated from other land masses by more than 1000 km and have never been historically connected to any continent or archipelago. Consequently, the flora is very disharmonic and a high proportion of endemism has been recognized by various authors (e.g., Tuyama, 1970; Yamazaki, 1970; Ono, 1991). According to our recent survey (Kobayashi & Ono, 1987), out of 460 taxa (including 120 recent introductions) of vascular plants growing on the islands, 140 are considered endemic (30.4%). This high level of endemism is less than that of Hawaii (94% angiosperms, 65% ferns; Carlquist, 1974) but comparable to the Galapagos Islands (32%; Wiggins & Porter, 1971).

Each island of the Bonin archipelago has a very small area and very low elevation, unlike the archipelagos mentioned above: the Hawaiian Islands are about 16 000 km^2 with the highest peak at *c.* 4100 m; the Galapagos are *c.* 7700 km^2 with the highest peak at about 2000 m. Conversely, even the largest of the Bonin Islands, Chichijima, has an area of only 24 km^2 and the highest mountain on the islands is at only 498 m (Mount Chibusa on Hahajima). The diversity of available habitats on the Bonin Islands is therefore much lower than in the other archipelagos, no doubt contributing to a lower level of diversity of plant taxa.

Endangered plant species of the Bonin Islands

In general, isolated oceanic islands such as the Bonin, Galapagos or Hawaiian Islands, have comparatively small areas and lower elevations than continents. Consequently, the diversity of available habitats for plants and animals is less than that found in continents. In such a situation, when environmental conditions change, plants and animals cannot survive by taking refuge into other habitats. As a result, many endemic taxa become threatened, endangered or even extinct. Such environmental change could be caused by volcanic explosions, climatic disasters such as hurricanes or typhoons, or human impact including overgrazing and fire.

Species may also be threatened by biological factors, because if population sizes drop below critical levels, the chances of cross-pollination become so greatly reduced that extinction becomes almost inevitable in a medium to long-term situation (Melville, 1979).

For the past two centuries, many examples of extinction of both plant and animal taxa have been reported from various oceanic islands (Carlquist, 1965). Besides the famous examples of wingless birds, e.g., the Dodo (extinct in Mauritius) and Moa (in New Zealand), a number of plant species endemic to St Helena had already become extinct before the first botanical expedition by Burchell during 1805–10; since then 11 additional endemic species have been lost (Lucas, 1979).

On the Hawaiian Islands, out of 11 endemic species of *Tetramolopium* (Compositae), four species are considered to have become extinct in this century (Lowrey & Crawford, 1985). A great number of endemic plant taxa surviving on oceanic islands are àt present threatened or critically endangered by rapid changes of environment caused mainly by humans. On the Canary Islands, for example, according to an assessment of the endemic flora made in 1972, of 514 species and subspecies, 75% were assigned to one or another of the categories recognized by the IUCN, and 70 (14%) were regarded as endangered (Melville, 1979).

As for the flora of the Bonin Islands, after a period of colonization for nearly a century, entire islands were left without human impact for 25 years after World War II. During this period many introduced weeds as well as escaped crops and trees grew widely on the islands, invading and destroying the original vegetation. The following species are representatives of such invaders: *Pinus lutchuensis* Mayer (Pinaceae), *Casuarina equisetifolia* Foster (Casuarinaceae), *Acacia confusa* Merrill, *Derris eliptica* Bentham, *Leucaena leucocephala* de Wit (Leguminosae), *Bishofia javanica* Blume (Euphorbiaceae) and *Psidium cattleyanum* Sabine (Myrtaceae) (Ono,1991).

In 1968, the Bonin Islands were restored to Japan after the military occupation by the USA. Subsequently, former inhabitants began returning to the islands, and the population has increased gradually; now about 2400 people live on the two large islands, Chichijima and Hahajima. Consequently, natural vegetation has been forced to survive under heavy pressures from human activities, not only by direct urban development but also by reduced water supplies, competition with naturalized weeds and crops, and by naturalized goats, etc.

According to our survey made during 1979–80 (approximately ten years after restoration and recolonization), out of 456 taxa of vascular plants indigenous to the islands (except recently naturalized plants), 143 were considered endemic (Kobayashi & Ono, 1987). Among them, 21 taxa of ferns and flowering plants were recognized as threatened (Ono, Kobayashi & Kawakubo, 1986). At that time six species including three orchids were already extinct and another three species had only one (*Melastoma tetramerum*) or a few (*Pittosporum parvifolium, Rhododendron boninense*) individual(s). In addition to these, 64 taxa of ferns and flowering plants were listed as R (rare) or V (vulnerable) in accordance with the IUCN conservation categories (Ono, Kobayashi & Kawakubo, 1986).

Factors leading to endangerment of endemic plant taxa

On the Bonin Islands, the most important factors which are causing endangerment of plant taxa are as follows:

Naturalized goats

Damage caused by naturalized goats can be observed throughout the islands. Except on Hahajima, where apparently no goats have ever been successfully introduced, many goats have proliferated since the last century, especially after World War II. They eat almost all plants, even roots (if they are particularly hungry), but they are especially fond of members of the Compositae (e.g., *Cirsium, Crepidiastrum*), Campanulaceae (*Lobelia*), Verbenaceae (*Callicarpa*), as well as some orchids.

African agate snail

The African agate snail (*Achantia furica*) was introduced to the Bonin Islands before World War II (Yoshikawa,1972). Although the purpose of this introduction is not clearly known, the snail has increased and spread over all the islands, causing serious damage on both natural (indigenous) vegetation and cultivated vegetables. Particularly on Hahajima, rare orchids such as *Calanthe hoshii* S. Kobayashi, the endemic evergreen

Fig. 8.1. Representative endemic vascular plants of the Bonin Islands. A, *Melastoma tetramerum*; B, *Pittosporum parvifolium*; C, *Metrosideros boninensis*; D, *Crepidiastrum linguaefolium*; E, *Calanthe hattorii*; F, *Calanthe hoshii*; G, *Malaxis boninensis*.

tree *Claoxylon centenarium* Koidz. (Euphorbiaceae) and even several species of ferns have been eaten by this snail.

For the past five years, however, the agate snail has suddenly become reduced in number on all islands. The reason for this abrupt decrease is uncertain, but it is extremely positive from the viewpoint of conservation of indigenous plants on the Bonin Islands.

Naturalized rats

In general, damage to endemic plant taxa by naturalized rats is not serious on the Bonin Islands. However on Hahajima, one rare endemic species of *Piper*, *P. postelsianum* Maxim. (Piperaceae) has suffered severe damage due to rats; two of its three remaining populations were completely eaten.

Damage by drought

Even though the Bonin Islands are located in a subtropical region, the climate is comparatively dry, having an annual precipitation level of only 1200 mm. Moreover, during the past 30 years, the precipitation level seems to have been decreasing (official statistics). Especially in 1973 and 1980, very severe drought occurred resulting in fatal damage to small surviving populations of several endangered species, such as *Pittosporum parvifolium* Hayata (Pittosporaceae), *Symplocos kawakamii* Hayata (Symplocaceae) and *Callicarpa nishimurae* Koidz. (Verbenaceae).

Influences of road construction and urban development

The human population of the Bonin Islands is more than 2400 (in 1997, official statistics of the Village). Road construction has caused a big impact on the natural vegetation. The dense, sclerophyllous, evergreen shrub forest, which is the most characteristic plant formation on the islands and once occurred widely on Chichijima and Anijima, has been separated and restricted by paved roads. As a consequence, the forest has become drier and more windy, decreasing moist habitats for small epiphytic or parasistic plants such as ferns, orchids or *Schiaphila* species (Triuridaceae).

Furthermore, the local government of the islands intends to build a civil airport on the central part of Anijima where the sclerophyllous, evergreen shrub forest remains large and intact.

Illicit collection of some rare orchids or ferns

Some rare orchids, e.g., *Calanthe hattorii* Schlch., *C. hoshii* S. Kobayashi, *Cirrhopetalum boninense* Schlch., *Corymborchis subdensa* Fukuyama, and *Malaxis boninensis* Nackejima, are the target of collectors for illicit cultivation and commerce on the black market. Certain species of epiphytic ferns (e.g., *Asplenium nidus* L., *Diplazium longicarpum* Kodama and *Ophioderma pendulum* Presl.) are also the target of collectors for ornamental use. Because the islands are so small, the impact of such illicit collections is great and can force these plant taxa, as well as associated small animals living in the forest, into an endangered status or even extinction.

Present status and conservation efforts

In 1993, about 35 species of flowering plants and ferns were considered endangered. The current status of several are described as follows. Conservation efforts have been made for these species over the past few years by local conservation authorities, with the co-operation of the botanical garden of the University of Tokyo (Shimozono & Iwatsuki, 1986; Iwatsuki, Shimozono & Hirai, 1992), other botanical investigators and local volunteers of the islands.

Species in critical status

Malaxis boninensis Nackejima (Orchidaceae). One endemic species of *Malaxis (M. boninensis)* was considered extinct in the early 1980s (Ono, Kobayashi & Kawakubo, 1986). Subsequently, one individual was found in 1991 on a steep slope in Chichijima, but it was lost within half a year, perhaps eaten by a naturalized goat. Another individual was rediscovered in 1994. This living stock is now under careful observation and a programme for its artificial propagation is in preparation.

Piper postelsianum Maxim. (Piperaceae). There were two populations of this species in the doline (limestone sink) of Mount Sekimon, Hahajima Island. In the latter half of the 1980s, however, all these plants were eaten by naturalized rats, resulting in a seriously endangered species. There is only one small population left, consisting of a few individuals newly found near the old destroyed populations. As this species is dioecious, only low levels of fertile seed can be expected from this small population.

Morus boninensis Koidz. (Moraceae). *Morus boninensis* is also considered as endangered, with only 20 mature trees remaining. Before World

War II, a closely related species, *M. australis* Poiret was introduced from the Japanese mainland for silkworm breeding. Subsequent natural hybridization between the endemic *M. boninensis* and this introduced species has occurred widely (Itow, 1991). Recently, however, a small population of about 20 individuals of *M. boninensis* has been discovered in Otohtojima, on the neighbouring island of Anijima. These newly discovered individuals have been documented as authentic *M. boninensis* by analysis of allozyme polymorphisms (Itow, 1991).

Pittosporum parvifolium Hayata (Pittosporaceae). Four endemic species of *Pittosporum* grow on the Bonin Islands. One species (*P. boninense* Koidz.) has wide distribution throughout the islands. The other three species are restricted to one island and the adjacent small islets: *P. chichijimense* Nakai ex Tuyama and *P. parvifolium* Hayata grow only on Chichijima, while *P. beecheyi* Tuyama is distributed on Hahajima and its neighbouring islets. Although individuals of these three species are limited in number, the status of *P. parvifolium* is the most critical. According to our survey made in 1994, only two plants have survived on Chichijima. These two plants are not expected to produce mature seeds by themselves because the species is considered to be functionally dioecious (Kobayashi, 1982).

Melastoma tetramerum Hayata (Melastomaceae). Since our initial survey in 1980 (Ono, Kobayashi & Kawakubo, 1986), the status of the species has changed a little. A new population consisting of about 200 individuals has been found on Chichijima, the same island on which only one individual of the original population survived until 1995. The new population, however, has some individuals showing a slight difference in morphology. For instance, flowers of the original plant are tetramerous (as in the type specimen), but several individuals in the new population have pentamerous flowers, a key character for the variety on Hahajima Island (var. *hahajimense*). The genetic vs. environmental components of this feature, both within the new population and between it and the original one, should be investigated.

Rhododendron boninense Nakai (Ericaceae). The status of *R. boninense* has recently become very critical: only one individual survives. Fortunately, it blooms almost every year and is still producing mature seeds. Efforts to grow young stocks from these seeds and return them to the original habitat are now being undertaken by the Tokyo Metropolitan Government in collaboration with the botanical garden of the University of Tokyo.

Callicarpa nishimurae Koidz. (Verbenaceae). There are three species of *Callicarpa* endemic to the Bonin Islands: *C. glabra* Koidz., *C. nishimurae* Koidz. and *C. subpubescens* Hook. et Arn. Of the three species, *C. nishimurae* is seriously endangered. This species is a shrub with hairy, thick leaves and is restricted to the semiarid zones of Chichijima and Anijima. On both of these islands, the species has been damaged by naturalized goats and as a result only a few individuals are surviving. Unless stringent conservation measures are taken, this species is very likely to be extinct in the near future.

Other endangered species

Metrosideros boninensis Tuyama (Myrtaceae). *Metrosideros boninensis* was also considered endangered in our previous survey (Ono, Kobayashi & Kawakubo, 1986). During the past 15 years, many additional individuals (more than 100) have been found by a local investigator on Chichijima at various localities. Despite this positive development, the species is still considered rare.

Three endemic species of *Crepidiastrum* (Compositae). There are three species of *Crepidiastrum* endemic to the Bonin Islands: *C. ameristophyllum* Nakai, *C. grandicollum* (Koidz.) Nakai and *C. linguaefolium* (A.Gray) Nakai. All of them are semisucculent, woody, perennial herbs related to *Lactuca* or *Ixeris*. The distribution of each species is restricted and all of them are now very rare due to damage by naturalized goats, the introduced agate snail (*Achantia furica*) and agriculture.

Several species of orchids. About 14 species of orchids have been described from the Bonin Islands, all of which are endemic and very rare. Other than *Malaxis boninensis* mentioned above, almost all of them are more or less endangered because of illicit collection by amateur collectors for garden cultivation.

Two additional species of orchids, *Calanthe hoshii* S. Kobayashi and *C. hattorii* Schlch., have a particularly critical status. The former species is endemic to Hahajima and only a few individuals are surviving under protection at a site near its original habitat. The latter species, on the other hand, has a relatively wide distribution throughout Chichijima and Anijima. As it has showy yellow flowers, however, it continues to be a target for illicit collectors both for cultivation and sale. Furthermore, in one of the habitats of the species, Anijima, the local government is intending to construct a civil airport. If this plan is realized, it could cause serious damage not only to this rare orchid, but also to other endangered or threatened plants. Efforts are being made to prevent execution of this plan, by recommending change of site to another island. Appeals are based on the importance of conservation of the sclerophyllous, evergreen shrub forest on the plateau of Anijima, where many endangered or rare endemic plants and small animals survive.

A few species of Pteridophyta. There are a few species of Pteridophyta which are now seriously endangered. The most critical one is *Ophioderma pendulum* (L.) Presl. This epiphytic species usually grows upon *Asplenium nidus* L. which itself is an epiphytic fern growing on tall evergreen trees in mist forests at higher parts of the island. However, in Chichijima such a mist forest is no longer present and therefore no large individuals of *Asplenium* have been found for the past 20 years. Only in the mist forest of Mount Sekimon on Hahajima has this *Ophioderma* species survived upon a few large individuals of *Asplenium*. A strong typhoon in 1983 destroyed the canopy of the mist forest and caused serious damage to various epiphytes, including several species of fern and orchid. As no recent information is available on this rare species of *Ophioderma*, it is possible that it may be extinct.

A second fern, *Diplazium longicarpum* Kodama is also very likely to have become extinct during the past ten years. According to T. Yasui (personal communication) a local botanist living in Chichijima, the habitat of this fern is restricted to a small valley in the northern part of Hahajima in a semimoist microclimate. Recently, however, the upper part of this valley was buried by road construction, resulting in the loss of the only known population of this rare endemic fern.

A third example of ferns with a critical status on the Bonin Islands is *Asplenium tenerum* L., a very rare endemic species restricted to Hahajima. It has been observed only in the mist forest on the central mountain ridge of Hahajima. Although the fern is not an epiphyte, it grows on mossy rocks or stumps and requires a very high moisture level. Also according to T. Yasui, the single known small population of this fern suddenly disappeared several years ago. The reason for this loss has not yet been determined, but it may have been stolen by an illicit collector. No conspicuous change of the habitat has occurred but rare ferns are sometimes the target of hobby collectors in Japan.

Acknowledgements

I wish to express my cordial thanks to my colleagues who have been engaged in surveying the flora of the Bonin Islands for the past 25 years, since the restoration of the islands.

Financial support for these surveys has been provided by the Tokyo Metropolitan Government, the Environment Agency of the Japanese Government, and the Ministry of Education, Science and Culture.

Literature cited

Carlquist, S. (1965). *Island Life: A Natural History of the Islands of the World.* New York: Natural History Press.

Carlquist, S. (1974). *Island Biology.* New York: Columbia University Press.

Itow, M. & Ono, M. (1990). Allozyme diversity and the evolution of *Crepidiastrum* (Compositae) on the Bonin (Ogasawara) Islands. *Botanical Magazine Tokyo*, **103**, 449–59.

Iwatsuki, K., Shimozono, F. & Hirai K. (1992). Program of propagation and returning to the native habitat of some endangered plant species of the Bonin Islands. *Annual Report of Ogasawara Research*, **16**, 78–96. [In Japanese.]

Kobayashi, S. (1982). A taxonomical note on *Pittosporum tobira* and its allied species. *Journal of Japanese Botany*, **57**, 70–80.

Kobayashi, S. & Ono, M. (1987). A revised list of vascular plants indigenous and introduced to the Bonin (Ogasawara) and the Volcano (Kazan) Islands. *Ogasawara Research*, **13**, i–iv, 1–55.

Lowrey, T. K. & Crawford, D. J. (1985). Allozyme divergence and evolution in *Tetramolopium* (Compositae: Astereae) on the Hawaiian Islands. *Systematic Botany*, **10**, 64–72.

Lucas, G. (1979). The threatened plants committee of IUCN and island floras. In *Plants and Islands*, ed. D. Bramwell, pp. 423–30. London: Academic Press.

Melville, R. (1979). Endangered island floras. In *Plants and Islands*, ed. D. Bramwell, pp. 361–77. London: Academic Press.

Nobushima, F. (1992). Distribution and ecology of *Metrosideros boninensis* Tuyama, an endemic plant species of the Bonin Islands. *Annual Report of Ogasawara Research*, **16**, 37–77. [In Japanese.]

Ono, M. (1985). Speciation and distribution of *Pittosporum* in the Bonin Islands. In *Evolution of Diversity in Plants and Plant Communities*, ed. H. Hara, pp. 7–17. Tokyo: Academia Scientific Book Co.

Ono, M. (1991). Flora of the Bonin Islands: endemism and dispersal mode. *Aliso*, **13**, 95–105.

Ono, M. (1993). Discovery of a new population of an endangered plant species. *Melastoma tetramerum* Hayata in Chichijima. *Annual Report of Ogasawara Research*, **17**, 74–7. [In Japanese.]

Ono, M., Kobayashi, S. & Kawakubo, N. (1986). Present situation of endangered plant species in the Bonin (Ogasawara) Islands. *Ogasawara Research*, **12**, 1–32.

Ono, M., Kobayashi, S. & Yasui, T. (1991). The vascular plant flora of the Bonin Islands and their present situation. In *Report of the Second General Survey on*

Natural Environment of the Bonin (Ogasawara) Islands, ed. M. Ono, pp. 1–12. Tokyo: Tokyo Metropolitan University. [In Japanese.]

Shimozono, F. & Iwatsuki, K. (1986). Botanical gardens and the conservation of an endangered species in the Bonin Islands. *Ambio*, **15**, 19–21.

Tuyama, T. (1970). Plants of the Bonin Islands. In *The Nature of the Bonin Islands*, ed. T. Tuyama & S. Asami, pp. 109–141. Tokyo: Hirokawa Shoten. [In Japanese.]

Wiggins, I. & Porter, D. (1971). *Flora of the Galapagos Islands*. California: Stanford University Press.

Yamazaki, T. (1970). *The vascular plants in the Bonin and Volcano Islands*. Tokyo: Culture Agency, Ministry of Education, Science and Culture.

Yoshikawa, K. (1972). On agate snail in the Bonin Islands – ecology of invaders. *Annual Report of Ogasawara Research*, **1**, 49–56. [In Japanese.]

9

Preliminary observations on the evolution of endemic angiosperms of Ullung Island, Korea

BYUNG-YUN SUN AND TOD F. STUESSY

Abstract

Ullung Island is located 150 km east of Korea in the East Sea at 37° N latitude. The island is of volcanic origin, 73km^2 in area and approximately 1.8 million years old. Ullung Island contains 700 species of vascular plants of which 37 angiosperms are endemic. Closest phytogeographic ties of the island flora are with South Korea and Central Japan. This chapter offers preliminary observations on relationships of Ullung Island endemics with presumptive source-area relatives. Initial assessments of morphological changes suggest a tendency towards increased stature in island taxa as well as a loss of pubescence and prickles (e.g., in *Rubus takesimana*). Detailed morphological comparisons using principal components analysis of the endemic *Hepatica maxima* and close congeners in northeastern Asia support further the observed trends. New chromosome counts are presented for 48 taxa of Ullung Island, including first reports for eight species, representing 40% of the native and endemic angiosperms. These data, in consort with previously published cytological information regarding presumptive relatives, suggest that few aneuploid or euploid changes have occurred during speciation of the angiosperm flora on this island. The endemic angiosperms of Ullung Island appear mostly to have evolved by simple anagenesis, there being little evidence of intra-island cladogenetic speciation events.

Ullung Island is a small island 150 km east of mainland Korea and 300 km west from Japan (Fig. 9.1), extending from 37°27' to 37°33' N and from 130°47' to 130°56' E. The total area of the island is about

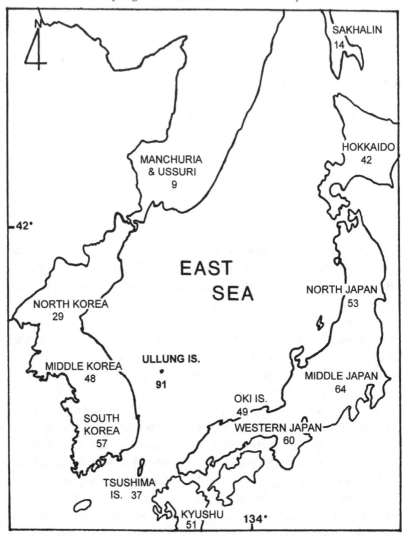

Fig. 9.1. Map showing the location of Ullung Island, and the number of native woody plant species on the island (91) and in neighbouring areas. (After Nakai, 1919.)

73 km^2 and the highest peak (Seongin-bong), which occupies the centre of the island, is 984 m above sea level (Fig. 9.2). The island is mountainous, composed mainly of five ridges and valleys extending down from the central peak. The average depth of the sea around the island is 2000 m and most of the seashore is composed of steep cliffs. Along the coast

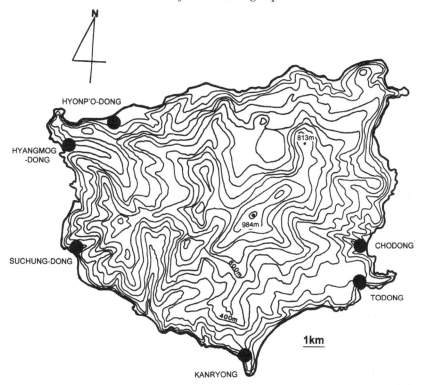

Fig. 9.2. Map of Ullung Island showing elevation and principal settlements. Contour interval 100 m.

there are several small villages (Fig. 9.2) and the total population of the island is approximately 20 000 people.

Of volcanic origin, Ullung Island has never been connected to any other land mass. Past volcanic activity on Ullung island has been examined by potassium-argon dating methods (Kim, 1985) and it is estimated that the first volcanic eruption under the sea occurred about 2.7 million years ago. The geological history of this island is divided into five stages based on the mode of volcanic activity and the nature of the volcanic products (Kim, 1985). The age of stage I is about 1.8 million years. The present geomorphology of the island was formed at stage III. The maximum age of the island during which it would have been available for colonization by diaspores, therefore, is 1.8 million years. In comparison with other oceanic islands, Ullung Island is very young, and this is an important factor in explaining evolutionary patterns and processes within the island.

Table 9.1. *List of endemic angiosperms of Ullung Island, Korea*

ACERACEAE
  Acer okamotoanum Nakai
  A. takesimense Nakai
ARACEAE
  Arisaema takesimense Nakai
CAMPANULACEAE
  Campanula takesimana Nakai
CAPRIFOLIACEAE
  Abelia insularis Nakai
  Lonicera insularis Nakai
  Sambucus seiboldiana var. *pendula* (Nakai) T. Lee
CARYOPHYLLACEAE
  Silene takesimensis Uyeki et Sakata
COMPOSITAE
  Chrysanthemun zawadskii var. *lucidum* (Nakai) T. Lee
CRASSULACEAE
  Sedum takesimense Nakai
CRUCIFERAE
  Arabis takesimana Nakai
CYPERACEAE
  Carex blepharicarpa var. *insularis* Nakai
  C. takesimensis Nakai
FAGACEAE
  Fagus multinervis Nakai
FUMARIACEAE
  Corydalis filistipes Nakai
GRAMINEAE
  Poa takeshimana Honda
LABIATAE
  Lamium takeshimense Nakai
LILIACEAE
  Lilium hansonii Leichtl.
OLEACEAE
  Ligustrum foliosum Nakai
  Syringa velutina var. *venosa* (Nakai) T. Lee
PHYTOLACCACEAE
  Phytolacca insularis Nakai
ROSACEAE
  Cotoneaster wilsonii Nakai
  Physocarpus insularis Nakai
  Potentilla dickinsii var. *glabrata* Nakai
  Prunus takesimensis Nakai
  Rubus takesimensis Nakai
RANUNCULACEAE
  Hepatica maxima Nakai
RUTACEAE
  Phellodendron insulare Nakai

Table 9.1. (*cont.*)

SALICACEAE
  Salix ishidoyana Nakai
SCROPHULARIACEAE
  Scrophularia takesimensis Nakai
  Veronica insulare Nakai
TILIACEAE
  Tilia insularis Nakai
UMBELLIFERAE
  Bupleurum latissimum Nakai
  Dystaenia takesimana (Nakai) Kitagawa
VALERIANACEAE
  Valeriana officinalis L. var. *integra* Nakai
VIOLACEAE
  Viola insularis Nakai
  V. takesimana Nakai

Modified after Lee & Yang (1981)

The flora of Ullung Island contains approximately 700 vascular plant species of which about 500 are native or endemic (Lee & Yang, 1981). Some 37 angiosperm taxa are endemic, representing 25 families and 34 genera (Table 9.1). Nakai (1919), who was the pioneer of floristic research on Ullung Island, examined the occurrence of 91 native woody plant species of the island in adjacent regions, including Korea, Japan, Manchuria, Russia and the Sakhalin Peninsula, and suggested that the island has closest floristic relationship with South Korea and Central Japan (Fig. 9.1). This has been analysed more recently for the entire flora by Kim (1988) in which the percentage of species in common between Ullung Island and Korea, Japan and Manchuria is 94, 86 and 59%, respectively.

One of the most important characteristics of the endemic vascular flora of Ullung Island is that only single endemic species have evolved from their immigrant progenitors (i.e., without subsequent cladogenesis). Exceptions might appear to be in *Acer, Carex* and *Viola* that contain two endemic species (Table 9.1), but close examination reveals that these congeners are more closely related to different extra-island relatives than to each other. Ullung Island, therefore, is an oceanic island where it appears that the evolution of new angiosperm species has been only by simple anagenetic change. Furthermore, the endemic taxa occur in many diverse angiosperm families. This is an excellent system, therefore, in which to address questions regarding the initial stages of island plant evolution.

Our present research involves investigating the origin and evolution of endemic angiosperms of Ullung Island, Korea, focusing on the understanding of changes taking place during anagenetic (phyletic) speciation. The specific steps involved are: (1) confirming ties of endemic species with relatives on other land masses, some of which have been suggested previously (e.g., Nakai, 1919; Lee, 1966; Lee, 1980, 1984); (2) investigating morphological, cytological, genetic and breeding system differences between extra-island progenitors and island derivatives; (3) comparing morphololgical and genetic variation within and between populations of island and continental taxa; and (4) summarizing patterns and offering suggestions on processes of evolution. The purpose of this chapter is to present preliminary observations on morphological and chromosomal patterns in endemic (and some native) angiosperms of Ullung Island, and between these taxa and close relatives in surrounding source areas.

Morphological variation of endemic vascular taxa

Important in interpreting modes of evolution of endemic angiosperms of Ullung Island is the determination of source-area relatives. Only by clarification of these relationships can the types of character state changes during evolution be assessed. A few of the more interesting genera are discussed below, with detailed comments on *Hepatica*.

Two endemic species of *Acer* (Aceraceae) exist on Ullung Island: *A. okamotanum* and *A. takesimense*. Preliminary examination of morphology of these two species suggests that they have evolved from different ancestral species. This has been confirmed recently by ITS sequences (Cho *et al.*, 1996). *Acer okamotoanum* is very closely related morphologically to *A. mono* and it probably originated from the latter species either from Korea or Japan. *Acer okamotoanum* can be distinguished from *A. mono* by the former's much larger leaves, flowers and fruits, and it seems likely that an increase in the size of various organs has occurred during speciation of *A. okamotoanum*. Another species, *A. takesimense*, however, is morphologically related to *A. pseudosieboldianum* rather than *A. mono*. Chang & Giannasi (1991), based on flavonoid data, have even suggested that *A. takesimense* and *A. pseudosieboldianum* might best be treated as conspecific, but Park, Oh & Shin (1993) point to useful morphological discriminators and underscore specific status (the debate continues, however: Chang, 1994; Park, Oh & Shin, 1994). There is no ITS sequence divergence between the two taxa (Cho *et al.*, 1996). We believe that *A. takesimense* may have speciated from *A. pseudosieboldianum*, perhaps by immigration from the Korean Peninsula. *Acer takesimense* differs most obviously morphologically from *A. pseudosieboldianum* by its more deeply divided leaves with usually 13 or more lobes.

The genus *Abelia* (Caprifoliaceae) consists of about 25 species worldwide and the centre of diversification is the eastern Asiatic region including Korea, Japan, China and the far eastern region of Russia (Chung & Sun, 1984). *Abelia insularis*, an endemic species of Ullung Island, belongs to sect. *Zabelia* series *Biflorae*. Besides *A. insularis*, this series includes three other species, *A. biflora, A. coreana* and *A. integrifolia. Abelia insularis* is closely related to *A. coreana* and hence sometimes is treated as a subspecies of it (Paik & Lee, 1989). Whatever the taxonomic rank of *A. insularis*, it differs in having broader, almost glabrous leaves. In addition, the flowers of *A. insularis* are succulent and sometimes have additional corolla lobes.

Dystaenia (Umbelliferae), sometimes included in the genus *Ligusticum* (Hiro & Constance, 1958), has two species worldwide (Kitagawa, 1937). *Dystaenia takesimana* is endemic to Ullung Island and the other species, *D. ibukiensis*, is endemic to Japan. *Dystaenia takesimana* is larger in stature than its Japanese relative.

The Ullung Island endemic *Rubus takesimensis* (Rosaceae), is distinct from its close relative, *R. crataegifolius*, from Korea and Japan. Changes occurring during the speciation of *R. takesimensis* from this putative ancestor seem to have involved an increase in the size of leaves and flowers, and a loss of prickles.

Corydalis filistipes (Fumariaceae), another endemic species, is enigmatic because its closest relative is still unknown. *Corydalis filistipes* is usually found above 800 m, and it has a single spherical tuber and rhizome. It belongs, therefore, to sect. *Pes-gallinus*. There are approximately ten species of sect. *Pes-gallinus* in Korea (Oh, 1986), but *C. filistipes* has no close morphological affinities to any of them. The island taxon has tripinnately compound leaves, with leaflets finely divided into small segments. In addition, flowers are white with a very short spur.

Among endemic taxa of Ullung Island, *Hepatica maxima* (Ranunculaceae) shows the most dramatic morphological changes from its putative relatives. The northeastern Asiatic region is very important for *Hepatica* because it is here that the highest level of diversification has occurred, with three species and two varieties among 11 taxa worldwide. *Hepatica asiatica* is distributed throughout the entire Korean Peninsula

including Manchuria, *H. insularis* occurs in southernmost Korea including Cheju Island, *H. maxima* is confined to Ullung Island, and *H. nobilis*
var. *japonica* and *H. nobilis* var. *pubescens* are distributed in Japan.

To understand the patterns of variation and relationships of *H. maxima* with congeners in the northeastern Asiatic region, preliminary analyses of leaf characters using principal components analysis were
employed. Emphasis was placed on vegetative features because they are
more diagnostic for the recognition of these taxa than are reproductive
characters. A total of 253 specimens was selected for numerical analysis
of morphological variation, including 88 specimens of *H. asiatica*, 57 of
H. insularis, 22 of *H. maxima* and 86 of *H. nobilis* var. *japonica* (list of
vouchers available from B. Y. Sun upon request). From these specimens,
22 leaf characters were scored for morphological analyses, two of which
were used for calculating ratios only, and six of which were derived ratios
(Table 9.2). Although ratios have been criticized for not being particularly helpful in removing the negative effects of scaling data (Atchley,
Gaskins & Anderson, 1976; Atchley & Anderson, 1978), we have used
them here because in our opinion they offer a more complete comparison
of morphological features among the taxa sampled; careful use of ratios
has also been supported by other workers (e.g., Corruccini, 1977;
Albrecht, 1978; Hills, 1978).

Table 9.3 shows loadings and cumulative percentages of the first three
principal components of leaf features in northeastern Asian species of
Hepatica. These accounted for 44.5, 12.2 and 10.6% of the total variance,
respectively, and subsequent components contributed less than 9% each.
The first principal component mainly reflects high positive loadings of
absolute length or width of the leaves, including the length of the central
lobe (character 4), width at the widest point of the central lobe (character
3), width at the base of the central lobe (character 2), length of the lateral
lobe (character 6), width at the widest point of the lateral lobe (character
9), width at the base of the lateral lobe (character 8), length of the entire
leaf (character 14), and width of the entire leaf (character 12).

Figure 9.3 shows the plot of individuals of *Hepatica* taxa in the northeastern Asian region projected on the first two principal components. The
analysis reveals the presence of four major groups. The right grouping (*)
represents individuals of the Ullung Island endemic, *H. maxima*. In this
plot, individuals of *H. maxima* are well separated from the rest, with little
overlap in the first principal component (relating to size factors such as
length or width of leaf parts). This species tends to have larger leaves and
helps underscore its recognition as a distinct taxon. The relative position

Table 9.2. *The 22 leaf characters used in numerical analyses of* Hepatica
species of northeastern Asia

1, apex angle of central lobe (degree); 2, width at base of central lobe (mm); 3,
width at widest point of central lobe (mm); 4, length of central lobe (mm); 5,
length from apex to the widest point of central lobe (mm); 6, length of lateral lobe
(mm); 7, apex angle of lateral lobe (degree); 8, width at base of lateral lobe (mm);
9, width at widest point of lateral lobe (mm); 10, angle between blade length and
width of lateral lobe (degree); 11, length from apex to widest point of lateral lobe
(mm); 12, width of leaf (mm); 13, base angle of leaf (degree); 14, length of leaf
(mm); 15, pubescence of adaxial surface of leaf (+, −); 16, duration of leaf
(annual, biennial); 17, ratio character 4/character 5; 18, ratio character 3/
character 2; 19, ratio character 6/character 11; 20, ratio character 12/character 4;
21, ratio character 9/character 6; 22, ratio character 14/character 12.

Table 9.3. *Loadings of the first three principal components in numerical
analyses of* Hepatica *species from northeastern Asia*

Character number (cf. Table 9.2)	Component		
	1	2	3
1	0.1701	0.1947	0.1326
2	0.3242	−0.0617	−0.0930
3	0.3322	−0.0151	−0.2420
4	0.3204	0.0394	−0.0216
6	0.3189	0.0368	−0.1393
7	0.1842	0.1766	0.0775
8	0.3091	−0.0672	−0.2009
9	0.3313	−0.0034	−0.0666
10	0.0781	0.0226	0.4298
12	0.3204	−0.0015	−0.1233
13	0.1123	−0.0340	0.5247
14	0.3199	0.0408	−0.1375
15	−0.0966	0.5237	−0.1727
16	0.0966	−0.5237	0.1727
17	0.0387	0.3451	0.0869
18	0.1359	0.2246	0.2896
19	0.0317	0.1775	−0.0971
20	−0.0706	−0.2385	−0.4697
21	0.2079	−0.1251	0.1241
22	0.0332	0.3067	−0.1086
Eigenvalue	8.8964	2.4457	2.1235
Cumulative % of Eigenvalue	44.5	56.7	67.3

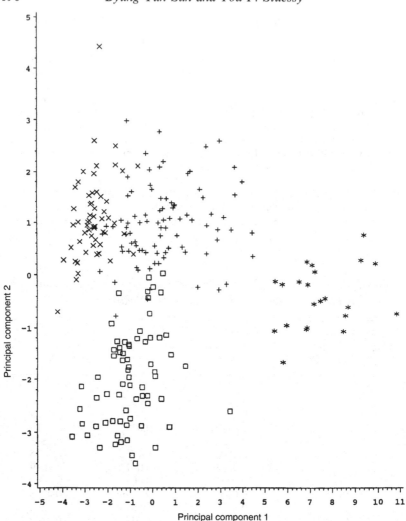

Fig. 9.3. Plot of the first two principal components showing relationships based on morphological leaf characters among taxa of *Hepatica* in northeastern Asia. * = *H. maxima*; × = *H. insularis*; + = *H. asiatica*; □ = *H. nobilis* var. *japonica*.

of the other three taxa on the first principal component suggest that *H. maxima* is most closely related (based on leaf characters) to *H. asiatica* (+). Considering the range of variation of each taxon in these analyses, *H. maxima* shows a limited range of variation compared to the other three taxa, perhaps due to founder effect and environmental uniformity over the island; this tendency is clearest in the second principal component.

A general survey of gross morphological differenes between 32 endemic vascular taxa of Ullung Island and putative relatives either on mainland Korea or Japan, reveals that 12 species are without hair or prickles and 15 species are larger in stature; only four species are smaller. Based on these observations, it can be suggested that one trend of morphological change during anagenesis of endemic angiosperm taxa on Ullung Island may be an increase in overall stature and loss of hairs or prickles. Detailed quantitative observations are obviously needed to confirm these initial impressions. Any adaptive explanations for these suggested changes in morphology are even more speculative, but one suggestion is that increased stature might relate to relatively low levels of light on this island. The average number of sunny days is only 56 per year (Anon., 1988); the other days are either cloudy, rainy or snowy. Loss of prickles in *Rubus takesimensis* may be related to the absence of herbivores on Ullung Island. This case could be similar to that of *Rhaphithamnus venustus* (Verbenaceae), a species endemic to the Juan Fernandez Islands, Chile, in which loss of spines has occurred during evolution from its well-armed ancestor, *R. spinosus* (Sun *et al.*, 1996*b*).

Chromosomal survey

For effective studies on the evolution of island endemics, it is important to understand chromosomal relationships among island taxa, and between them and source-area relatives. Having this information can do much to help suggest hypotheses on modes of speciation in the evolutionary development of the flora. Toward these ends we have completed a preliminary chromosomal survey of the native and vascular plants of Ullung Island, including 40% of the native and endemic species. Table 9.4 lists 49 chromosome counts for native or endemic taxa, 45 of which are new reports. Among them, first counts are reported for eight taxa: *Acer takesimense, Abelia insularis, Gymnadenia camtschatica, Lonicera insularis, Rubus takesimensis, Sambucus sieboldiana* var. *pendula, Smilax riparia* var. *ussuriensis* and *Spiraea blumei*. A total of 38 additional counts are for taxa reported previously, two of which are new chromosome numbers. Our discussion focuses on counts for the 14 endemic taxa (following alphabetical order of families in Table 9.4).

Acer takesimense is reported here for the first time as $n = 13$. Many species of the genus have been examined and most are also

Table 9.4. *Chromosome numbers of angiosperms from Ullung Island, Korea*
Most are new reports. Vouchers are deposited in the Herbarium of Chonbuk
National University.

Taxon	Meiotic chromosome number (*n*)	Voucher or reference
Aceraceae		
* *Acer takesimense* Nakai	13	Sun *et al.* 1472
Berberidaceae		
Berberis amurensis Rupr.	14	Sun *et al.* 1014, 1458
Campanulaceae		
Campanula takesimana Nakai	17	Sun *et al.* 1131
Caprifoliaceae		
* *Abelia insularis* Nakai	54	Sun *et al.* 1459
* *Lonicera insularis* Nakai	9	Sun *et al.* 1010
* *Sambucus sieboldiana* var. *pendula* (Nakai) T. Lee	18	Sun *et al.* 1526
Caryophyllaceae		
Silene takesimensis Uyeki et Sakata	12	Sun *et al.* 1158
Compositae		
Aster spathulifolius Max.	9	Sun *et al.* 1139
Chrysanthemum zawadskii var. *lucidum* (Nakai) T. Lee	2*n* = 36	Lee (1967)
Erigeron annuus (L.) Pers.	27	Sun *et al.* 1143
Hieracium umbellatum L.	27	Sun *et al.* 1154
Solidago virga-aurea L. var. *gigantea* Nakai	9	Sun *et al.* 1153
Sonchus brachyotus Roem. et Schult.	18	Sun *et al.* 1106, 1498
Convolvulaceae		
Calystegia soldanella (L.) Roem. et Schult.	11	Kim, T. 1145
Cornaceae		
Cornus macrophylla Wallich	11	Sun *et al.* 1129
Cruciferae		
Wasabia japonica (Miq.) Matsum.	14	Sun *et al.* 1032
Fumariaceae		
Corydalis speciosa Max.	8	Sun *et al.* 1468
Leguminosae		
Sorbus aurensis Koehne	17	Sun *et al.* 1463
Liliaceae		
+ *Allium victorialis* var. *platyphyllum* (Hulten) Makino	8	Sun *et al.* 1025, 1046
Lilium hansonii Leichtl.	12	Sun *et al.* 1133
Maianthemum dilatatum (Wood) Nelsons et Macbr.	18	Sun *et al.* 1019, 1467
Smilax riparia var. *ussuriensis* Hara et Koyama	16	Sun *et al.* 1132

Table 9.4. (*Cont.*)

Taxon	Meiotic chromosome number (*n*)	Voucher or reference
Moraceae		
Morus bombyis Koidz	14	Sun *et al*. 1454
Orchidaceae		
* *Gymnadenia camtschatica* (Cham.) Miyabe et Kudo	20	Sun *et al*. 1026
Orobanchaceae		
Orobanche coerulescens Steph.	19	Sun *et al*. 1136
Papaveraceae		
Chelidonium majus L. var. *asiaticum* (Hara) Ohwi	5	Sun *et al*. 1029
Phytolaccaceae		
Phytolacca insularis Nakai	$2n = 72$	Lee (1972)
Plantaginaceae		
Plantago asiatica Dcne.	12	Sun *et al*. 1135
Ranunculaceae		
Hepatica maxima Nakai	7	Sun *et al*. 1036
Rosaceae		
* *Rubus takesimensis* Nakai	7	Sun *et al*. 1011
* *Spiraea blumei* G. Don	18	Sun *et al*. 1135, 1461
Rubiaceae		
+ *Asperula odorata* L.	11	Sun *et al*. 1024, 1473, 1479, 1504
Saxifragaceae		
Tiarella polyphylla D. Don	7	Kim 1423
Scrophulariaceae		
Veronica insulare Nakai	16	Sun *et al*. 1136
Ulmaceae		
Zelkova serrata (Thunb.) Makino	14	Sun *et al*. 1460
Umbelliferae		
Anthriscus sylvestris (L.) Hoffm.	8	Sun *et al*. 1455
Bupleurum latissimum Nakai	$2n = 16$	Lee (1967)
Dystaenia takesimana (Nakai) Kitagawa	22	Sun *et al*. 1150
Valerianaceae		
Valeriana fauriei Briq.	28	Sun *et al*. 1400
Violaceae		
Viola takesimana Nakai	$2n = 20$	Lee (1967)
Viola verecunda A. Gray	12	Sun *et al*. 1041

* = First report for taxon
+ = New reported number for taxon

$n = 13$ ($2n = 26$: e.g., Foster, 1993; Mehra & Khosla, 1969; Mehra, Khosla & Sareen, 1972). *Acer pseudoplatanum*, however, is tetraploid at $n = 26$ (Mehra, 1976; Ferakova, 1978). The close relative of *A. takesimense* is *A. pseudosieboldianum* (Pax) Kom. (Park, Oh & Shin, 1993), for which counts of $2n = 26$ are available (Foster, 1933; Gurzenkov, 1973). Because the island endemic is at the same chromosomal level as its close ally, as well as most of the rest of the genus (Fedorov, 1969; Mehra, Khosla & Sareen, 1972; Mehra, 1976; Santamour, 1988), it is unlikely that it has undergone a chromosome number change during its phyletic evolutionary history.

Campanula takesimana is known cytologically as $n = 17$ (our report) and $2n = 34 + 2B$ (Lee, 1969*b*). Its close relative is *C. punctata* (Nakai, 1919; Yoo, 1995; T. Lammers, personal communication), also known chromosomally as $n = 17$ (Fedorov, 1969; Probatova & Sokolovskaya, 1989). It seems clear, therefore, that no change in chromosomal level has occurred in the origin of this endemic taxon.

Abelia insularis is counted here for the first time as $n = 54$. This is an extremely interesting count, because it is the first report of what appears to be the duodecaploid ($12x$) level in the genus. Previous counts of $n = 9$, 16 and 18 encompassing four species (*A. engleriana, A. schumannii, A. trifolia* and *A. uniflora*: Sax & Kribs, 1930; de Poucques, 1949; Mehra, 1976; Bedi, Bir & Gill, 1981, 1982) reveal only diploid ($n = 9$), tetraploid ($n = 18$) and an apparent aneuploid at the tetraploid ($n = 16$) levels. None of these taxa is closely related to the endemic *A. insularis*. The closest relative of the island taxon is *A. coreana* of sect. *Zabelia* (treated at the generic level by Fukuoka, 1968, 1972) series *Biflorae* (Chung & Sun, 1984) and it is known as $2n = 36$ from a population in Manchuria (Probatova & Sokolovskaya, 1988). A new count has also been obtained recently from a population in mainland Korea as $n = 54$ (Sun *et al.*, 1996*a*). Hence, it seems likely that no change in chromosome level has occurred in the evolution of the endemic *A. insularis*. The $n = 36$ report needs to be investigated further, however.

Diploid chromosome counts for $2n = 18$ are abundant in the genus *Lonicera* (Fedorov, 1969; Rüdenberg & Green, 1969). It is not surprising, therefore, that the endemic *L. insularis* also occurs at the same chromosomal level (Table 9.4). The closest relative appears to be *L. morrowii* A. Gray (Nakai, 1919), reported on several occasions also as $2n = 18$ (Fedorov, 1969). It is a reasonable hypothesis, therefore, that no change in chromosome number occurred during the evolution of *L. insularis*.

The endemic taxon, *Sambucus sieboldiana* var. *pendula*, is counted here as $n = 18$, consistent with other known counts in the genus (Fedorov, 1969). Previous counts are available for *S. sieboldiana* at $2n = 36$ (Funabiki, 1958*a*) and $2n = 38$ (Hounsell, 1968; Ourecky, 1970; Benko-Iseppon, 1992). As $2n = 38$ also occurs in a few other taxa of the genus (e.g., *S. buergeriana* and *S. racemosa*: Funabiki, 1958*b*; Fedorov, 1969), further sampling here on the island as well as of other relatives needs to be completed before chromosomal alterations during evolution of this taxon can be understood.

The relationships of the Ullung Island endemic *Silene takesimensis* to other taxa in the genus are unclear. The count of $n = 12$ confirms the previous count for the species of $2n = 24$ (Lee, 1972) that is consistent with most other counts in the genus (Fedorov, 1969) and is the lowest recorded number. It is most likely, therefore, that the endemic island species evolved from a relative at the same chromosomal level, but further studies are obviously needed.

Chrysanthemum zawadskii var. *lucidum* is endemic to Ullung Island, whereas the species occurs in both Korea and Japan. The published counts of $2n = 36$ for this infraspecific taxon (Lee, 1969*a*, 1975) correspond to other counts of $2n = 36$ for *C. zawadskii* (Harn & Lee, 1968; Lee, 1967, 1975). Other counts in the species are $2n = 54$ and 72 (Shimotomai, 1937; Fedorov, 1969; Lee, 1967, 1969*a*, *b*, 1975; Watanabe, Nishii & Tanaka, 1972). The putative tetraploid chromosomal level of the island endemic raises interest as to its origin from relatives in Asia, but the existing counts in the parental species at the same level may suggest that no chromosomal change has accompanied divergence of this island taxon.

Nakai (1919) hypothesized that the endemic *Phytolacca insularis* is closely allied to *P. esculenta*. The former is known cytologically from a previously reported count of $2n = 72$ (Lee, 1972), the octoploid level within the genus (Fedorov, 1969). *Phytolacca esculenta* van Houtt is also known at this level as is *P. japonica* Makino (Oginuma, Tanaka & Suzuki, 1980). Still other species have been reported as $2n = 36$ (Davis, 1985) and one is $2n = 18$ (*P. sessilifolia*; Sugiura, 1936, 1939). Further detailed studies are obviously of fundamental importance here, but the preliminary viewpoint suggests that no change in chromosome number has occurred during anagenesis.

Hepatica maxima is closely related to *H. asiatica* Nakai, as documented earlier in this chapter. Park (1974), in fact, treated the former as a variety of the latter. The two taxa are both known now as $n = 7$ ($2n = 14$ in *H.*

asiatica; Probatova & Sokolovskaya, 1988) and hence there has been no change in chromosomal level during anagenesis of the island endemic. *Hepatica insularis* from South Korea has also been reported as $n = 7$ ($2n = 14$; Okada & Tamura, 1979). Most other counts in the genus are also $n = 7$ (Kurita, 1957; Baumberger, 1970; Okada & Tamura, 1979).

The endemic *Rubus takesimensis* is now known cytologically for the first time as $n = 7$. The closest relative is *R. crataegifolius* Bunge, which is also $n = 7$ (as $2n = 14$; Iwatsubo & Naruhashi, 1992) thus suggesting no chromosomal change during evolution of the island endemic. Another recent count of $2n = 21$ in *R. crataegifolius* also exists, however, (Li, Song & Chen, 1993).

Veronica insulare is counted here as $n = 16$. The taxon is sometimes placed in the segregate genus *Pseudolysimachion* (Lee & Yamazaki, 1983). Whatever the generic perspective, the closest relative appears to be *V. schmidtiana* Regel from Japan and known counts are $2n = 34$ (Sakai, 1935; Yamazaki & Tateoka, 1959). Within the genus, $n = 17$ is the most common number reported (Fedorov, 1969), with the lesser occurrence of $n = 16$. It seems in this case, therefore, that aneuploid reduction may have occurred in the evolution of the endemic *V. insulare* from its Japanese relative, *V. schmidtiana*.

The island endemic *Bupleurum latissimum* is known chromosomally as $n = 8$ ($2n = 16$; Lee, 1967). It appears to be most closely related to *B. longiradiatum* Turcz., known from both the Korean Peninsula (Choi *et al.*, 1996) as well as Japan (Hiroe & Constance, 1958). Most species of the genus are $2n = 16$ (Fedorov, 1969), including one count at this level for the close relative *B. longiradiatum* ($n = 8$; Bell & Constance, 1957). This latter species, however, has more often been reported as $2n = 12$ (or $n = 6$; Gurzenkov & Gorovoy, 1971; Arano & Saito, 1977, Cauwet-Marc, 1979; Pan *et al.*, 1985; Gorovoy & Volkova, 1987; Kim & Yoon, 1990; Daushkevich, Vasil'eva & Pimenov, 1993). A report of $n = 16$ has also been recorded for this species (Cauwet-Marc, 1978). In view of the cytological variation within *B. longiradiatum*, therefore, more study clearly seems warranted before comments can be made regarding chromosomal change in the evolution of the island endemic.

The $n = 22$ chromosome report for *Dystaenia takesimana* is the same ploidy level as the $2n = 44$ reported for the close relative from Japan, *D. ibukiensis* (Arano & Saito, 1980). That this species is the closest relative of the island taxon seems clear because the genus is usually regarded as bitypic (Kitagawa, 1937; Ohwi, 1965). It should be noted, however, that Hiroe & Constance (1958) submerged *Dystaenia* into *Ligusticum*, a genus of 20–25 species of circumboreal distribution. Arano & Saito (1980) also suggested a close relationship with *Angelica*.

The endemic *Viola takesimana* is most likely allied to *V. grypoceras* (Nakai, 1919; Lee & Yang, 1981) of Korea and Japan. Lee (1967) reported a chromosome number for *V. takesimana* of $2n = 20$. *Viola grypoceras* is also known as $2n = 20$ (Miyaji, 1913, 1927, 1929, 1930) and thus reveals no chromosomal change in this phyletic line.

These results indicate, based on preliminary interpretations, that most of the endemic angiosperm species of Ullung Island show no aneuploid or euploid evolutionary relationships with close relatives in source-area regions, with one exception. *Veronica insulare* appears to have speciated with corresponding aneuploid chromosomal reduction. Most of the results from Ullung Island, however, remain consistent with the concept that evolution involving chromosome number alterations has not played an important role in the evolution of endemic species of oceanic islands. Similar results have been revealed from Hawaii (Carr, 1978, 1985; see Chapter 1), the Bonin islands (Ono, 1991) and the Juan Fernandez Islands (Sanders, Stuessy & Rodríguez R., 1983; Spooner *et al.*, 1984; Sun, Stuessy & Crawford, 1990). On oceanic islands in general, reticulate evolution seems a very rare phenomenon; such chromosomal alterations may play a disruptive and negative role for survival in these isolated archipelagos (Sun, Stuessy & Crawford, 1990; see detailed discussions in Chapters 1 and 12). Ullung Island is geologically youthful and is not isolated from major source-areas (even less so during Pleistocene sea-level lowering), however, which could have resulted in differing evolutionary patterns and processes.

Acknowledgements

Appreciation is expressed to the National Science Foundation (USA; to Tod Stuessy) for the support of one field excursion to Ullung Island for the preliminary collection of plant materials, and to the Korea Science and Engineering Foundation for research support for Byung-Yun Sun (Grant no. 911-0409-068-2). Thanks also go to T. Lammers for help with literature regarding Campanulaceae, Larissa Menon for preparing labels for the figures, and curators of the following herbaria for the loan of specimens of *Hepatica*: (BM, E, GH, KYO, L, LE, M, OS, PR, PRC, TI).

Literature cited

Albrecht, G. H. (1978). Some comments on the use of ratios. *Systematic Zoology,* **27**, 67–71.

Anon. (1988). *Ullung County Report.* Korea: Ullung County.

Arano, H. & Saito, H. (1977). Cytological studies in family Umbelliferae. II. Karyotypes in some genera of *Bupleurum, Spuriopimpinella* and *Pimpinella. Kromosomo,* **2**, 178–85.

Arano, H. & Saito, H. (1980). Cytological studies in the family Umbelliferae. V. Karyotypes of seven species in subtribe Seselinae. *Kromosomo* (ser. 2) **17**, 471–80.

Atchley, W. R. & Anderson, D. (1978). Ratios and the statistical analysis of biological data. *Systematic Zoology,* **28**, 71–8.

Atchley, W. R., Gaskins, C. T. & Anderson, D. (1976). Statistical properties of ratios. I. Empirical results. *Systematic Zoology,* **25**, 137–48.

Baumberger, H. (1970). Chromosomenzahlbestimmungen und Karyotypanalysen bei den Gattungen *Anemone, Hepatica* und *Pulsatilla. Berichte Schweizerischen Botanischen Gesellschaft,* **80**, 17–96.

Bedi, Y. S., Bir, S. S. & Gill, B. S. (1981). IOPB chromosome number reports. LXIII. *Taxon,* **30**, 843.

Bedi, Y. S., Bir, S. S. & Gill, B. S. (1982). Cytological studies in certain woody members of family Caprifoliaceae. *Journal of Tree Science,* **1**, 27–34.

Bell, C. R. & Constance, L. (1957). Chromosome numbers in Umbelliferae. *American Journal of Botany,* **44**, 565–73.

Benko-Iseppon, A. M. (1992). Karyologische Untersuchung der Caprifoliaceae s.l. und möglicher verwandter Familien. Ph.D. thesis, University of Vienna.

Carr, G. D. (1978). Chromosome numbers of Hawaiian flowering plants and the significance of cytology in selected taxa. *American Journal of Botany,* **65**, 236–42.

Carr, G. D. (1985). Additional chromosome numbers of Hawaiian flowering plants. *Pacific Science,* **39**, 303–6.

Cauwet-Marc, A. M. (1976). Biosytématique des espèces vivaces de *Bupleurum* L. (Umbelliferae) du bassin Méditerranéen occidental. Fasicule II. Connaissances actuelles sur la genre. Ph.D. thesis, University of Languedoc.

Cauwet-Marc, A. M. (1978). IOPB chromosome number reports. LXI. *Taxon,* **27**, 385–6.

Chang, C. S. (1994). A critique of 'Reexamination of vascular plants in Ullung Island, Korea II: taxonomic identity of *Acer takesimense* Nakai (Aceraceae)'. *Korean Journal of Plant Taxonomy,* **24**, 279–84.

Chang, C. S. & Giannasi, D. E. (1991). Foliar flavonoids of *Acer* sect. *Palmata* series *Palmata. Systematic Botany,* **16**, 225–41.

Cho, H. J., Kim, S., Suh, Y. & Park, C. W. (1996). ITS sequences of some *Acer* species and phylogenetic implication. *Korean Journal of Plant Taxonomy,* **26**, 271–91.

Choi, H. K., Kim, H. J., Shin, H. & Kim, Y. (1996). Phylogeny and ribosomal DNA variations of *Bupleurum* (Umbelliferae). *Korean Journal of Plant Taxonomy,* **26**, 219–33.

Chung, Y. H. & Sun, B. Y. (1984). Monographic study on the endemic plants of Korea. IV. Taxonomy and interspecific relationships of the genus *Abelia. Korean Journal of Plant Taxonomy,* **14**, 137–52.

Corruccini, R. S. (1977). Correlation properties of morphometric ratios. *Systematic Zoology,* **6**, 211–14.

Daushkevich, Y. V., Vasil'eva, M. G. & Pimenov, M. G. (1993). Chromosome numbers and their variation in some *Bupleurum* species (Umbelliferae). *Botaniceskij Zurnal (Moscow & Leningrad)*, **78** (11) 93–100. [In Russian.]

Davis, J. I. (1985). Introgression in Central American *Phytolacca* (Phytolaccaceae). *American Journal of Botany*, **72**, 1944–53.

de Poucques, M. L. (1949). Recherches caryologiques sur les Rubiales. *Revue Generale de Botanique*, **56**, 5–27, 74–96, 97–138, 172–188.

Fedorov, A. A. (1969). *Khromosomnye Chisla Tsvetkovykh Rastenii [Chromosome Numbers of Flowering Plants]*. Leningrad: Acad. Sciences USSR.

Ferakova, V. (1978). Index of chromosome numbers of Slovakian flora (part 6). *Acta Facultatis Rerum Naturalium Universitatis Comenianae, Botanica*, **26**, 1–42.

Foster, R. C. (1933). Chromosome number in *Acer* and *Staphylea*. *Journal of the Arnold Arboretum*, **14**, 386–93.

Fukuoka, N. (1968). Phylogeny of the tribe Linnaeeae. *Acta Phytotaxonomica Geobotanica*, **23**, 82–94.

Fukuoka, N. (1972) Taxonomic study of the Caprifoliaceae. *Memoires Faculty of Science Kyoto University*, ser. *Biology*, **6**, 15–58.

Funabiki, K. (1958*a*). Distribution and polyploidy of angiosperms. I. *Kromosomo*, **37–38**, 1253–67.

Funabiki, K. (1958*b*). Distribution and polyploidy of angiosperms. II. Northern flora of Japan. *Kromosomo*, **37–38**, 1268–75.

Gorovoy, P. G. & Volkova, S. A. (1987). A morphological, geographical and karyological study of East Asian *Bupleurum longiradiatum* Turcz. and *B. sachalinense* Fr. Schmidt. *Feddes Repertorium Specierum Novarum Regni Vegetabilis*, **98**, 838–9.

Gurzenkov, N. N. (1973). Studies of chromosome numbers of plants from the south of the Soviet Far East. *Komarov Lectures*, **20**, 47–61.

Gurzenkov, N. N. & Gorovoy, P. G. (1971). Chromosome numbers of Umbelliferae of the Far East. *Botaniceskij Zurnal*, **56**, 1805–15. [In Russian.]

Harn, C. Y. & Lee, M. S. (1968). Studies on the putative parents of cultivated *Chrysanthemum*. IV. *Korean Journal of Botany*, **11**, 33–7.

Hills, M. (1978). On ratios – a response to Atchley, Gaskins and Anderson. *Systematic Zoology*, **27**, 61–2.

Hiroe, M. & Constance, L. (1958). Umbelliferae of Japan. *University of California Publications in Botany*, **30**, 1–144.

Hounsell, R. W. (1968). Cytological studies in *Sambucus*. *Canadian Journal Genetics and Cytology*, **10**, 235–47.

Iwatsubo, Y. & Naruhashi, N. (1992). Cytotaxonomical studies of *Rubus* (Rosaceae). I. Chromosome numbers of 20 species and 2 natural hybrids. *Journal of Japanese Botany*, **67**, 270–5.

Kim, J. W. (1988). The phytosociology of forest vegetation on Ulreung-do, Korea. *Phytocoenologia*, **16**, 259–81.

Kim, Y. K. (1985). Petrology of Ulreung volcanic island, Korea. 1. Geology. *Journal of Japanese Association of Minerologists, Petrologists and Economic Geologists*, **80**, 128–35. [In Japanese.]

Kim, Y. S. & Yoon, C. Y. (1990). A taxonomic study on the genus *Bupleurum* in Korea. *Korean Journal of Plant Taxonomy*, **20**, 209–42.

Kitagawa. M. (1937). Miscellaneous notes on Apiaceae (Umbelliferae) of Japan and Manchuria. II. *Botanical Magazine (Tokyo)*, **51**, 805–12.

Kurita, M. (1937). Chromosome studies in Ranunculaceae. I. Karyotypes of the subtribe Anemoninae. *Reports Biological Institute Ehime University*, **1**, 11–17.

Lee, T. B. (1980). *Illustrated Flora of Korea.* Seoul: Hyangmoonsa.

Lee, T. B. (1984). Outline of Korean endemic plants and their distribution. *Korean Journal of Plant Taxonomy,* **14**, 21–32.

Lee, T. B. & Yamazaki, T. (1983). A revision of the Scrophulariaceae in Korea. *Bulletin Kwanak Arboretum. Korea,* **4**, 34–70.

Lee, W. T. & Yang, I. S. (1981). The flora of Ulreung and Dogdo Island. In *A Report on the Scientific Survey of the Ulreung and Dogdo Islands,* ed. Anon., pp. 61–5. Seoul: The Korean Association for Conservation of Nature.

Lee, Y. N. (1966). Taxonomic studies on *Fagus multinervis, Fagus japonica* and *Fagus crenata. Journal Korean Research Institute Ewha Women's University,* **10**, 373–7.

Lee, Y. N. (1967). Chromosome number of flowering plants in Korea. 1. *Journal Korean Research Institute Ewha Women's University,* **11**, 455–78.

Lee, Y. N. (1969a). A cytotaxonomic study on *Chrysanthemum zawadskii* complex in Korea. 2. Polyploidy. *Korean Journal of Botany,* **12**, 35–48.

Lee, Y. N. (1969b). Chromosome number of flowering plants in Korea. 2. *Journal Korean Research Institute for Better Living,* **2**, 141–5.

Lee, Y. N. (1972). Chromosome number of flowering plants in Korea. 4. *Journal Korean Research Institute for Better Living,* **8**, 41–51.

Lee, Y. N. (1975). Taxonomic study on white flowered wild *Chrysanthemum* in Asia. *Journal Korean Research Institute for Better Living,* **14**, 63–79.

Li, X. Y., Song, W. Q. & Chen, R. Y. (1993). Studies on karyotype of some berry plants in North China. *Journal Wuhan Botanical Research,* **11**, 289–92.

Mehra, P. N. (1976). *Cytology of Himalayan Hardwoods.* Calcutta: Sree Saraswaty Press.

Mehra, P. N. & Khosla, P. K. (1969). IOPB chromosome number reports. XX. *Taxon,* **18**, 213–21.

Mehra, P. N., Khosla, P. K. & Sareen, T. S. (1972). Cytological studies of Himalayan Aceraceae, Hippocastanaceae, Sapindaceae and Staphyleaceae. *Silvae Genetica,* **21**, 96–102.

Miyaji, Y. (1913). Untersuchungen über die Chromosomenzahlen bei einigen *Viola*-Arten. *Botanical Magazine (Tokyo),* **27**, 443–60.

Miyaji, Y. (1927). Untersuchungen über die Chromosomenzahlen bei einigen *Viola*-Arten. *Botanical Magazine (Tokyo),* **41**, 262–8.

Miyaji, Y. (1929). Studien über die Zahlenverhältnisse der Chromosomen bei der Gattung *Viola. Cytologia,* **1**, 28–58.

Miyaji, T (1930). Betrachtungen über die Chromosomenzahlen von *Viola,* Violaceen und verwandten Familien. *Planta,* **11**, 631–49.

Nakai, T. (1919). Report on the Vegetation of the Island Ooryongto or Dagelet Island, Corea. Seoul: The Government of Chosen. [In Japanese.]

Oginuma, K., Tanaka, R. & Suzuki, K. (1980). Karyomorphological studies on three species of *Phytolacca* of Japan. *Chromosome Information Service,* **29**, 6–8.

Oh, B. U. (1986). A taxonomic study of the genus *Corydalis* in Korea. Ph.D. thesis, Korea University.

Ohwi, J. (1965). *Flora of Japan,* ed. F. G. Meyer & E. H. Walker. Washington DC: Smithsonian Institution Press.

Okada, H. & Tamura, M. (1979). Karyomorphology and relationship in the Ranunculaceae. *Journal of Japanese Botany,* **54**, 65–77.

Ono, M. (1991). Flora of the Bonin Islands: endemism and dispersal mode. *Aliso,* **13**, 95–105.

Ourecky, D. K. (1970). Chromosome morphology in the genus *Sambucus*. *American Journal of Botany*, **57**, 239–44.

Paik, W. K. & Lee, W. T. (1989). A taxonomic study of the genus *Abelia* in Korea. *Korean Journal of Plant Taxonomy*, **19**, 139–55.

Pan, Z., Chin, H., Wu, Z., Yuan, C. & Liou, S. (1985). A report on the chromosome numbers of Chinese Umbelliferae. *Acta Phytotaxonomica Sinica*, **23**, 97–102. [In Chinese.]

Park, C. W., Oh, S. H. & Shin, H. (1993). Reexamination of vascular plants in Ullung Island, Korea. II. Taxonomic identity of *Acer takesimense* Nakai (Aceraceae). *Korean Journal of Plant Taxonomy*, **23**, 217–32.

Park, C. W., Oh, S. H. & Shin, H. (1994). Character analysis and taxonomic identity of *Acer takesimense* Nakai. *Korean Journal of Plant Taxonomy*, **24**, 285–94.

Park, M. K. (1974). *Keys to the Herbaceous Plants in Korea (Dicotyledonae)*. Seoul: Jungumsa.

Probatova, N. W. & Sokolovskaya, A. P. (1988). Chromosome numbers in vascular plants from Primorye Territory, the Amur River basin, north Koryakia, Kamchatka and Sakhalin. *Botaniceskij Zurnal*, **73**, 290–3. [In Russian.]

Probatova, N. W. & Sokolovskaya, A. P. (1989). Chromosome numbers in vascular plants from Primorye Territory, the Amur region, Sakhalin, Kamchatka and the Kuril Islands. *Botaniceskij Zurnal*, **74**, 120–3. [In Russian.]

Rüdenberg, L. & Green, P. S. (1969). A karyological survey of *Lonicera*. II. *Journal of the Arnold Arboretum*, **50**, 449–61.

Sakai, K. (1935). Studies on the chromosome number in alpine plants. II. *Japanese Journal of Genetics*, **11**, 68–73. [In Japanese.]

Sanders, R. W., Stuessy, T. F. & Rodríguez R., R. (1983). Chromosome numbers from the flora of the Juan Fernandez Islands. *American Journal of Botany*, **70**, 799–810.

Santamour, F. S., Jr. (1988). New chromosome counts in *Acer* (maple) species, sections *Acer* and *Goniocarpa*. *Rhodora*, **90**, 127–31.

Sax, K. & Kribs, D. A. (1930). Chromosomes and phylogeny in Caprifoliaceae. *Journal of the Arnold Arboretum*, **11**, 147–53.

Shimotomai, N. (1937). Chromosomenzahlen bein einigen Arten von *Chrysanthemum. Zeitschrift Induktion Abstammungs und Verebungslehre*, **74**, 30–3.

Spooner, D. M., Stuessy, T. F., Crawford, D. J. & Silva O., M. (1984). Chromosome numbers from the flora of the Juan Fernandez Islands. II. *Rhodora*, **89**, 351–6.

Sugiura, T. (1936). A list of chromosome numbers in angiospermous plants. II. *Proceedings of the Imperial Academy of Tokyo*, **12** (5), 144–6.

Sugiura, T. (1939). Studies on the chromosome numbers in higher plants. III. *Cytologia*, **10**, 205–12.

Sun, B-Y., Park, J. H., Kwak, M. J., Kim, C. H. & Kim, K. S. (1996*a*). Chromosome counts from the flora of Korea with emphasis on Apiaceae. *Journal of Plant Biology*, **39**, 15–22.

Sun, B-Y., Stuessy, T. F., Humaña, A. M., Riveros G., M. & Crawford, D. J. (1996*b*). Evolution of *Rhaphithamnus venustus* (Verbenaceae), a gynodioecious hummingbird-pollinated endemic of the Juan Fernandez Islands, Chile. *Pacific Science*, **50**, 55–65.

Sun, B-Y., Stuessy, T. F. & Crawford, D. J. (1990). Chromosome counts from the flora of the Juan Fernandez Islands, Chile. III. *Pacific Science,* **44**, 258–64.

Watanabe, K., Nishii, Y. & Tanaka, R. (1972). Anatomical observations on the high frequency callus formation from another culture of *Chrysanthemum*. *Japanese Journal of Genetics,* **47**, 249–55.

Yamazaki,T. & Tateoka, T. (1959). Cytotaxonomic studies in *Veronica* and related genera. I. *Annual Report National Institute of Genetics (Japan)*, **9**, 54.

Yoo, K. O. (1995). Taxonomic studies on the Korean Campanulaceae. Ph.D. thesis, Kangweon National University.

10

Evolution in *Crossostylis* (Rhizophoraceae) on the South Pacific Islands

HIROAKI SETOGUCHI, HIDEAKI OHBA AND
HIROSHI TOBE

Abstract

Crossostylis (Rhizophoraceae) is one of the inland genera of Rhizophoraceae. Its distribution is on the South Pacific Islands, encompassing New Caledonia, the Fiji Islands, Vanuatu, the Solomon Islands, Samoa, the Society Islands and the Marquesas Islands. Species of this genus show morphological diversity especially of the inflorescences, flowers and seed coats. Based on cladistic analyses of morphological characters and restriction site variation of chloroplast DNA (cpDNA), we discuss phylogeny, speciation and evolutionary trends of morphological characters in *Crossostylis*. Cladistic analyses based on morphological and DNA data yield the same topological cladogram. Phylogenetic analyses suggest that species of *Crossostylis* are divided into two pronounced monophyletic groups: one comprising six species distributed on the Solomon Islands, Vanuatu and the Fiji Islands, and the other comprising four species distributed on New Caledonia and Polynesia. Monophyly of the former group agrees with general assumptions of phytogeographical affinities, but the monophyletic Polynesia–New Caledonian species group is not coincident with general floristic affinities for the region. All species endemic to the Fiji Islands are very closely related and this suggests speciation from a single ancestor on the islands. Some morphological characters were optimized on the cpDNA tree obtained and examined for evolutionary trends. Many of the floral morphological features support macromolecular clades, but all seed coat characters are homoplasious. Perhaps this reflects high adaptive significance, such as is involved with seed dispersal.

Crossostylis J. R. Forst. et G. Forst., one of the inland genera of Rhizophoraceae, is distributed on the major islands of the South Pacific. Since J. R. Forster and G. Forster described the first species in the genus in 1776, 16 species (including synonyms) have been described from the Fiji Islands, New Caledonia, Vanuatu, the Solomon Islands, the Samoa Islands, the Society Islands and the Marquesas Islands. These islands are scattered over 6500 km from east to west in the South Pacific Ocean (Fig. 10.1).

Crossostylis consists of trees up to 30 m tall that grow in inland habitats from 10 to 1000 m above sea level. Species of *Crossostylis* occasionally develop abundant aerial roots at the base of the stem (Fig. 10.2), like stilt roots of *Rhizophora*. Rhizophoraceae, a mangrove family, contains about 15 genera and 135 species (Juncosa & Tomlinson, 1988a). *Crossostylis* belongs to the tribe Gynotrocheae along with *Carallia*, *Gynotroches* and *Pellacalyx* (see van Vliet, 1976, for a review of infrafamilial classification in Rhizophoraceae), which are mainly distributed in South East Asia. The distribution of *Crossostylis* only overlaps with other genera of the family on the Solomon Islands (with *Carallia* and *Gynotroches*; Ding Hou & van Steenis, 1963). *Crossostylis* is clearly distinguished from allied genera by several obvious morphological

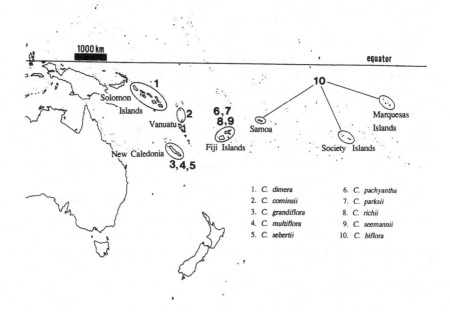

Fig. 10.1. Geographical distribution of species of *Crossostylis*.

Fig. 10.2. Aerial roots of *Crossostylis cominsii*. Vanua Lava Island, Vanuatu.

characters, such as semi-inferior ovaries (vs. superior or inferior in allied genera), capsular fruits (vs. baccate) and arillate seeds (vs. non-appendaged). Capsular fruits of *Crossostylis* are indehiscent, with pericarp detaching at the base and with a naked white arillate sarcocarp. Seeds are fleshy-arillate and dispersed by birds and/or ants.

Although 16 species of *Crossostylis* have been reported from the South Pacific Islands, species delimitation and relationships are obscure. Morphological characters for all species have been presented in original descriptions and in local floras, and Juncosa (1988) and Juncosa & Tomlinson (1988*b*) have provided information on the morphology of several species of *Crossostylis*. Information on details of morphology of most species, however, is still lacking. The necessity for comprehensive studies on various morphological characters and monographic work has been pointed out by Ding Hou & van Steenis (1963) and Smith (1981).

In this chapter, we first report the distribution of each species of *Crossostylis* in the South Pacific. Second, we present the morphological diversity, especially of inflorescences, flowers and seed coats. Third, we discuss phylogeny, speciation and evolutionary trends of morphological

Table 10.1. *Species of* Crossostylis *and their distributional ranges*

Species	Distribution range
Crossostylis biflora J. R. et G. Forst.	Samao, Society Islands, Marquesas Islands
C. cominsii Hemsl.	Vanuatu, Banks Group, Vanua Lava Island Solomon Islands, Santa Cruz Group
C. dimera Ding Hou	Solomon Islands (from Bougainville Island to San Cristobal Island)
C. grandiflora Brongn. et Gris	New Caledonia, Grande Terre Island
C. multiflora Brongn. et Gris	New Caledonia, Grande Terre Island
C. pachyantha A. C. Sm.	Fiji Islands, Viti Levu Island, Vanua Levu Island
C. parksii (Gillespie) A. C. Sm.	Fiji Islands, Viti Levu Island
C. richii (A. Gray) A. C. Sm.	Fiji Islands, Viti Levu Island, Vanua Levu Island
C. sebertii (Pancher) ex Brongn. et Gris	New Caledonia, Grande Terre Island
C. seemannii (A. Gray) A. Schimp.	Fiji Islands, Viti Levu Island, Vanua Levu Island, Ovalau Island

characters in *Crossostylis* through cladistic analyses using morphological characters and restriction site variation of chloroplast DNA (cpDNA).

Distribution on South Pacific Islands

Setoguchi *et al.* (unpublished) have recently examined species delimitations in *Crossostylis* and have revised them into ten species (Table 10.1) endemic to the Fiji Islands, New Caledonia, Vanuatu, the Solomon Islands, the Samoa Islands, the Society Islands and the Marquesas Islands (Fig. 10.1).

New Caledonia

On New Caledonia, three endemic species, *C. grandiflora*, *C. multiflora* and *C. sebertii*, are allopatrically distributed. Their distributions clearly correspond to edaphic conditions: *C. multiflora* is distributed in substrate from michaschistes parental material; *C. serbertii* is in ultramaphic soils containing iron oxides with Cr, Ni and Mg in the southern part of New Caledonia; and *C. grandiflora* is commonly distributed in soil with leaf mould.

The flowers of each species are hermaphroditic but quite different in size: disc *c.* 1.5 cm diameter and stamens 3 cm long in *C. grandiflora* (Fig. 10.3A), *c.* 6 mm and 5 mm in *C. sebertii* (Fig. 10.3B), and *c.* 3 mm and 1.7 mm in *C. multiflora* (Fig. 10.3C). They also exhibit substantial variation in floral and seed coat structures as described below.

The Fiji Islands

Four endemic species occur on the Fiji Islands: *C. pachyantha*, *C. parksii*, *C. richii* and *C. seemannii*. Contrary to New Caledonia, morphological characters of the Fijian species are relatively similar. In all species the trees are dioecious. Taxonomically they can be distinguished by small differences in the number of stamens and carpels, shapes of leaves, and number of secondary veins. Size of flowers and number of stamens vary continuously among species. Floral and seed coat structure is also uniform (described in more detail below).

Samoa, the Society Islands and the Marquesas Islands

Only one species, *C. biflora*, is distributed on Samoa, the Society Islands and the Marquesas Islands. The only structural differences that have been observed between populations on these different islands are in the staminodia (Setoguchi *et al.*, unpublished). Staminodia are hairy in populations from Samoa, but glabrous in those from the Marquesas, and populations consisting of individuals with hairy staminodia and those with glabrous ones are found on the Society Islands. This morphological differentiation might be genetically based, but precise molecular analyses will be needed to confirm (or reject) this hypothesis.

Fig. 10.3. Flowers of *Crossostylis* on New Caledonia. A, *C. grandiflora*; B, *C. sebertii*; C, *C. multiflora*. Scale bars equal 1 cm.

Vanuatu and the Solomon Islands

Crossostylis cominsii is distributed on Vanua Lava Island of Vanuatu and Vanikoro Island of the Solomon Islands. This species is very similar to species on the Fiji Islands in floral and inflorescence features; seed coat structure is quite different, however.

Crossostylis dimera is the only species distributed on the Solomon Islands. It is characterized by having diplostemonous stamens and diplocarpelous carpels.

Morphological diversity

Inflorescence

Inflorescences, which are always born in leaf axils, are basically dichasia (Fig. 10.4A) but show various configurations according to the degree of branching of axes as well as the position of suppressed flowers and pedicels (Fig. 10.4A–G). Quaternary axes are the highest order of branches (Fig. 10.4D); terminal flowers are sometimes lacking in any order of branches. The number of flowers per inflorescence varies from two (*C. biflora*, Fig. 10.4C) to 12 (*C. multiflora*, Fig. 10.4D).

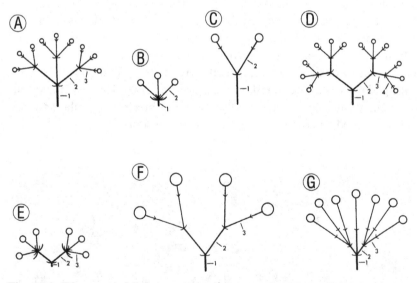

Fig. 10.4. Diagrams of inflorescences of *Crossostylis*. Numerals 1, 2, 3 and 4 indicate positions of primary, secondary, tertiary and quaternary axes. A, *C. dimera*; B, *C. cominsii, C. parksii, C. richii* and *C. seemannii*; C, *C. biflora*; D, *C. multiflora*; E, *C. pachyantha*; F, *C. grandiflora*; G, *C. sebertii*.

Floral sexuality

Flowers are hermaphroditic in four species: *C. biflora, C. grandiflora, C. multiflora* (Fig. 10.5A, B) and *C. sebertii*. They are unisexual (trees dioecious) in six species: *C. cominsii* (Fig. 10.5C–F), *C. dimera, C. pachyantha, C. parksii, C. richii* and *C. seemannii*. In dioecious species, female flowers have sterile stamens (Fig. 10.5D, F) which remain on the flower up to later fruiting stages (Fig. 10.5G, H; seen easily on herbarium materials).

Number and arrangement of floral parts

Both uni- and bisexual flowers in *Crossostylis* have sepals, petals, stamens and a gynoecium. Sepals vary from four to six and are fused at the base, alternating with the petals (of equal number). Stamens are arranged in a whorl, varying from eight to 28. Figure 10.6 presents floral diagrams that were drawn on the basis of flowers of each species of *Crossostylis*, illustrating number and arrangement of floral parts. Table 10.2 gives floral formulae of the ten species. Three of the ten species (*C. grandiflora, C. biflora* and *C. sebertii*; Fig. 10.6F, G and H, respectively) have a whorl of staminodia positioned between the whorl of stamens and the gynoecium. The gynoecium is multicarpellate with a short undivided style; number of carpels varies from three to 26 (Table 10.2).

In summary, the number of sepals and petals vary between four and six, and the number of stamens (and staminodia) and carpels are about two to five or more times greater than the perianth parts. Looking at individual species, most have a varying number of floral elements. In *C. sebertii* and *C. multiflora* (Table 10.2; Fig. 10.6H and I, respectively) the number of sepals and petals is always four, and the number of stamens (and staminodia) and carpels always eight. In these two species the number of stamens is always twice as many as the petals, with episepalous stamens alternating with epipetalous ones. In addition, in *C. biflora* and *C. grandiflora* the number of sepals and petals is always four, whereas the number of stamens, staminodia and carpels is variable (Table 10.2).

Morphology of floral parts

The free parts of sepals (i.e., calyx lobes) are thick and triangular. In fruiting stages, calyx lobes are reflexed in *C. cominsii, C. dimera, C. pachyantha, C. parksii, C. richii* and *C. seemannii* (Fig. 10.7A), spreading

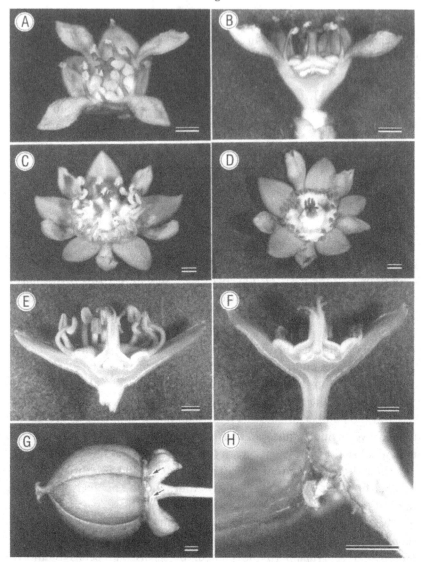

Fig. 10.5. Floral sexuality in *Crossostylis*. Hermaphroditic flower of *C. multiflora* (A, upper view; B, longitudinal section). Male flower of *C. cominsii* (C, upper view; E, longitudinal section). Female flower of *C. cominsii* (D, upper view; F, longitudinal section). Fruit of *C. parksii* showing short sterile stamens (G and H). Scale bars equal 1 mm.

Table 10.2. *Floral formulae of species of* Crossostylis

K = calyx; C = corolla; A = androecium; S = staminodia; G = gynoecium. Calyx and gynoecium are always connate, the numbers given in parenthesis. Floral elements invariant within species are enclosed by rectangles.

Species	Floral formula
C. biflora	K(4) C4 A20–28 S20–28 G(11–18)
C. cominsii	K(4–6) C4–6 A19–31 G(8–11)
C. dimera	K(5–6) C5–6 A10–12 G(5–6)
C. grandiflora	K(4) C4 A23–28 S23–28 G(20–26)
C. multiflora	K(4) C4 A8 G(8)
C. pachyantha	K(4–6) C4–6 A19–25 G(8–13)
C. parksii	K(4–5) C4–5 A10–14 G(4–5)
C. richii	K(4–6) C4–6 A14–20 G(3–6)
C. sebertii	K(4) C4 A8 S8 G(8)
C. seemannii	K(4–6) C4–6 A14–20 G(6–9)

Source: From Setoguchi, Ohba & Tobe (1996)

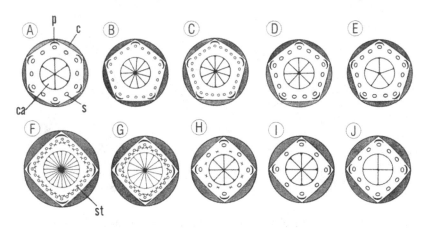

Fig. 10.6. Floral diagrams of *Crossostylis* showing variation in the number of sepals (calyx lobes), petals, stamens, staminodia and carpels. Abbreviations: c = calyx member; ca = carpel; p = petal; s = stamen; st = staminodium. A, *C. dimera*; B, *C. pachyantha*; C, *C. cominsii*; D, *C. seemannii*; E, *C. richii*; F, *C. grandiflora*; G, *C. biflora*; H, *C. sebertii*; I, *C. multiflora*; J, *C. parksii*.

in *C. biflora* (Fig. 10.7B), *C. multiflora* and *C. sebertii*, and erect or straight in *C. grandiflora* (Fig. 10.7C).

The petals are keeled and have entire margins (Fig. 10.7D). They are always somewhat undulate near the apex in *C. sebertii* (Fig. 10.7E) and sometimes so in *C. biflora* and *C. grandiflora*. The apex of the petals is pointed in *C. biflora*, *C. grandiflora* (Fig. 10.7F, G) and *C. sebertii*, but is not pointed in the other seven species (Fig. 10.7D). The outer surface of the petals is hairy in all species of *Crossostylis*; the inner surface is glabrous in *C. biflora* and *C. grandiflora*, and hairy in the eight other taxa.

In five of the ten species, *C. biflora, C. grandiflora, C. multiflora, C. pachyantha* (Fig. 10.8A) and *C. sebertii*, a membranous, webbed structure is observed between the base of the filaments. Such a membranous web is lacking in the other five species (Fig. 10.8B, *C. cominsii*). Among rhizophoraceous genera, a similar structure is reported in *Sterigmapetalum* (Steyermark & Liesner, 1983). As in *Sterigmapetalum*, the membranous

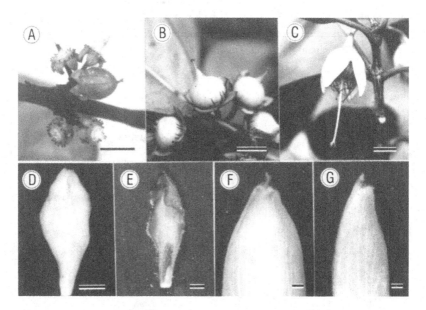

Fig. 10.7. Flowers and fruits of *Crossostylis* showing details of floral parts. A, reflexed calyx lobes in *C. seemannii*; B, spreading calyx lobes in *C. biflora*; C, ascending calyx lobes in *C. grandiflora*; D, petal of *C. pachyantha*; E, petal with undulate margin in *C. sebertii*; F and G, abaxial and lateral views (respectively) of petal with a pointed apex in *C. grandiflora*. Scale bars equal 1 cm in A–C, 1 mm in D–G.

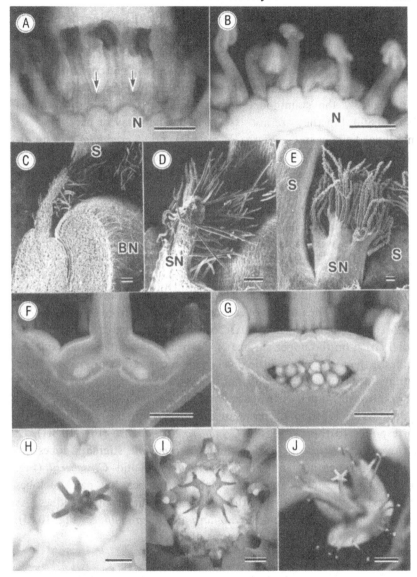

Fig. 10.8. Flowers of *Crossostylis* showing details of floral parts. A, stamens with membranous webs (arrows) between filament bases in *C. pachyantha*; B, stamens without membranous webs in *C. cominsii*; C, longitudinally dissected stamino-dium and nectary in *C. biflora*; D, staminodium with unicellular hairs in *C. biflora*; E, staminodium with multicellular hairs in *C. sebertii*; F, longitudinally dissected ovary of *C. cominsii* showing ovarian locules with little interior space; G, longitudinally dissected ovary of *C. biflora* showing ovarian locules with much interior space; H, stigma with separate lobes in *C. cominsii*; I and J, stigmas with fused lobes in *C. sebertii* and *C. grandiflora*, respectively. Abbreviations: BN = bulge of nectary; N = nectary; S = stamen; SN = staminode. Scale bars equal 1 mm in A and B; 100 μm in C–E; 1 mm in F–I; and 5 mm in J.

web of *Crossostylis* may have been derived by the fusion of filament bases.

Staminodia are found in three species: *C. biflora, C. grandiflora* and *C. sebertii*. They are short-filiform or subulate, and are present on the inner side of the stamens and alternate with them (Fig. 10.6F, G, H). Histologically the staminodia appear similar to nectariferous tissues (Fig. 10.8C), but they are not vascularized. They may or may not have unicellular hairs as in *C. biflora* (Fig. 10.8D) and *C. grandiflora*, or multicellular hairs as in *C. sebertii* (Fig. 10.8E). Juncosa (1988) and Juncosa & Tomlinson (1988b) used the term 'androecious appendages' for the staminodia of *C. grandiflora* and they regarded these structures as representing an autapomorphy for the genus. However, since the staminodia are restricted only to three of the ten species, it seems inappropriate to regard their presence as an autapomorphy for the entire genus. It is uncertain whether the staminodia of *Crossostylis* have any function.

The ovary is semi-inferior in all species of *Crossostylis*. It is multicarpellate with two ovules in each carpel and with imperfect septa between ovarian locules. The number of carpels, which varies from three (*C. richii*) to 26 (*C. grandiflora*; Table 10.2), can be determined by counting the radial striae on the upper surface of the ovary. Placentation is axial. At anthesis, the locules have little open space because they are largely filled with mature ovules (Fig. 10.8F); in *C. biflora* (Fig. 10.8G) and *C. grandiflora*, however, much open space remains.

Stigmas are lobed in all species of *Crossostylis* and the number of lobes corresponds to the carpels. In *C. dimera, C. cominsii* (Fig. 10.8H), *C. pachyantha, C. parksii, C. richii* and *C. seemannii*, stigmatic lobes are separate throughout. In *Crossostylis multiflora* and *C. sebertii* (Fig. 10.8I), stigmatic lobes are always eight in number, but pairs of lobes are fused at the base so that the stigma has four groups of lobes. As many as 20 lobes are found in *C. biflora* and *C. grandiflora* (Fig. 10.8J), which are also fused into four groups of lobes.

Floral nectary

The floral nectary of *Crossostylis*, like that of other Rhizophoraceae, is annular and located inside the androecial whorl. In *C. biflora, C. grandiflora, C. multiflora* (Fig. 10.9A) and *C. sebertii*, a part of the nectary is swollen inwardly from the staminal side. Such swollen nectariferous tissues are not observed in the six other species of *Crossostylis* (e.g., *C. cominsii*; Fig. 10.9B), nor are they found in other Rhizophoraceae. In

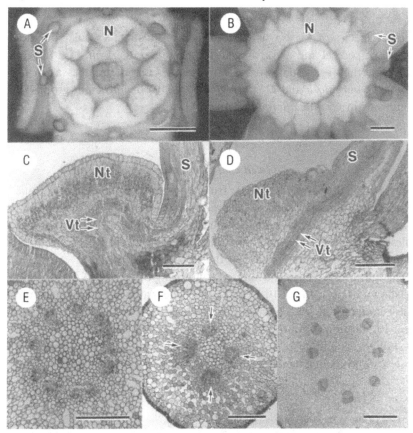

Fig. 10.9. Flowers of *Crossostylis* showing details of floral parts and anatomy. A and B, upper views of the nectary of *C. multiflora* and *C. cominsii*, respectively. Note that the nectary of *C. multiflora* has swollen tissues; C and D, longitudinal sections of nectaries of *C. multiflora* and *C. cominsii*, respectively; E–G, transverse sections of the pedicel of *C. richii*, *C. sebertii* and *C. grandiflora*, respectively, showing the number and arrangement of vascular bundles. Arrows indicate bunched vascular bundles. N = nectary; Nt = nectariferous tissue; S = stamen; Vt = vascular tissue. Scale bars equal 1 mm in A and B; 500 μm in C–E; and 100 μm in F and G.

the nectary with swollen tissues, a staminal vascular bundle deeply penetrates and curves into it (e.g., *C. multiflora*; Fig. 10.9C). Such a vascularization may suggest that the nectary was derived from the base of the filament. In the nectary without swollen tissues, however, the staminal vascular bundle does not enter the nectary (e.g., Fig. 10.9D).

Vasculature in the pedicel

In most species of *Crossostylis*, more than ten distinct vascular bundles run in the pedicel and are arranged in a ring (e.g., *C. richii*; Fig. 10.9E). However, they are usually fused into four groups of bundles as in *C. biflora*, *C. multiflora* and *C. sebertii* (Fig. 10.9F) or into eight groups of bundles as in *C. grandiflora* (Fig. 10.9G).

Seed coat structure

Seeds, which are borne in capsular fruits and range in number per capsule from one (*C. multiflora* and *C. parksii*) to 30 (*C. grandiflora*), are ellipsoid to ovate and 1.6–4.1 mm long (Setoguchi, Tobe & Ohba, 1992), and have a fleshy and membranaceous aril that grows from the tip of the outer integument and covers nearly half the seed (Fig. 10.10).

The mature seed coat and surface sculpturing of species of *Crossostylis* are shown in Figs. 10.11, 10.12 and 10.13. The mature seed coat structure consists mainly of both a well-developed exotesta and a fibrous exotegmen. In addition, both the thin, mostly vestigial, mesotesta and the thin crystalliferous endotesta are observed. Species differ in the thickness of exotesta and exotegmen, the shape of exotestal cells, the presence and

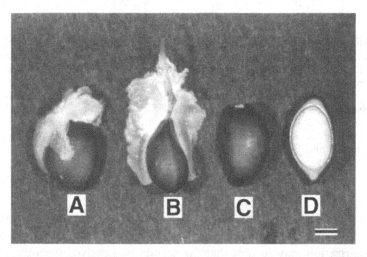

Fig. 10.10. Mature seeds of *Crossostylis parksii*. Seeds viewed from dorsal side (A, C, D), and from raphal side (B). In C and D, an aril has been removed from the seed. Scale bar equals 1 mm.

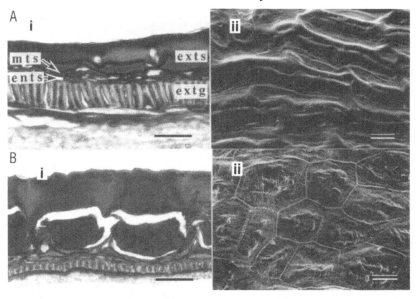

Fig. 10.11. Seed coat structure (transverse section, i) and surface sculpturing (SEM, ii) of species of *Crossostylis* from the Solomon Islands and Vanuatu. A, *C. cominsii*; B, *C. dimera*. Exts = exotesta; mts = mesotesta; ents = endotesta; extg = exotegmen. Scale bars equal 50 µm.

absence of persistent mesotesta, tanniferous deposits in exotestal cells, and seed surface sculpturing. On the basis of the position of the characteristic mechanical layer, we classified the seed coat as 'exotestal' or 'exotestal-exotegmic' after Corner (1976) and Schmid (1986).

Exotestal seed coats were observed in seven species of *Crossostylis*, such as *C. dimera* (Fig. 10.11B), *C. grandiflora* (Fig. 10.13A), *C. sebertii* (Fig. 10.13C) and all four species on the Fiji Islands (*C. pachyantha, C. parksii, C. richii* and *C. seemannii*, Fig. 10.12A–D, respectively). Exotestal cells are enlarged and densely tanniferous with an extremely thick cuticle in these species. In *C. grandiflora*, exotestal cells are longitudinally elongated, but are radially enlarged and cuboidal in the six other species.

On the other hand, exotestal-exotegmic seed coats were observed in *C. cominsii* (Fig. 10.11A), *C. multiflora* (Fig. 10.13B), and *C. biflora* (Fig. 10.13D). The exotestal cells in these species are flattened, and longitudinally elongate in *C. cominsii* and *C. multiflora*, but not elongated in *C. biflora*. In the two species, *C. multiflora* and *C. biflora*, exotestal cells lack tanniferous deposition.

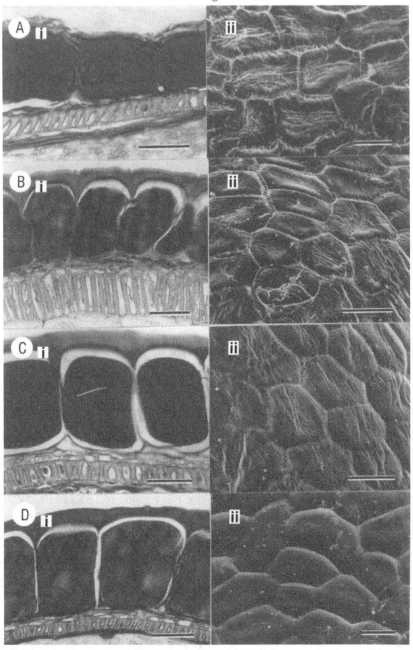

Fig. 10.12. Seed coat structure (transverse section, i) and surface sculpturing (SEM, ii) of species of *Crossostylis* from the Fiji Islands. A, *C. pachyantha*; B, *C. parksii*; C, *C. richii*; D, *C. seemannii*. Scale bars equal 50 μm.

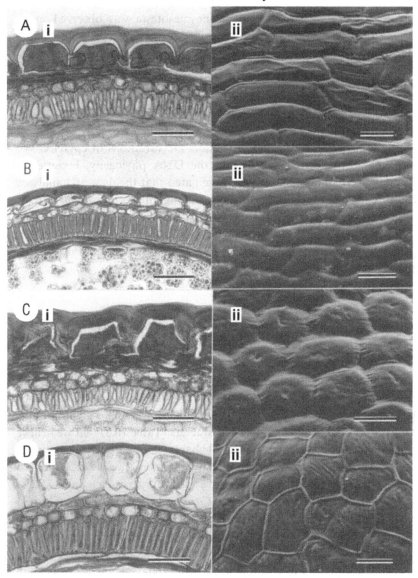

Fig. 10.13. Seed coat structure (transverse section, i) and surface sculpturing (SEM, ii) of species of *Crossostylis* from New Caledonia and the islands of Samoa and French Polynesia. A, *C. grandiflora*; B, *C. multiflora*; C, *C. sebertii*; D, *C. biflora*. Scale bars equal 50 μm.

A persistent layer (or layers) of the mesotesta was observed in *C. biflora*, *C. grandiflora* and *C. sebertii*. Mesotestal cells were almost crushed in the remaining taxa.

Phylogenetic analysis and trends in character evolution

We cladistically elucidated the phylogeny within *Crossostylis* based on morphological characters and restriction site variations of cpDNA. We were unable to add *C. dimera* into the DNA phylogeny, because no materials were obtainable. It was anticipated that the data and analyses would reveal the evolutionary lineages among all species of *Crossostylis*, as well as serving to re-evaluate the infrageneric classification. Also, the evolutionary trends of the various morphological characters in *Crossostylis* were evaluated by their parsimonious distribution on the cpDNA tree.

Phylogeny inferred from morphological characters

To determine evolutionary relationships among the ten species of *Crossostylis*, we conducted cladistic analyses using parsimony (PAUP version 3.1.1; Swofford, 1993). To get rooted trees, *Carallia* and *Gynotroches*, which are considered sister groups of *Crossostylis* (Tobe & Raven, 1988; Juncosa & Tomlinson, 1988*b*), were used as outgroups. We have chosen 24 morphological characters (Table 10.3) that we believe are cladistically informative. Of the 24 characters, 17 are floral characters (nos. 1–17 in Table 10.3), five seed coat characters (nos. 18–22) and two vegetative characters (nos. 23, 24). We added four characters (nos. 25–28) which distinguish *Crossostylis* from *Carallia* and *Gynotroches* (Corner, 1952; Ding Hou, 1958; Geh & Keng, 1974; Juncosa & Tomlinson, 1988*a*,*b*; Tobe & Raven 1988; H. Setoguchi, personal observation). Character state evaluation was made on the basis of outgroup comparison; each character was coded 0 or 1 when its character state occurred in *Carallia* and/or *Gynotroches*. We used Wagner parsimony (Farris, 1970) of PAUP 3.1.1.

The results, using the branch and bound option, yielded seven equally parsimonious trees. These were 46 steps long with a consistency index (CI) of 0.65 and a retention index (RI) of 0.76. Six of the seven parsimonious trees indicate that the species of *Crossostylis* are divided into two monophyletic groups (Fig. 10.14). One comprises *C. biflora*,

Table 10.3. *Characters and states used in the morphological cladistic analysis of* Crossostylis

Character	Character state
FLOWERS AND INFLORESCENCES	
1. Flower	0 = hermaphroditic; 1 = unisexual (trees dioecious)
2. Central buds in inflorescences	0 = present; 1 = absent
3. Merosity in flowers	0 = 4–6; 1 = 4
4. Calyx lobes in fruiting stage	0 = spreading to ascending; 1 = reflexed
5. Outer surface of calyx	0 = glabrous; 1 = hairy
6. Inner surface of calyx lobes	0 = glabrous; 1 = hairy
7. No. stamens vs. calyx lobes	0 = twice; 1 = more than twice
8. Apex of petals	0 = pointed; 1 = not pointed
9. Inner surface of petals	0 = glabrous; 1 = hairy
10. Base of filaments	0 = glabrous; 1 = hairy
11. Staminodia	0 = absent; 1 = present
12. Stigmatic lobes	0 = free; 1 = fused into four groups
13. Ovary surface	0 = glabrous; 1 = hairy
14. Ovarian locules at anthesis	0 = with much space; 1 = without much space
15. Swollen tissue on nectary	0 = absent; 1 = present
16. Vascularization at flower base	0 = not fasciculated; 1 = fasciculated
17. Staminal vascular bundle in nectary	0 = absent; 1 = present
SEED COAT	
18. Seed coat type	0 = exotestal; 1 = exotestal-exotegmic
19. Tanniniferous deposition in exotestal cells	0 = present; 1 = absent
20. Shapes of exotestal cells	0 = radially enlarged, cuboidal; 1 = slightly flattened, irregularly cuboidal; 2 = extremely flattened
21. Elongation of exotestal cell	0 = non-elongated; 1 = elongated
22. Mesotesta	0 = persistent; 1 = vestigial
VEGETATIVE MORPHOLOGY	
23. Lower surface of leaves	0 = glabrous; 1 = pilose
24. Shoot	0 = glabrous; 1 = pilose
CHARACTERS DISTINGUISHING *CROSSOSTYLIS* FROM OTHER GENERA	
25. Ovary	0 = superior; 1 = inferior
26. Nectary at filament bases	0 = not annular, discontinuous; 1 = annular, continuous
27. Seed appendages	0 = absent; 1 = present
28. Filamentous appendages on petal	0 = present; 1 = absent

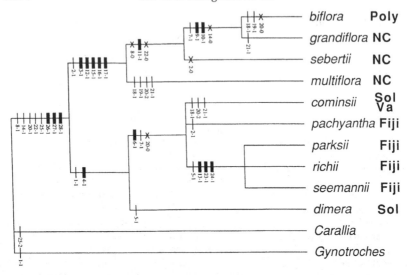

Fig. 10.14. The 50% majority rule consensus tree of seven equally most parsimonious trees of *Crossostylis* inferred from morphological characters. Length = 46; CI = 0.65; RI = 0.76. Character numbers correspond to those in Table 10.3. Thick bars indicate characters with consistency indices of 1.00; thin bars characters less than 1.00; crosses indicate reversals. Poly = Polynesia; NC = New Caledonia; Sol = Solomon Islands; Va = Vanuatu; Fiji = Fiji Islands. (After Setoguchi, Ohba & Tobe, 1996.)

C. grandiflora, *C. multiflora* and *C. sebertii*, and the other *C. cominsii*, *C. dimera*, *C. pachyantha*, *C. parksii*, *C. richii* and *C. seemannii*.

Phylogeny inferred from cpDNA restriction site variations

Nine species of *Crossostylis* (excepting *C. dimera* that could not be obtained) were used for cpDNA restriction site variation analysis. For rooting of the phylogenetic tree, we chose *Gynotroches* as the most closely related outgroup among genera of Gynotrocheae, because preliminary experiments revealed that many cpDNA restriction sites in *Crossostylis* were also found in this genus. As the leaf-powder suspended solution was too sticky to purify total DNA with the usual CTAB (hexadecyl trimethyl ammonium bromide) extraction method (Hasebe & Iwatsuki, 1990), we used an SDS extraction which involved washing leaf powder in HEPES (2-hydroxyethyl piperazine-N′-2-ethanesulfonic acid) buffer (Setoguchi & Ohba, 1995).

DNA was digested with 19 endonucleases with 6 bp recognition sequences. DNA fragments were separated by electrophoresis using agarose gels and blotted to nylon membranes. The probes used in Southern hybridization were from the clone bank of tobacco cpDNA (Sugiura *et al.*, 1986). Restriction site data were analysed using Wagner parsimony of PAUP 3.1.1. The most parsimonious tree was obtained using the branch and bound option, and bootstrap analysis (Felsenstein, 1985) of 1000 replicates was conducted to obtain estimates of reliability for each monophyletic group.

We detected 23 restriction site changes, 13 of which were shared by more than two species, i.e., cladistically informative. Length mutations were excluded from phylogenetic analysis in this study. A phylogenetic analysis using Wagner parsimony resulted in a single most parsimonious tree of length 14, CI of 0.93 and RI of 0.98, as shown in Fig. 10.15. Only one mutation was homoplasious.

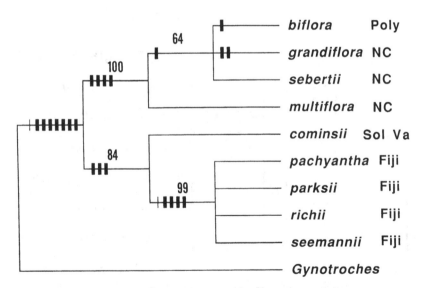

Fig. 10.15. The most parsimonious phylogenetic tree inferred from cpDNA restriction site data for species of *Crossostylis* and outgroup (*Gynotroches*) generated by Wagner parsimony of PAUP 3.1.1. Length = 14; CI = 0.93; RI = 0.98. Heavy bars indicate synapomorphic site mutations; thin bars indicate homoplasious site mutations. The number at each branch gives the confidence interval of each monophyletic group based on 100 bootstrap replicates using Wagner parsimony methods. Poly = Polynesia; NC = New Caledonia; Sol = Solomon Islands; Va = Vanuatu; Fiji = Fiji Islands. (After Setoguchi & Ohba, 1995.)

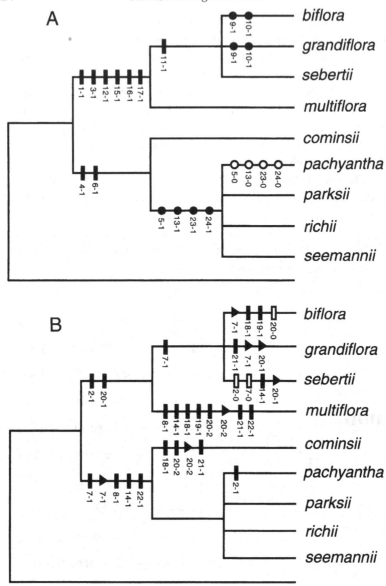

Fig. 10.16. Parsimonious replacement of various morphological characters (Table 10.3) with ACCTRAN and DELTRAN optimization on the most parsimonious phylogenetic tree based on cpDNA data (Fig. 10.15). A, parsimonious replacement of synapomorphic character states (solid bars) and those which cannot be evaluated as synapomorphic or homoplasious (solid circles). Open circles denote reversals. There is no difference in the character distribution between ACCTRAN and DELTRAN. B, parsimonious replacement of homoplasious characters with ACCTRAN (bars) and DELTRAN (triangles) optimization. Open bars denote reversals.

The tree obtained reveals that species of *Crossostylis* are divided into two monophyletic groups: group A, *C. biflora, C. grandiflora, C. multiflora* and *C. sebertii*; group B, *C. cominsii, C. pachyantha, C. parksii, C. richii* and *C. seemannii*. Confidence intervals of bootstrap in the monophyletic groups A and B were 100 and 84%, respectively. Moreover, three species, *C. biflora, C. grandiflora* and *C. sebertii* were monophyletic at 64% in group A; and four species, *C. pachyantha, C. parksii, C. richii* and *C. seemannii*, were monophyletic at 99% in group B.

In the present analysis, phylogeny inferred from cpDNA data is completely concordant with that generated from morphological data (Fig. 10.14). The topological coincidence between the two cladograms heightens confidence in the resolved phylogeny.

Character transformations of morphological elements

We parsimoniously distributed morphological character states on the cladogram based on cpDNA data to help reveal evolutionary trends. Figure 10.16 shows the parsimonious distribution of 24 characters (Table 10.3) in *Crossostylis* using two methods of optimization, accelerated transformation (ACCTRAN) and delayed transformation (DELTRAN). The distribution of each character was examined and its evolutionary trend interpreted.

As a result, 13 characters were revealed to be synapomorphic as shown in Fig. 10.16A. Many of the floral morphological characters were synapomorphic as sexuality and merosity in flowers, etc. Distribution and evolutionary trends of these characters according to ACCTRAN were concordant with those of DELTRAN and no discrepancy was recognized between them. On the other hand, all five seed coat characters were homoplasious along with several floral characters (Fig. 10.16B). In two characters (no. 7 and 20), the distribution pattern and evolutionary trends differed between ACCTRAN and DELTRAN. It could not be determined whether the remaining morphological characters (no. 9 and 10) were synapomorphic or homoplasious, because these characters occur within the multiple furcation (Fig. 10.16A).

Nine characters are clearly homoplasious among the 24 characters. Five of these are seed coat characters, which may point to adaptive significance. Seed coat characters, such as tanniferous deposits in the exotesta, and thickness of exotesta and exotegmen, may yield stronger seed coats to survive dispersal by birds. Form of inflorescence and number of stamens, also homoplasious, might be related to type of pollinator. The flowers of *C. grandiflora* are known to be bird-pollinated (Juncosa, 1988), and species in the Fiji Islands and Vanuatu have bee-pollinated flowers (H. Setoguchi, personal observation).

Speciation

The distribution of each species of *Crossostylis* in relation to evolutionary affinities was also examined (Fig. 10.14 and 10.15). Two monophyletic groups were revealed: (1) six species distributed on the Solomon Islands, Vanuatu and the Fiji Islands; and (2) four species on New Caledonia and Polynesia (Samoa, the Society Islands and the Marquesas Islands). These two groups are distributed allopatrically in the two different geographical regions. It is assumed, therefore, that speciation among ten species in *Crossostylis* has occurred independently in these two areas.

Smith (1955, 1981) and Parham (1972) reported that the flora of the Fiji Islands generally has close affinities with that of the Solomon Islands, Vanuatu, New Guinea and Malaysia. The present result in *Crossostylis* agrees with the general assumption of phytogeographical affinity among the Solomon Islands, Vanuatu and the Fiji Islands.

On the other hand, Morat, Veillon & Mackee (1986) estimated a very low coefficient of phytogeographical relationship between New Caledonia and Polynesia. The floristic relationship of rainforest genera between New Caledonia and Polynesia, including the Society Islands and the Marquesas Islands, was estimated as 0.5% (6.8–9.8%, however, among the Solomon Islands, Vanuatu and the Fiji Islands). Contrary to this general tendency, the monophyly of *C. biflora* (species distributed in Polynesia) and three New Caledonian species is supported in the present study. They share as synapomorphies five morphological characters and five restriction site changes. The distance between New Caledonia and Samoa is over 2500 km, and the Fiji Islands are located between them. We are uncertain whether the Fiji Islands could have functioned as a 'land bridge' in the past, but an unusual long-distance dispersal event in the South Pacific could also have led to speciation in Polynesian–New Caledonian *Crossostylis*.

On the Fiji Islands, the occurrence of four monophyletic species suggests that speciation occurred from a single ancestral introduction. Although we do not know if edaphic variation has affected the distribution of each species, they are currently allopatric on the Fiji Islands (Setoguchi *et al.*, unpublished). These four species of *Crossostylis* might have diversified by adaptation to different environments on the islands. Environmental factors, such as differences in pollinators and seed dispersers, might not effect changes in floral and seed coat structure. Uniform structures might be maintained from their common ancestry.

In New Caledonia, ecological factors might have been strong enough to bring about obvious changes in flowers, inflorescences and seed coats, as observed in *C. grandiflora, C. multiflora* and *C. sebertii.* For example, the flowers of *C. grandiflora* have been reported as bird-pollinated (Juncosa, 1988) whereas the other species are bee-pollinated (H. Setoguchi, personal observations). The flower of *C. grandiflora,* with its long stamens and gynoecium, can be interpreted as adaptive morphological changes to a bird pollination system. In the case of New Caledonia, ecological factors (including pollinators and seed dispersers) might have had a great effect on morphological divergence in *Crossostylis.*

Acknowledgements

We thank: Dr K. Oginuma, Kouchi Women's University, for his co-operation and advice on the present study; Dr P. Raven, Missouri Botanical Garden, for his co-operation in collecting seed materials; and Professor M. Ono and Professor T. Stuessy for giving us the opportunity to submit our article to this book. We are deeply indebted to the following people who kindly co-operated in the field collections: Dr S. Vodonaivalu (University of the South Pacific, Fiji); Dr J. Florence (ORSTOM in Tahiti); Mr C. Djeen and Mr E. Brotherson (Economie Rulae in Raiatea, French Polynesia); Mr M. Kilman and the late Ms C. Clunies-Ross (Department of Forestry, Vanuatu); Dr T. Jaffré and Dr J. M. Veillon (ORSTOM in Noumea); Ms C. Latreille (Department of Forestry in Province Nord); Mr M. Boulet (Direction du Developpement Rural, Province Sud); Ms M. L. Pierrez and the late Dr J. F. Cherrier (CTFT, New Caledonia); Dr S. Lum (Nanyang Technological University, Singapore); and Dr H. Doi and Dr H. Koyama (National Science Museum, Tokyo). The authors also thank A, BISH, BM, L, MO, K, NOU, NY, P, PAP, SUVA and US for allowing us to observe herbarium specimens. This study was supported in part (to Hiroaki Setoguchi) by the Dan Charitable Trust Fund for Research in the Biological Sciences and the Fujiwara Natural History Foundation.

228   *Hiroaki Setoguchi* et al.

Literature cited

Corner, E. J. H. (1952). *Wayside Trees of Malaya*, vol. 1, 2nd edn. Singapore: Government Printing Office.

Corner, E. J. H. (1976). *The Seeds of Dicotyledons*, vols. 1 & 2. Cambridge: Cambridge University Press.

Ding Hou (1958). Rhizophoraceae. *Flora Malesiana*, **5**, 429–93.

Ding Hou & van Steenis, C. G. G. J. (1963). *Crossostylis*. In *Pacific Plant Areas*, vol. 1, ed. C. G. G. J. van Steenis, pp. 290–1. Manila: Manila Bureau of Printing.

Farris, J. S. (1970). Methods for computing Wagner trees. *Systematic Zoology*, **19**, 83–92.

Felsenstein, J. (1985). Confidence limits on phylogenies: an approach using the bootstrap. *Evolution*, **39**, 783–91.

Geh, S. Y. & Keng, H. (1974). Morphological studies on some inland Rhizophoraceae. *Garden Bulletin Straits Settlement*, **27**, 183–220.

Hasebe, M. & Iwatsuki, K. (1990). *Adiantum capillus-veneris* chloroplast DNA clone bank: as useful heterologous probes in the systematics of the leptosporangiate ferns. *American Fern Journal*, **80**, 20–5.

Juncosa, A. M. (1988). Floral development and character evolution in Rhizophoraceae. In *Aspects of Floral Development*, ed. P. Leins, S. C. Tucker & P. K. Endress, pp.83–101. Berlin: Gebruder Borntraeger.

Juncosa, A. M. & Tomlinson, P. B. (1988*a*). A historical and taxonomic synopsis of Rhizophoraceae and Anisophyleaceae. *Annals of the Missouri Botanical Garden*, **75**, 1278–95.

Juncosa, A. M. & Tomlinson, P. B. (1988*b*). Systematic comparison and some biological characters of Rhizophoraceae and Anisophylleaceae. *Annals of the Missouri Botanical Garden*, **75**, 1296–1318.

Morat, Ph., Veillon, J. M. & Mackee, H. S. (1986). Floristic relationships of New Caledonian rainforest phanerogams. *Telopea*, **2**, 631–79.

Parham, J. W. (1972). *Plants of the Fiji Islands*. 2nd edn. Suva: The Government Printer.

Schmid, R. (1986). On Cornerian and other terminology of angiospermous and gymnospermous seed coats: historical perspective and terminological recommendations. *Taxon*, **35**, 476–91.

Setoguchi, H. & Ohba, H. (1995). Phylogenetic relationships in *Crossostylis* inferred from restriction site variation of chloroplast DNA. *Journal of Plant Research*, **108**, 87–92.

Setoguchi, H., Ohba, H. & Tobe, H. (1996). Floral morphology and phylogenetic analysis in *Crossostylis* (Rhizophoraceae). *Journal of Plant Research*, **109**, 7–19.

Setoguchi, H., Tobe, H. & Ohba, H. (1992). Seed coat anatomy of *Crossostylis* (Rhizophoraceae): its evolutionary and systematic implications. *Botanical Magazine Tokyo*, **105**, 625–38.

Smith, A. C. (1955). Botanical studies in Fiji. *Smithsonian Report*, **1954**, 301–15.

Smith, A. C. (1981). *Flora Vitiensis Nova*, vol. 2. Honolulu: Pacific Tropical Botanical Garden.

Steyermark, J. A. & Liesner, R. (1983). Revision of the genus *Sterigmapetalum* (Rhizophoraceae). *Annals of the Missouri Botanical Garden*, **70**, 179–93.

Sugiura, M., Shinozaki, K., Zaita, N., Kusuda, M. & Kumano, M. (1986). Clone bank of tobacco (*Nicotiana tabacum*) chloroplast genome as a set of overlapping restriction endonuclease fragments: mapping of eleven ribosomal protein genes. *Plant Science*, **44**, 211–16.

Swofford, D. L. (1993). *PAUP: Phylogenetic Analysis Using Parsimony*. Mac Version 3.1.1 (Computer program and manual). Champaign: Illinois Natural History Survey.

Tobe, H. & Raven, P. H. (1988). Seed morphology and anatomy of Rhizophoraceae, inter- and infrafamilial relationships. *Annals of the Missouri Botanical Garden*, **75**, 1319–42.

van Vliet, G. J. C. M. (1976). Wood anatomy of the Rhizophoraceae. *Leiden Botanical Series*, **3**, 20–75.

Part four

General evolutionary patterns and processes on oceanic islands

Introduction

In this part of the book we turn our attention toward general patterns and processes of evolution, although these issues have been dealt with to some extent in other chapters. In Chapter 11, Bruce Bohm provides the first review of distributions of plant secondary products in island ecosystems. Some workers have suggested (e.g., Carlquist, 1980) that in view of their probable defensive functions from fungi and insects, many compounds have, in the more open (and certainly different) habitats of oceanic islands, possibly undergone loss of diversity. That is, relaxation of predation pressure would surely lead to loss of defensive chemicals over many generations. Bohm's survey suggests the contrary. No obvious loss of compounds is seen when comparing chemical arsenals in island plants with continental relatives.

In Chapter 12, the theme of chromosomal stability in endemic plants of oceanic islands is continued by Stuessy & Crawford, that was initiated in Chapter 1 by Gerald Carr for the Hawaiian Islands. This new chapter extends Carr's observations for the Hawaiian Islands into the endemic angiosperm floras of the Juan Fernandez, Bonin and Galapagos Islands, as well as comparisons with the Canary Islands and the Queen Charlotte Islands. The Canaries have a complex geological history, are older relative to the other archipelagos and are very close to the source area in continental Africa. Much more cytological variation in certain genera is seen. On the Queen Charlotte Islands, which are essentially part of the Canadian continent separated only by vagaries of sea level, typical continental levels of chromosomal variation are revealed.

Chapter 13 forms a summary of the present volume and the editors sketch protocols for continued work in oceanic archipelagos. The contributions in this book set a tone for approaching evolutionary patterns and processes in endemic plants of oceanic islands, and this last chapter

231

brings these perspectives into focus. Clearly we have come a very long way in marshalling the necessary morphological, cytological, molecular (micro- and macro-), phylogenetic and biogeographic data, and methods of analysis for more precise results and insightful syntheses. Future work must maintain these critical standards.

Literature cited

Carlquist, S. (1980). *Hawaii: A Natural History. Geology, Climate, Native Flora and Fauna above the Shoreline*, 2nd edn. Honolulu: Pacific Tropical Botanical Garden.

11

Secondary compounds and evolutionary relationships of island plants

BRUCE A. BOHM

Abstract

Studies of plants, endemic or otherwise, on islands have disclosed the presence of a wide variety of natural products, acetylenes, alkaloids, cyanogenic glycosides, flavonoids, terpenoids and others. In the vast majority of instances, the chemistry of the island species parallels the chemistry of related continental species with regard to the class or classes of natural product under consideration. In some cases, compound profiles are very similar to those of continental species; in some, simpler profiles are seen, while in others, enriched arrays of compounds occur. In a only a few well-documented cases, notably cyanogenic species on the Galapagos Islands, has loss of the compounds in question apparently occurred.

Many classes of naturally occurring compounds have been implicated as herbivore feeding deterrents and that they exert that activity because of their toxicity. I offer the alternative suggestion that their effectiveness arises from simply tasting bad. It has been suggested that deterrent compounds would be lost if the predator were removed. However, in the absence of herbivore pressure most island species appear to maintain their capacity to manufacture these putative defensive compounds. Therefore, I suggest that the compounds may serve other functions in the plant.

The aims of this chapter are to review the literature concerning secondary compounds reported from island species with emphasis on endemics and to comment on: (1) the use of these data to evaluate evolutionary relationships with mainland species and with each other; and (2) the operation of these compounds in defensive roles in the island environ-

ment. The reader will find information on acetylenic compounds, flavonoids (including condensed tannins), simple phenolic molecules, various
terpenoids, alkaloids, cyanogenic glycosides, histamine derivatives as well
as several other types of compounds that do not fit comfortably into the
above categories.

I have included island systems for which information was available to
me, especially the Hawaiian Islands, the Juan Fernandez Islands and the
Canary Islands. New Zealand and its associated islands also play a significant role. Some readers might find the inclusion of New Zealand as an
island system inappropriate as it is often considered to be a miniconti-
nent, albeit in two parts. There are several reasons for including New
Zealand in this review, among which are the existence of an impressive
endemic flora, the sizable amount of chemical work done on the endemic
species, and the fact that it is surrounded by water. I have also included
examples from Tasmania. The survey is not exhaustive, but instead representative. A difficulty encountered is the need to know something about
both the flora of an island, endemic or otherwise, and something about
the chemistry of its plants as well. I suspect there may be a literature
based primarily on pharmacologically driven research concerning island
plants; readers who know of such information can help by informing me
of such oversights.

Chemistry of island species
The Pacific Ocean
The Hawaiian Islands

No island system has attracted the attention of biologists as have the
Hawaiian Islands. From the early days of exploration of the Sandwich
Islands to the current all-out assault on the islands' unique biota with
modern techniques, more information has been, and is being, assembled
for the Hawaiian chain than for any other island system. The Hawaiian
Islands are the most isolated of oceanic systems lying in the Central
Pacific Ocean between 154°48′ and 178°20′ W longitude, and 18°56′ to
28°22′ N latitude. The archipelago consists of eight main islands (the
'tourist' group) plus a line of degraded volcanic islets and atolls that
stretches over 2580 km to the northwest ending at Midway. The
Hawaiian Islands exhibit one of the highest degrees of species endemism
known for any island system with well over 90% of the species found

nowhere else in the world (Wagner, Herbst & Sohmer, 1990). There are over 200 endemic genera on the islands but no endemic families.

Pioneering studies of natural products of Hawaiian plants involved investigations on alkaloids (Folkers & Koniuszy, 1939; Gorman *et al.*, 1957) and a survey of plants for antibacterial properties (Bushnell, Fukuda & Makinodan, 1950). A systematic examination of Hawaiian plant chemistry, however, had to await the collaboration at the University of Hawaii of P. J. Scheuer and H. St John of the Bishop Museum who examined the endemic Hawaiian flora for alkaloids (Swanholm, St John & Scheuer, 1959, 1960; Scheuer, Horigan & Hudgins, 1962). Results clearly showed the presence of many alkaloid-bearing species in the Hawaiian flora, to a very large extent in the families where they would be expected, e.g., Apocynaceae, Rutaceae. These and other studies from these two families are discussed in detail below.

Ochrosia, a member of Apocynaceae, consists of 23 (Mabberley, 1987) or about 30 (Wagner, Herbst & Sohmer, 1990) species that occur on Pacific islands, with, 'tentatively', four (Wagner, Herbst & Sohmer, 1990) in Hawaii. *Ochrosia sandwicensis* A. Gray ('holei'), one of the endemics in Hawaii, was shown to accumulate the indole alkaloids holeinine (Fig. 11.1 [1]) (Scheuer & Metzger, 1961), ochrasandwine and a hunterburnine methochloride (Jordan & Scheuer, 1965). The latter compound is also known from species of *Hunteria*, an African genus of the Apocynaceae (Bartlett *et al.*, 1963, and citations therein). Studies of *Fagara semiarticulata* (St John & Hosaka) Degener (now considered *Zanthoxylum kauaense* A. Gray) (Rutaceae) (Wagner, Herbst & Sohmer, 1990), restricted to the Koolau Mountains of Oahu, showed the presence of chelerythrine and dihydrochelerythrine (Fig. 11.1 [2]) (Scheuer, Horigan & Hudgins, 1962). Chelerythrine, widespread in the Papaveraceae, is also known from other members of *Zanthoxylum* and from *Toddalia* (Cannon *et al.*, 1953; Govindachari & Thyagarajan, 1956). Coumarins, triterpenes, the flavanone hesperidin, and the alkaloids canthin-6-one, chelerythrine, nitidine and tembetarine have been isolated from *Zanthoxylum dipetalum* H. Mann (Fish, Gray & Waterman, 1975; Fish *et al.*, 1976). A rare protopine alkaloid 'thalictricine' (Fig. 11.1 [3]) along with four acylhistamines (Fig. 11.1 [4–7]) were described from the same species by Arslanian *et al.* (1990). Thalictricine and acylhistamines (Fig. 11.1 [5] and [7]) occurred in plants from the islands of Hawaii and Kauai, whereas the plants from a population on Oahu lacked the alkaloid but had acylhistamines [4–6]. Variation of chemical characters in this species will be discussed again below.

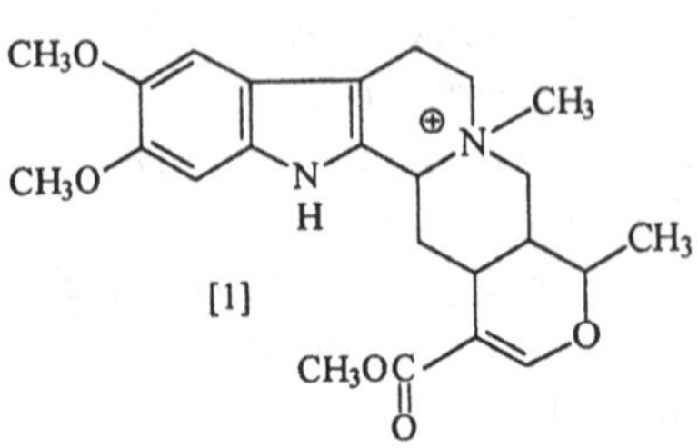

[1]

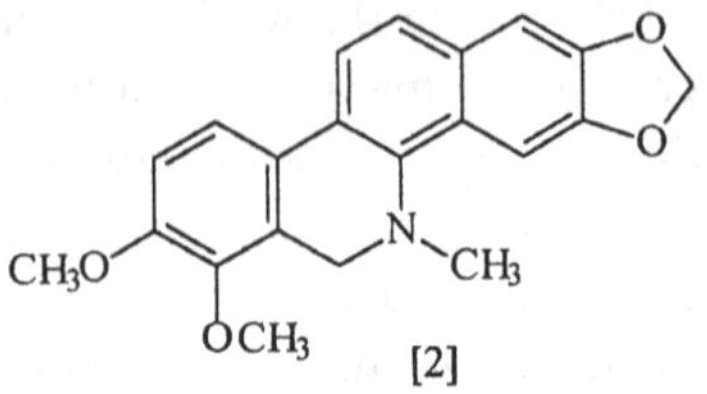

[2]

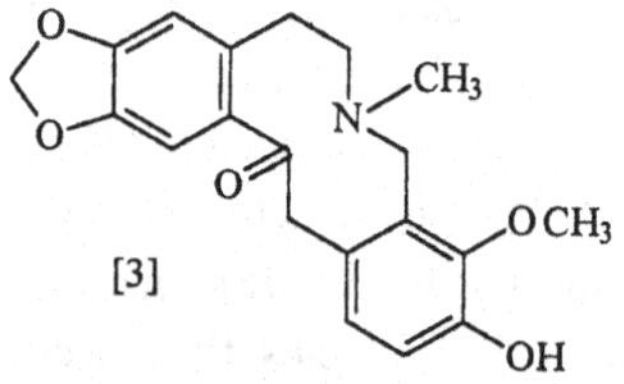

[3]

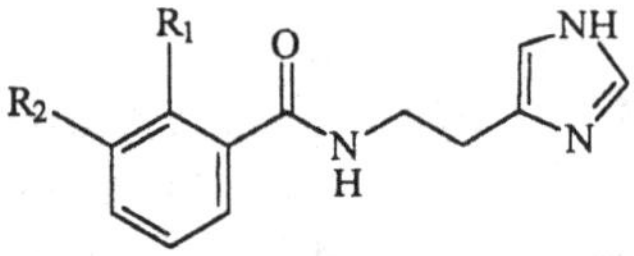

[4] $R_1 = R_2 = H$

[5] $R_1 = OCH_3$, $R_2 = H$

[6] $R_1 = R_2 = OCH_3$

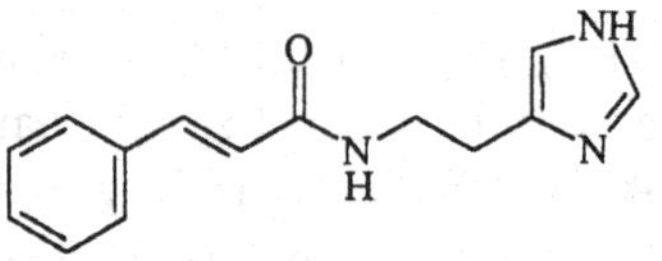

[7]

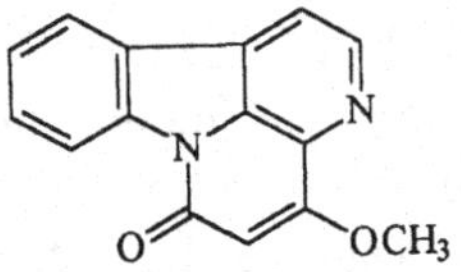

[8]

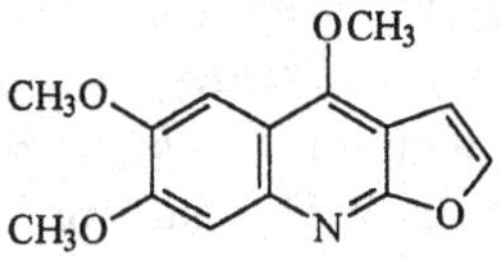

[9]

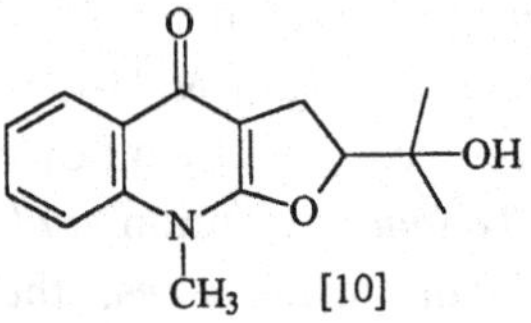

[10]

Fig. 11.1. Secondary plant compounds mentioned [by number] in the text.

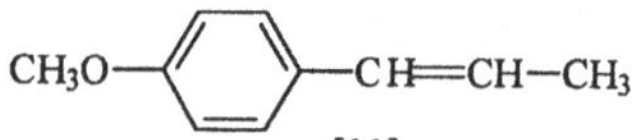

[11]

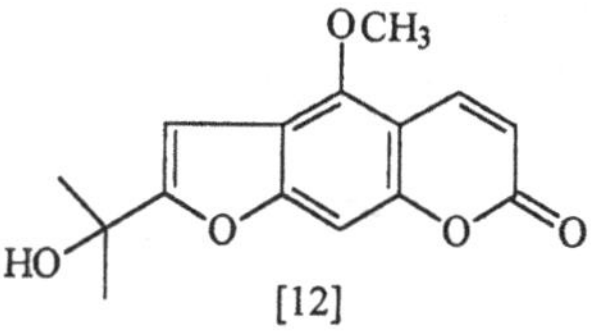

[12]

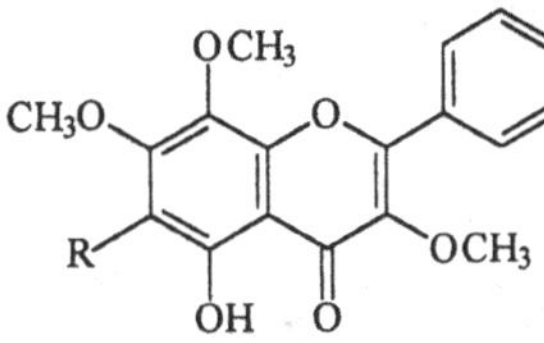

[13] R = H
[14] R = OCH₃

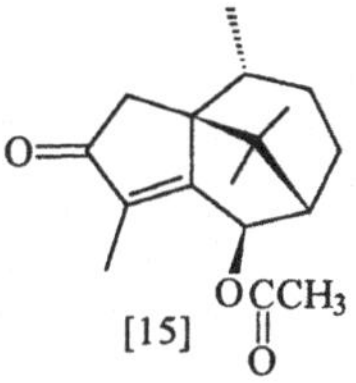

[15]

[16] CH₃ ... CH₃

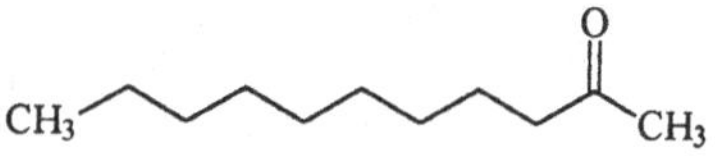

[17] CH₃ ... CH₃

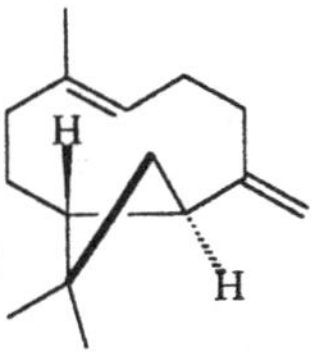

[18]

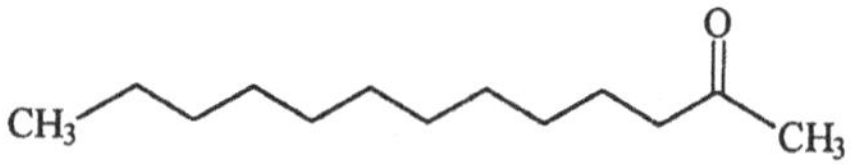

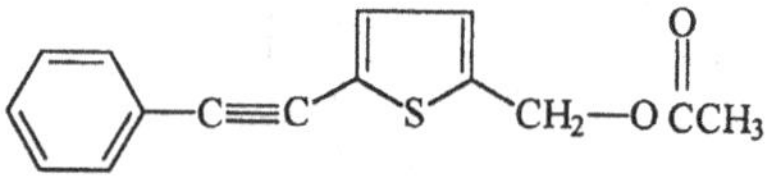

[19]

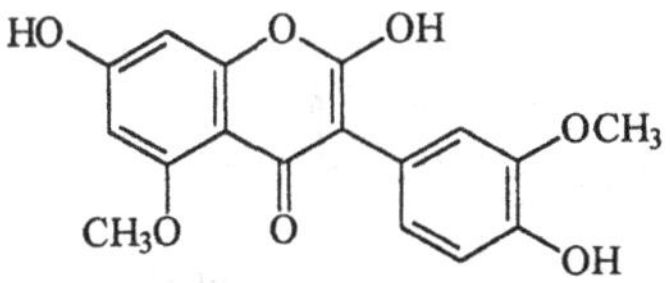

[20]

Fig. 11.1. continued

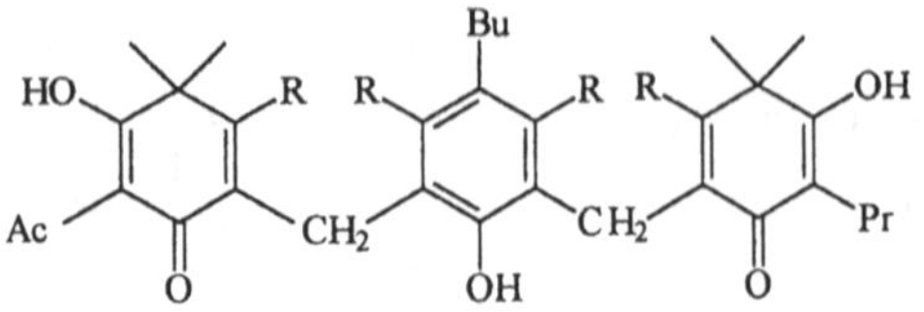

[21]

R = OH, Ac = Acetyl, Pr = Propionoyl, Bu = Butyroyl

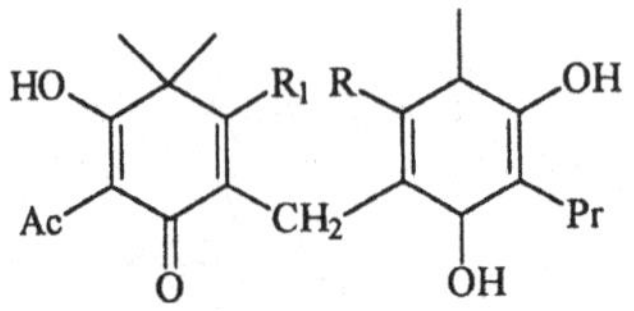

[22] R₁ = OH, R = OCH₃, Ac = Acetyl,
Pr = Propionyl

[23]

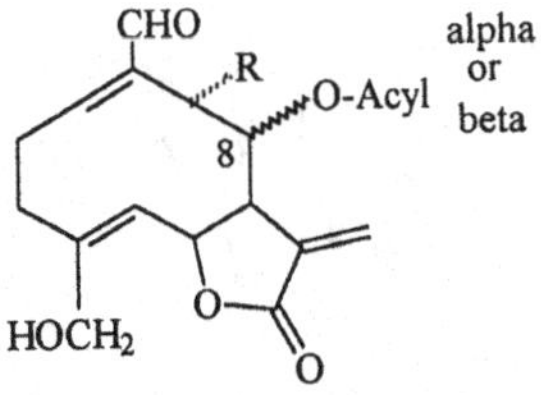

[24] R = H or Acetyl

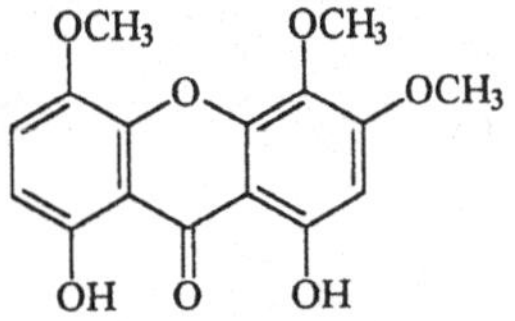

[25]

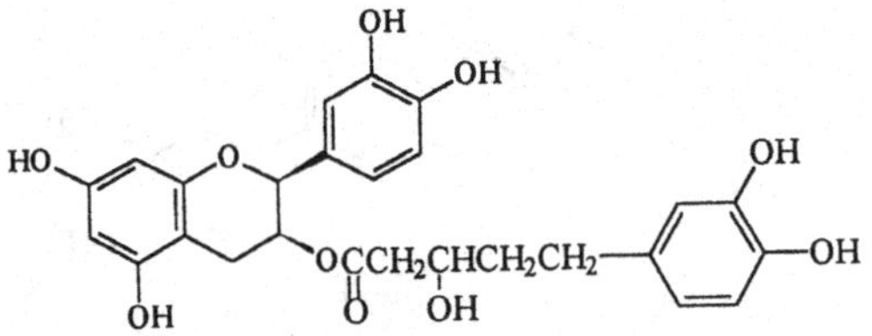

[26]

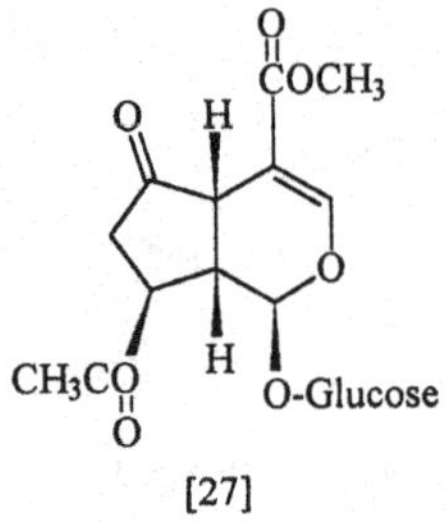

[27]

Fig. 11.1. continued

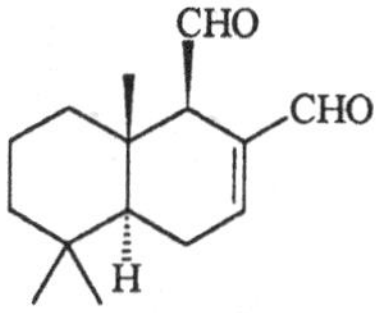

[28]

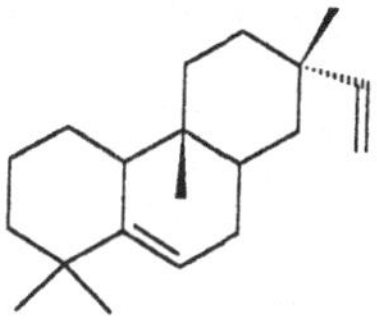

[29]

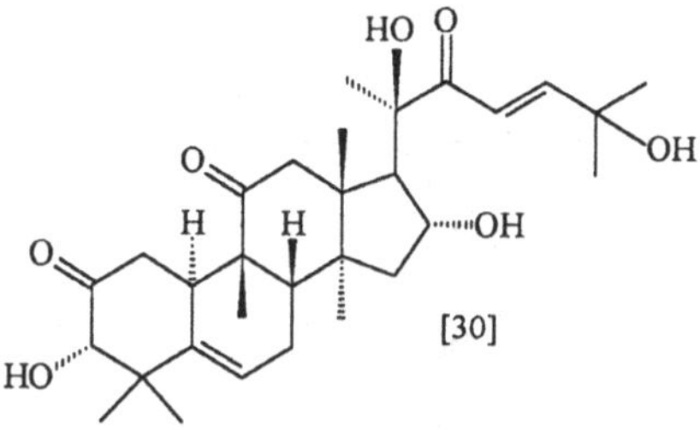

[30]

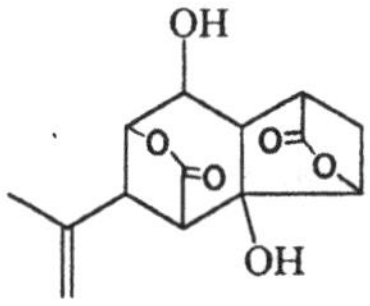

[31]

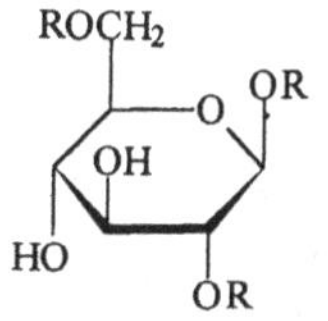

[32] R = $\overset{O}{\underset{\parallel}{C}}$CH₂CH₂NO₂

CH₂=CH-CH-C≡C-C≡C-CH-CH=CH-(CH₂)₆CH₃
| |
OH OH

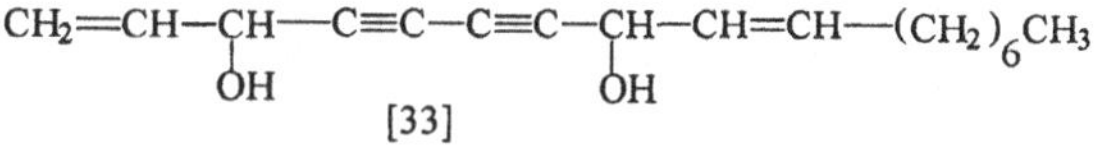

[33]

[34]

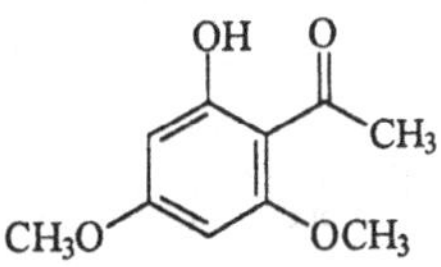

[35]

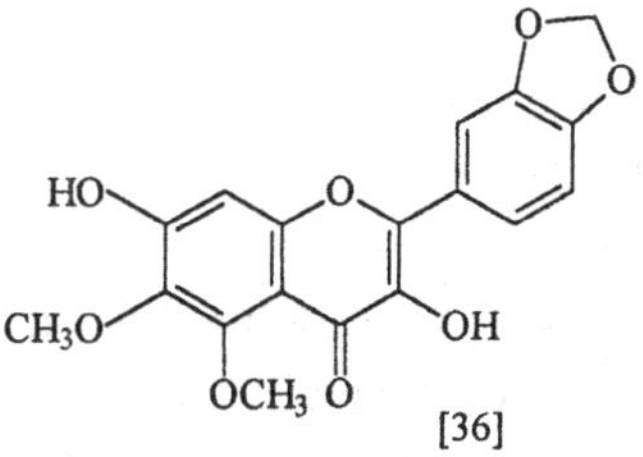

[36]

Fig. 11.1. continued

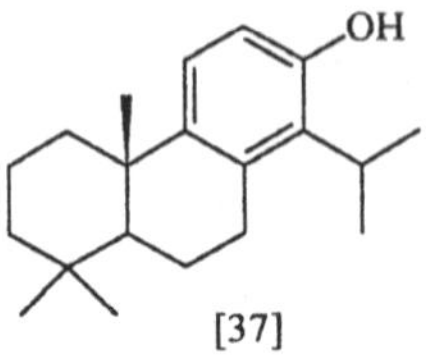

[37]

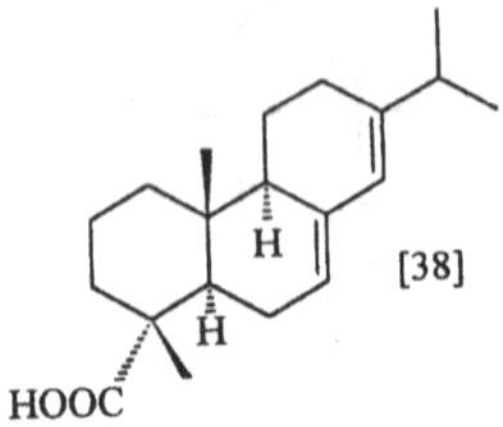

[38]

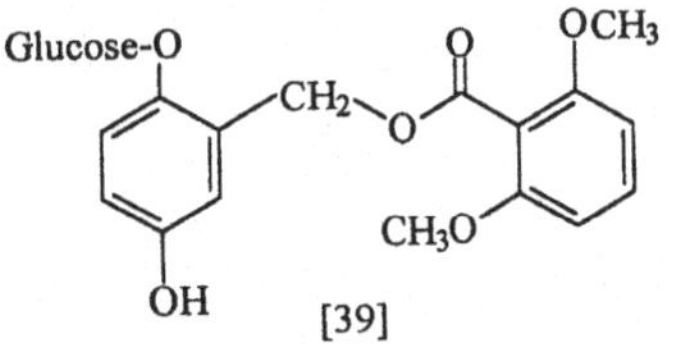

Glucose-O [39]

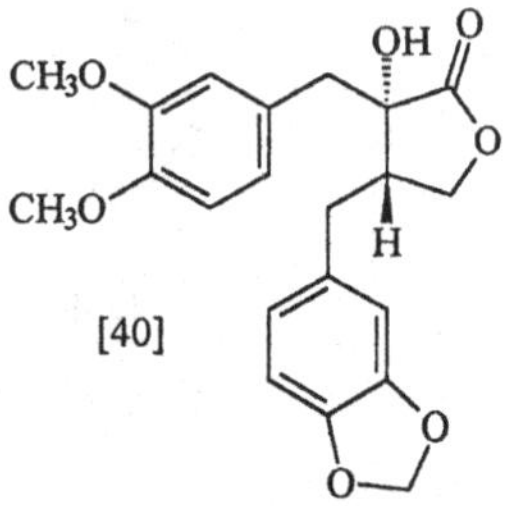

[40]

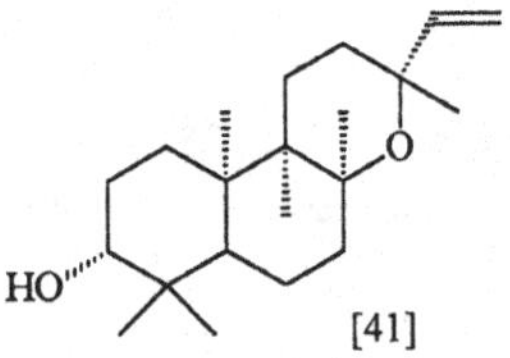

[41]

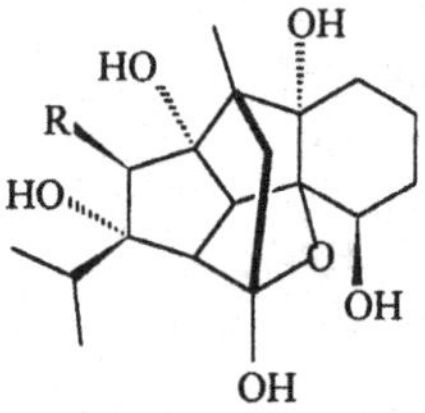

[42] R = OH
[43] R = H

Fig. 11.1. continued

Alkaloids, four furoquinolines and two quinolones, were isolated and identified (Werny & Scheuer, 1963) from *Platydesma campanulatum* H. Mann (now considered *P. remyi*). *Platydesma*, a member of Rutaceae, consists of four species all of which occur on the Hawaiian Islands (Wagner, Herbst & Sohmer, 1990).

Charpentiera obovata Gaud. (Amaranthaceae) afforded 4-methoxy-6-canthinone (Fig. 11.1 [8]) (Scheuer & Pattabhiraman, 1965). Similar compounds are known from Rutaceae and Simaroubaceae.

The genus *Pelea* (Rutaceae) has attracted the attention of many workers owing to its size and chemical riches. In a recent paper, Hartley &

Stone (1989) reduced all species of *Pelea* to *Melicope*. With the exception of *P. parviflora* Hillebr., which has become *M. mauii* Hartley & Stone, specific epithets have not been altered, except for minor orthographic changes. The genus is large and well represented on the Hawaiian Islands where endemism is high. A study of *P. barbigera* (A. Gray) Hillebr. showed the presence of the furoquinoline alkaloid kokusaginine (Fig. 11.1 [9]), a quinolone alkaloid isoplatydesmine (Fig. 11.1 [10]), and edulinine, an o-quinolone which was shown to be an artifact (Higa & Scheuer, 1974*a*). *Pelea anisata* H. Mann is endemic to the island of Kauai. The plant, known to the Hawaiians as 'mokihana', has a pleasant, anise-like aroma, as its specific epithet implies. Its capsules have been and continue to be used in the making of leis which, as we will see below, can be hazardous. Both gas (Hudgins & Scheuer, 1964) and thin layer chromatographic techniques (Floyd, 1979) have been applied to this taxonomically troublesome genus. Detailed chemical studies have shown that the steam volatile substances of *P. anisata* consist principally (*c.* 40%) of p-methoxypropenylbenzene (anethole) (Fig. 11.1 [11]) with lesser amounts of methyleugenol, limonene, methylisoeugenol and estragole (Scheuer, 1955; Scheuer & Hudgins, 1964). An unpleasant aspect of wearing a lei made of *P. anisata* is the tendency for the wearer to develop blistering of the skin where it comes in contact with the plant. Marchant *et al.* (1985) identified the coumarins psoralen, 8-methoxypsoralen, pimpinellin and isopimpinellin. Psoralen derivatives are well known for their capacity to cause dermatitis in the presence of sunlight. Higa & Scheuer (1974*b*) had already identified scopoletin, psoralen, marmesin and the newly encountered coumarin, 2-(1-hydroxy-1-methylethyl)-4-methoxyfuro[3,2-g][1]benzopyranon-7-one (Fig. 11.1 [12]) from *P. barbigera*. 5-Hydroxy-3,7,8-trimethoxy-3′,4′-methylenedioxyflavone (Fig. 11.1 [13]) and 5-hydroxy-3,6,7,8-tetramethoxy-3′, 4′-methylenedioxyflavone (Fig. 11.1 [14]) were also described as components of *P. barbigera* in that paper.

The Hawaiian Islands are also home to a chemotype of *Cyperus rotundus* L., the purple nutsedge, a weedy plant native to India, that has become widely established in southeastern and eastern Asia, including Japan (Holm, 1971; Komai, Iwamura & Ueki, 1977; Wagner, Herbst & Sohmer, 1990). The sesquiterpene chemistry and geographical distribution of purple nutsedge have been examined in detail by Komai, Iwamura & Ueki (1977) and Komai & Ueki (1981). Earlier work resulted in the recognition of three chemotypes, the H-type in Japan, the M-type in China, Hong Kong, Japan, Taiwan and Vietnam, and the O-type in

Japan, Taiwan, Thailand and the Philippines. More recently, Komai & Tang (1989) discovered a fourth chemotype in plants from Kauai which they called the K-type. Some 29 populations of *C. rotundus* from Hawaii were analysed, of which 19 were K-type, five were similar to the Asian O-type and five, all from the Kalapana area on the southeastern coast of Hawaii, were most similar to the Japanese O-type. The new Hawaiian K-chemotype is most closely related to the Japanese O-type plants and differs from them essentially in the presence of petchoulenyl acetate and sugeonyl acetate (Fig. 11.1 [15]) compounds either absent or seen only in trace amounts in the other chemotypes. The presence of these two compounds requires the activity of at least two enzymes, one a reductase and the other an acetyltransferase. It is useful to point out that both Japanese O-type and H-type plants grown on Hawaii retained their respective sesquiterpene profiles indicating genetic rather than environmental control of their biosyntheses. Komai, Tang & Nishimoto (1991) have shown that the essential oils from the different chemotypes of *C. rotundus* can inhibit germination of lettuce and oats. The biological significance of this behaviour under field conditions remains to be established.

Literature reports of insecticidal secondary compounds from members of the large, mainly tropical, genus *Zanthoxylum* prompted Marr & Tang (1992) to investigate three endemic Hawaiian members: *Z. dipetalum* H. Mann, *Z. hawaiiense* Hillebr. and *Z. kauaense* A. Gray. Using a fruit fly assay, they were able to determine the relative toxicities of a number of aliphatic ketones, terpenoids and phenylpropanoids. Results were variable. Of 47 *Z. kauaense* trees sampled only 12 showed biological activity. Of 12 *Z. dipetalum* trees tested only one gave a toxic extract, while none of 21 trees of *Z. hawaiiense* sampled showed any toxicity. Ovicidal extracts of *Z. kauaense* contained primarily 2-undecanone (Fig. 11.1 [16]) and 2-tridecanone (Fig. 11.1 [17]). The dominant compounds present in the toxic extracts of *Z. dipetalum* were anethole and estragole, both phenylpropanoids, and the sesquiterpene caryophyllene (Fig. 11.1 [18]). The authors stressed the importance of establishing limits of variation in natural systems. This is an extremely important consideration in assessing the ecological significance of the compounds in question, or when using the compounds as systematic indicators (Bohm, 1987).

The genus *Bidens* on Hawaii has attracted a good deal of attention, including efforts to reassess the taxonomy of the genus (Ganders & Nagata, 1983). With a clearer understanding of species limits within the genus it was possible to address questions concerning its origin(s) and the stages by which its constituent species have come to be established in the wide variety of habitats in which they are found. Two detailed studies of secondary compounds of most of the taxa have been described, one involving polyacetylenes, the other flavonoids. The study of acetylenes involved analysis of roots and leaves of 19 species and six subspecies (Marchant *et al.*, 1984). Roots of all species yielded acetylenes while leaves from only six species were positive. In all, 18 compounds were identified including four aromatic derivatives, four thiophenes and a tetrahydropyran. Most species exhibited a characteristic array of acetylenes but the most significant observation, from a biosynthetic point of view, was the finding of thiophene derivatives, e.g. 2-(2-phenylethyne-1-yl)-5-acetoxymethyl thiophene (Fig. 11.1 [19]), unique to the Hawaiian species. This observation is consistent with other information suggesting that the island species originated from a single ancestral colonizer. The biosynthetic steps by which the compounds unique to the islands were formed have not been investigated.

In a subsequent study, Marchant & Towers (1987) showed that there was no correlation between phylloplane fungi of *Bidens* species and the nature of the corresponding polyacetylenic profile.

Some 19 species and eight subspecies of *Bidens* were subjected to a detailed examination of floral and leaf flavonoids (Ganders, Bohm & McCormick, 1990). Kaempferol and quercetin 3-O-glycosides were present in all but two taxa. In the case of *B. macrocarpa* (A. Gray) Sherff, they were replaced by apigenin and luteolin 7-O-glycosides. The second exceptional taxon, *B. sandvicensis* Less. subsp. *confusa* Nagata & Ganders, does not accumulate flavonols or apigenin and luteolin glycosides, but does have, uniquely within the island taxa, diosmetin (luteolin 4'-methyl ether). Aurone and chalcone (anthochlor) glycosides were common in all species. Maritimetin, okanin, butein and sulfuretin glycosides occurred in a variety of combinations. The chemosystematic value of flavonoids is seriously compromised by high levels of interpopulational variation. An example, just noted, is *B. sandvicensis* subsp. *confusa*, which lacks flavonols, as compared to subsp. *sandvicensis*, which possesses a kaempferol and two quercetin 3-O-glycosides. These two taxa also have very different arrays of anthochlors; subsp. *confusa* has only maritimetin 6-O-glucoside while subsp. *sandvicensis* has a set of maritimetin glycosides and a set of okanin glycosides. The level of flavonoid variation observed in this study of *Bidens* is higher than for any other system reported to date from the Hawaiian Islands and is amongst the highest so far recorded for any other closely related group of species (Bohm

1987). Obviously, little specialization in flavonoid biosynthetic capacities has occurred during speciation and adaptive radiation on the islands.

Although the flavonoid chemistry of island *Bidens* species does not lend itself to analysis of local problems of relationships, we can say with reasonable certainty that the Hawaiian taxa show an overall flavonoid chemistry characteristic of the genus at large (see Ganders, Bohm & McCormick, 1990, for a discussion).

As famous as the Hawaiian Islands are as a natural laboratory for the study of evolution, so the silverswords must be one of the most frequently cited examples of adaptive radiation on an island (e.g., Carlquist, 1970; Carr & Kyhos, 1981; Carr, 1985, 1987; Witter & Carr, 1988). The silverswords themselves, members of the genus *Argyroxiphium*, are discussed from several points of view in this book (Chapters 1 and 2). The silverswords belong to Madiinae, a subtribe of Asteraceae represented in western North America by some ten or 11 genera. They are collectively known as the 'tarweeds'. On the Hawaiian Islands, Madiinae are represented by three genera: *Argyroxiphium* DC., silverswords and greenswords, five extant species restricted to the Islands of Hawaii and Maui; *Wilkesia* A. Gray, two species restricted to western Kauai; and *Dubautia* Gaud., 26 species scattered on all the major islands.

Studies of the flavonoids of several North American tarweed genera, including *Achyrochaena*, *Adenothamnus*, *Blepharipappus*, *Calycadenia*, *Holocarpha*, *Lagophylla*, *Layia*, *Madia* and *Raillardella* (Crins & Bohm, 1989, and citations therein; Bohm, Crins & Wells, 1994), and *Hemizonia* (Proksch *et al.*, 1984; Tanowitz *et al.*, 1987), provided a basis for assessing changes that may have occurred in the flavonoid profile of the tarweeds as the new species evolved. The tarweeds are so called because of the sticky glandular exudate that covers some or all of their aerial surfaces. Most of the flavonoid studies cited above dealt with the non-polar flavonoids present in leaf exudates. Both species of *Wilkesia*, both subspecies of *Argyroxiphium sandwicense* and several species of *Dubautia* yielded O-methylated flavonoids that resembled closely the profiles established for mainland members of the subtribe. The main flavonoid difference lies in the existence of a group of genera that make 8-substituted flavonoids, as opposed to those that only have extra oxygenation at C-6. The 8-oxygenated flavonoids occur in *Argyroxiphium* and *Wilkesia* but have not yet been detected in *Dubautia*. On the mainland, 8-oxygenated flavonoids have been reported from *Calycadenia*, *Hemizonia* and *Madia* (see Crins & Bohm, 1989, for discussion). This suggests that the capacity for 8-hydroxylation was present in the ancestor of the island species. Its apparent absence from *Dubautia* may indicate a loss, but further sampling of *Dubautia* is required before this possibility can be examined. It ought to be pointed out, though, that about half the species of *Dubautia* are eglandular so that glandular flavonoids would, obviously, not be available for study.

Flavonoids were used successfully to determine the parentage of hybrids between *Dubautia scabra* and *D. ciliolata* (Crins, Bohm & Carr, 1988) on the island of Hawaii. In a few instances, the flavonoid profiles of hybrid individuals showed compounds characteristic of both parent species. In several cases the flavonoid profiles of hybrids were more similar to one parent than they were to the other suggestive of introgression. The availability of synthetic F_1 hybrids (made by G. D. Carr) provided valuable controls for this study.

Vacuolar flavonoids of island tarweeds have been examined in only two species of *Dubautia* and provide no additional insights (Crins, Bohm & Carr, 1988).

Lipochaeta (Asteraceae) presents a particularly intriguing evolutionary problem. The genus, a moderately large one by Hawaiian standards, consists of 25 endemic species (29 taxa in all). Gardner (1976*a, b*) studied the genus using chemical, cytological and morphological data. Both diploid ($n = 15$) and tetraploid ($n = 26$) species exist. Diploid species have five-merous flowers while tetraploid flowers are four-merous. Flavonoid data show a similar correlation with ploidy level. Diploid species appear to make only quercetin glycosides (three reported) while the tetraploids make a complex array of flavone glycosides based upon apigenin and luteolin as well as two quercetin glycosides, which are different from the ones accumulated by the diploids. The seven tetraploid taxa whose flavonoids were reported exhibited taxon-specific arrays whereas the diploid profiles were much more uniform. Gardner (1976*b*) speculated on the possible role of *Wedelia* as an ancester of *Lipochaeta*. The closest mainland species that might have contributed propagules is the Latin American species *W. biflora* (L.) DC. This species is characterized by having achenes with thick, corky walls which would aid flotation. Although most species of *Lipochaeta* have achenes with thin walls, two species, *L. integrifolia* (Nutt.) A. Gray and *L. succulenta* (Hook. & Arn.) DC., both of which are strand plants, also have corky achenes. Gardner (1976*b*) stated that *L. biflora* '. . . has a chemical profile (i.e., flavonols only) virtually identical to the diploid species of *Lipochaeta*'. More recent work on the flavonoid chemistry of *Wedelia* would suggest that the situation may be more complex than that. Although Wagner *et al.* (1986)

reported apigenin from *W. calendulacea*, Miles *et al.* (1993) identified 5,7,4′-trihydroxy-3,3′-dimethoxyflavone (quercetin 3,3′-dimethyl ether), 3,5,4′-trihydroxy-7,3′-dimethoxyflavone (quercetin 7,3′-dimethyl ether) and the very unusual 2,7,4′-trihydroxy-5,3′-dimethoxyisoflavone (Fig. 11.1 [20]). Isoflavones are known from only a few other genera in the Asteraceae, isoflavones with a 2- hydroxyl group from no others. A re-examination of the flavonoid chemistry of this system, *Lipochaeta* and *Wedelia*, might well repay the efforts. See comments also in Chapter 1 on the likely biphyletic origin of *Lipochaeta*.

Scaevola L. (Goodeniaceae) is represented on the Hawaiian Islands by seven diploid endemic species. A close systematic relationship between these species is supported by similar flavonoid profiles, which consist of an apigenin glycoside and several quercetin glycosides (Patterson, 1984). As was pointed out in that paper, the only unusual compound encountered was quercetin 3-O-galactoside-7-O-rhamnoside which was found only in *S. coriacea* Nutt.

Three endemic species of *Vaccinium* (Ericaceae) occur on the Hawaiian Islands: *V. calycinum* Smith, *V. dentatum* Smith and *V. reticulatum* Smith. A study of *V. reticulatum* and *V. calycinum* for flavonoids, including proanthocyanidins (condensed tannins), has been carried out using material of *V. reticulatum* from the Kilauea Volcano area of the island of Hawaii and *V. calycinum* from Kauai (Bohm & Koupai-Abyazani, 1994). The flavonoid profiles of the two species were qualitatively identical consisting of quercetin 3-O-monoglycosides, quercetin 3-O-methyl ether, isorhamnetin and epicatechin. The proanthocyanidin fraction consisted solely of procyanidin with *cis*-stereochemistry. Extension and terminal units and the average molecular weight of the oligomers were determined. Neochlorogenic acid was a major phenolic acid component of both species as well. Analysis of population samples of *V. reticulatum* failed to show qualitative variation in any of the phenolic compounds seen in this study (visual observation of 2-D chromatograms). The procyanidin oligomers also appeared to be identical in the two species. Recent work from our group (Yang, Page & Bohm, unpublished data) using high performance liquid chromatography (HPLC) has shown minor quantitative differences in flavonoid composition among individuals of all three species as well as the presence of trace amounts of other flavonol glycosides.

Two species of the widespread fern genus *Dryopteris* Adans. endemic on the Hawaiian Islands have been investigated by Patama & Widén (1991) as part of the latter author's ongoing study of phloroglucinol derivatives in pteridophytes. *Dryopteris fusco-atra* (Hillebrand) W. Robinson contains filixic acid (Fig. 11.1 [21]) and flavaspidic acid which are characteristic of section *Fibrillosae* to which these two species are thought to belong. *Dryopteris hawaiiensis* (Hillebrand) W. Robinson, on the other hand, does not contain these compounds, but contains instead p-aspidin (Fig. 11.1 [22]), margaspidin and phloraspidinol. The phloroglucinol derivative profiles of the two species are mutually exclusive. Other than pointing out that *D. hawaiiensis* is unique with regard to its chemical profile, no suggestions were offered as to the evolutionary or systematic significance of these findings.

The Galapagos Islands

The Galapagos Islands (Archipiélago de Colón), Ecuador, lie essentially on the equator at *c*. 90° W longitude. This island group gained biological attention through the visit of Charles Darwin. The many varied forms of plant and animal species, different from, but still similar enough to mainland taxa to suggest close relationships, played a major role in the development of his ideas on speciation. The presentation of the concept of adaptive radiation is often introduced to students using examples from these islands.

Despite the uniqueness of the Galapagos Islands and their diverse flora, only a few studies of secondary metabolites of endemic species appear to have been undertaken. The impact of the available information is significant, however. Adsersen, Adsersen & Brimer (1988) surveyed a large fraction of the flora of the Galapagos Islands for cyanogenic compounds. The aim of the study was to determine whether species that had originated on the islands had a lower frequency of occurrence of cyanogenesis, compared to the mainland, as might be expected for plants evolving in the absence of former herbivores. Their study involved two groups of plants, those endemic to the islands and those found on both the islands and on mainland South America. Indeed, they showed that cyanogenesis is more common among mainland plants than among island endemics, although their results using fresh vs. herbarium material are not consistent. Using fresh material, 45% of endemic species were positive for hydrogen cyanide (HCN) compared to 62% of species also known from the mainland. When herbarium specimens were tested, however, 17.5% of endemic species were positive compared to 18.6% for the non-endemic species. Degradation of cyanogenic glycosides in older, dried specimens may account for the latter set of values. Data from fresh material, however, give a clear indication that cyanogenesis is reduced among island endemic species.

In a recent study (Adsersen *et al.*, 1993), cyanogenesis in the Galapagos endemic *Passiflora colinvauxii* Wiggins was investigated. This species, restricted to Isla Santa Cruz, is morphologically similar to the Ecuadorian species *P. biflora* Lam. and *P. punctata* L. Passibiflorin [(1S,4R)-1-D-glucopyranosyloxy)-4-(6-deoxy-β-D-gulopyranosyloxy) 2- cyclopentene-1-carbonitrile] (Fig. 11.1 [23]) was identified in the island endemic and in *P. biflora*. The occurrence of this unusual diglycoside in both species was taken as evidence of a close relationship between the two. Whether the island endemic species arose directly from the mainland species or whether there was/is a common ancestor is not known.

Lecocarpus (Asteraceae) is a genus of three species endemic to the Galapagos Islands. *Lecocarpus pinnatifidus* Decne. is further restricted to Isla Santa Maria, while *L. lecocarpoides* Cronquist & Stuessy, is known from Isla Española and Gardner. The third species is *L. darwinii*. In the first of two papers, Macias and co-workers (Macias & Fischer, 1992) described the existence of melampolides (ten-carbon carbocyclic sesquiterpene lactones) in *L. pinnatifidus* most of which were shown to possess structures identical to compounds known from other members of the Melampodiinae. One unique feature, however, was noted. Two of the melampolides have different stereochemistry at C-8 (Fig. 11.1 [24]). Normally one sees the hydroxyl group (and its esters) in the β-orientation; the unusual pair have their substituents at C-8 in the α-orientation. This is a seemingly small difference but it could indicate some subtle alteration of the biosynthetic enzyme(s) that allows a different approach of the substrate. Could this be the sort of change one might anticipate from simple mutations?

Information relevant to this question comes from the alkaloid biosynthesis literature. Tropinone is a branchpoint for the biosynthesis of tropane alkaloids through reduction to the corresponding alcohols. Tropinone reductase-I (TR-I) forms tropine in which the resulting alcohol function lies in the α-configuration, while tropinone reductase-II (TR-II) yields the β-diastereoisomer. The two enzymes from *Datura stramonium* have been studied and shown to have 64% sequence homology in 260 amino acids. The authors (for discussion see Hashimoto & Yamada, 1994) conclude by stating that '. . . a relatively small number of amino acid substitutions is sufficient to alter the stereospecificity of enzyme reactions'.

In the second paper, Macias *et al.* (1993) described melampolides and *cis, cis*-germacranolides (also ten-carbon carbocyclic sesquiterpene lactones) from *L. lecocarpoides*. A 'volunteer' compound was encountered in the isolation of the lactones from both species and identified as 5,4'-dihydroxy-3,6,7-trimethoxyflavone (penduletin), which is a widely distributed flavonoid in the Asteraceae.

The Juan Fernandez Islands

The Juan Fernandez Island group consists of three islands, two large and one small, located about 667 km off the coast of Chile at 80° W, 33° S. The two larger islands are Masatierra (Isla Robinson Crusoe), which has been dated at 3.7–4.4 million years), and Masafuera (Isla Alejandro Selkirk), dated at 1–2 million years, which lies farther to the west at a distance of about 181 km from Masatierra. The third island, Santa Clara, lies off the southwestern corner of Masatierra and is likely to have arisen from the latter through erosion. Geological information on the islands, including comments on their volcanic origin, has been reviewed by Stuessy *et al.* (1984).

A high level of endemism characterizes the flora of these islands (Stuessy *et al.*, 1992). Figures presented in that paper tell us that, in the vascular flora, the islands harbour one endemic family (Lactoridaceae), 12 endemic genera and 126 endemic species. This amounts to 11% endemism at the generic level and 60% at the species level. The comparatively high level of endemism, coupled with the need to assess the status of this unusual flora for conservation efforts, prompted workers at Ohio State University and the Universidad de Concepción to undertake a detailed study of the islands. Some of their results are summarized below (see also Chapters 3–5).

Asteraceae makes by far the largest contribution (29 species) to the list of endemic species on the Juan Fernandez Islands. *Erigeron* contributes six of these: *E. fernandezianus* (Colla) Harling, *E. ingae* Skottsb., *E. luteoviridis* Skottsb., *E. rupicola* Philippi, *E. stuessyi* Valdebenito and *E. turricola* Skottsb. *Erigeron fernandezianus* occurs on both islands; the other five species occur only on the younger island (Masafuera). Flavonoid profiles were determined for each of these as well as for five species from mainland South America (Valdebenito *et al.*, 1992*b*). The overall profiles show a high degree of similarity with glycosides of quercetin, quercetagetin, apigenin, luteolin and 6-C-glucosylacacetin occurring in various combinations although not all taxa, or indeed, all populations, exhibited all compounds. The authors were reluctant to place much emphasis on the flavonoid components because of the variation, although they did state that their data agreed, in general, with relationships based upon morphological data.

A combination of morphological data and the fact that five of the species occur only on Masafuera was taken as evidence that colonization of the younger island occurred first followed by adaptive radiation and then migration of *E. fernandezianus* to Masatierra. Flavonoid structural information points to a different scenario. The occurrence of luteolin 7-O-diglucoside and 6-C-glucosylacacetin 7-O-diglucoside play an important role in this. Except for the mainland species, *E. campanensis* Valdebenito, Lowrey & Stuessy, which exhibited the most depauperate flavonoid profile reported, these two compounds occur in at least some of the collections of the mainland species examined. They also were reported from at least one collection each of *E. ingae*, *E. luteoviridis* and *E. turricola*. It is impossible to judge the significance of the absence of these compounds in the single specimen of *E. stuessyi* available considering the variation exhibited throughout the genus otherwise. These two compounds apparently are not formed in *E. rupicola* and, significantly, in the Masafueran collections of *E. fernandezianus*. As pointed out by the authors (Valdebenito *et al.*, 1992*b*), the more parsimonius explanation of these results would involve colonization of Masatierra first, followed by dispersal to Masafuera with subsequent speciation. A few populations of *E. fernandezianus* lack one or both of the compounds in question and could conceivably have served as the source of propagules from which the species on Masafuera was derived. This would suggest that this species is the most primitive one of the island group, but this is not borne out by the morphological data. It is interesting to consider the biosynthesis of these two compounds in greater depth to see how many steps are involved which, in turn, would give an idea of genetic divergence between the two chemotypes. Three biosynthetic differences distinguish the two chemotypes: (1) C-glucosylation; (2) placement of the 'outer' glucose to establish the diglucoside derivatives; and (3) transfer of a methyl group to the 4'-OH to give the acacetin (apigenin 4'-methyl ether) structure. The formation of the 7-O-diglucosides probably is controlled by the same glucosyltransferase. The 4'-O-methylation occurs no where else in the array of flavonoids observed, although 6-O-methylation does. The O-methyltransferases are generally considered to be fairly selective in their position of action so it seems reasonable to suggest that the 4'-O-methyltransferase required for the formation of the acacetin derivative is unique within this group of species. By this reckoning, three enzyme systems are involved. If the direction of movement of the colonizer was from the mainland to the younger island and then to the older island, some thought must be given to the disappearance and reappearance of these enzymes. The loss and regaining of three enzyme systems seems drastic. Perhaps the genes responsible for these enzyme systems were not lost but were simply quiet during residence on the younger island. Perhaps in moving to the older island some subtle ecological factor enabled (or required?) the return to 'life' of the genes responsible for these steps with the observed chemistry re-emerging. Detailed molecular biological study of the enzymes and their controlling genes would be helpful in resolving situations of this sort. It would, of course, be of interest to know the influence of environmental factors on the expression of flavonoid features.

Dendroseris D. Don (Asteraceae, Lactuceae), endemic to the Juan Fernandez Islands (see Chapter 4), consists of 11 species arrayed in three subgenera. Eight species are restricted to the older island and three species, one from each subgenus, are restricted to the younger island. Flavonoid profiles were established for all of these taxa (Pacheco *et al.*, 1991*b*). Apigenin, luteolin and quercetin glycosides occurred in all taxa but the level of glycosylation of the aglycones varied. An unresolved problem surrounds the flavone glycosides. Three apigenin 7-O-monoglucosides and four luteolin 7-O-monoglucosides were reported. Hydrolysis of the individual glycosides yielded only the aglycone and glucose. Although the authors were aware of the possibility, they did not determine whether different isomeric forms of glucose, i.e., furanose vs. pyranose, were involved. If one also considers the possibility of α- and β-conformations, four isomeric glucosides are possible. This phenomenon is well known in the case of quercetin 3-O-arabinosides (Harborne, 1967, pp. 71–72). The three luteolin 7-O-monoglucosides could be explained in the same fashion. The suggestion by Pacheco *et al.* (1991*b*) that the multitude of glucosides could involve attachment of glucose via some position other than the anomeric carbon is not likely. Before one can assess the chemosystematic significance of the occurrence of this array of glucosides, it would be necessary to determine the structures rigorously and to establish that they were the products of specific enzyme reactions. The existence of isomeric diglucosides is less troublesome insofar as there are several ways in which the 'outer' glucose may be attached to the 'inner' glucose. Several flavonoid diglucosides are known; two common ones are sophorosides with the $(1'->2)$ linkage and gentiobiosides with $(1'->6)$.

Taken at face value, the array of flavonoid derivatives isolated from *Dendroseris* shows some differences between the subgenera. For example, subgenus *Phoenicoseris* lacks quercetin glycosides and has a much simpler array of apigenin glycosides than either of the other two subgenera. Subgenus *Rea* has a more complex array of luteolin glycosides and lacks quercetin 3-O-diglucoside. Despite intraspecific variation, which the reader will recall also plagued the *Erigeron* study, the distribution of flavonoids showed an overall correlation with subgeneric boundaries. Subgenus *Rea* is considered the most primitive on the basis of a cladistic analysis of morphological characters with the obvious suggestion that it most closely approximates the original colonizer. This subgenus exhibited the largest array of flavonoids. With the exception of *D. berteroana*, which possesses an apigenin 7-O-diglucoside not seen elsewhere in the island species, subgenera *Dendroseris* and *Phoenicoseris* exhibit flavonoid profiles that are subsets of the *Rea* profile. Reduction of flavonoid complexity can clearly be claimed as a companion of speciation in this group.

Attempts to locate a likely ancestor on the South American mainland using flavonoid characters pointed to species of *Hieracium* and *Hypochaeris* as possible candidates, with the choice of Pacheco *et al.* (1991*b*) going to a *Hieracium* species, possibly one from Chile. Apigenin and luteolin glycosides were identified in species from Chile and, although quercetin was not found in the Chilean species, quercetin glycosides are known from other members of the genus. Recent macromolecular studies (S. C. Kim *et al.*, unpublished; see also Chapter 4) suggest closer generic ties to *Sonchus* s.l., or perhaps *Embergeria* from New Zealand.

Robinsonia DC. is the second largest genus of the Asteraceae (Senecioneae) endemic to the Juan Fernandez Islands (see also Chapter 4). Pacheco *et al.* (1985) studied the flavonoid chemistry of all seven species [including *R. berteroi* (Hemsl.) Sanders, Stuessy & Marticorena, which was previously considered to comprise the monotypic genus *Rhetinodendron*]. Compounds identified included glycosides of the flavonols kaempferol, quercetin and isorhamnetin, the flavones apigenin and luteolin, the flavanones naringenin and eriodictyol, and the dihydroflavonol taxifolin. The array of flavonoids in *Robinsonia* offers more assistance in searching for relationships than was available in either *Erigeron* or *Dendroseris* because of a lower level of interpopulational variation and a more diverse array of flavonoid structural types. The pigment profiles are such that several questions of relationship may be addressed. It is worth while to reproduce the *Robinsonia* data in condensed form (Table 11.1). The species have been grouped according to their infrageneric taxonomy and interpopulational variation has been ignored.

From a strict phenetic point of view, comparison of flavonoid profiles offers support for several relationships suggested by morphological data. Recognition of the monotypic genus *Rhetinodendron* Meisner as *R. berteroi* is supported by its possession of a flavonoid profile, which consists solely of quercetin 3-O-mono- and digalactoside, identical to that of *R. macrocephala* Dcne., and similar to that of *R. gayana* Dcne. and *R. thurifera* Dcne., which make only quercetin glycosides. The close morphological similarity between *R. gayana* and *R. thurifera*, which comprise sect. *Robinsonia*, exhibit identical flavonoid profiles. The members of sect. *Eleutherolepis*, *R. evenia* Phil., *R. gracilis* Dcne. and *R. masafuerae* Skottsb., are characterized by unique assortments of flavonoids: *R. evenia* is the only species in the genus that accumulates a derivative of apigenin, for example, while *R. gracilis* is unique in accumulating isorhamnetin derivatives, both flavanones and the dihydro-

Table 11.1. *Distribution of flavonoids in* Robinsonia

	Flavonoids[a]												
	1	2	3	4	5	6	7	8	9	10	11	12	13
Subg. *Rhetinodendron*													
R. berteroi	+	+	−	−	−	−	−	−	−	−	−	−	−
Subg. *Robinsonia*													
sect. *Symphyochaeta*													
R. macrocephala	+	+	−	−	−	−	−	−	−	−	−	−	−
sect. *Robinsonia*													
R. gayana	+	+	+	−	−	−	−	−	−	−	−	−	−
R. thurifera	+	+	+	−	−	−	−	−	−	−	−	−	−
sect. *Eleutherolepis*													
R. evenia	+	−	+	−	−	−	−	−	+	+	−	−	−
R. gracilis	−	−	+	+	+	+	+	−	−	−	+	+	+
R. masafuerae	−	+	+	+	−	−	−	+	−	−	−	−	−

[a]1 = Q-3-Gal; 2 = Q-3-DiGal; 3 = Q-7-Glc; 4 = Q-3, 7-DiGlc; 5 = K-7-Glc; 6 = IR-3-Glc; 7 = IR-7-Glc; 8 = Lute-7-Glc; 9 = Lute-7-DiGly; 10 = Apig-7-DiGly; 11 = Erio-7-Glc; 12 = Nar-7-Glc; 13 = Tax-7-Glc (Q = quercetin; K = kaempferol; IR = isorhamnetin; Lute = luteolin; Apig = apigenin; Erio = eriodictyol; Nar = naringenin; Tax = taxifolin; Gal = galactose; Glc = glucose)

quercetin glycosides and is the only species that accumulates luteolin 7-O-monoglucoside.

Pacheco *et al.* (1985) present a detailed analysis of flavonoid biosynthetic events associated with the evolutionary history of the genus. They suggest that the likely ancestor was a Chilean species of *Senecio* that contained only flavonols. The formation of flavones and the 'sequestering' of flavanones and the dihydroflavonol are seen as derived traits based on altered enzymic capacity to utilize these precursors. Much of the rest of the structural variation seen involves the nature of glycosylation. I should like to comment on one of the assumptions made by these workers in their attempts to apply the flavonoid data to the problem of evolutionary relationships within *Robinsonia*. They assume that the different sugar substitutions observed have little phyletic value. They apparently took this position because the sugar substitution patterns do not agree with morphological data in this system. There is little information, so far as I am aware, as to whether the same O-glycosyltransferase works equally as well with glucose and galactose as substrates. On the other hand, there is sufficient information available on O-methylation and O-sulfation as well as O-glycosylation to suggest, at least, that different enzymes are required for different positions, the 3- and 7-positions in the present case.

The genus *Peperomia* (Piperaceae) is represented on the Juan Fernandez Islands by three endemic species: *P. margaritifera* Bert., restricted to Masatierra; *P. skottsbergii* C. DC., restricted to Masafuera; and *P. berteroana* Miq., which occurs on both islands. A fourth species, *P. fernandeziana* Miq., occurs on both islands and on mainland Chile. A complex array of flavonoids was described from all four island species plus four additional species from mainland South America (Valdebenito *et al.*, 1992*a*). The compounds identified included the flavones apigenin, acacetin, luteolin, luteolin 7-methyl ether and luteolin 4'-methyl ether (diosmetin), which occurred as aglycones in some cases, as well as various glycosides and as sulfates. In some instances direct sulfation of a phenolic hydroxyl was seen; in others, the sulfate moiety was found attached through a sugar hydroxyl. Flavonols identified were kaempferol 3,7-diglucoside, kaempferol 3-methyl ether 7-O-glucoside and quercetin 7-methyl ether 3-O-glucoside. The predominant compounds present were C-glycosylflavones, 19 of which were identified.

The entire flavonoid data table for *Peperomia* is too large to reproduce here, but flavonoid types and major structural features for the island species and ones from the mainland are summarized in Table 11.2. All

Table 11.2. *Occurrence of flavonoid structural features in* Peperomia *from the Juan Fernandez Islands and South American mainland*

Peperomia species	Origin[b]	Flavonoid Structural Feature[a]									
				C-Glys			Oxygen				
		F'one	Fl'ol	Mono	Di	Sulf	4'	3'4'	Aca	Dio	7-OMe
berteroana	MT	+	−	−	+	+	+	+	+	+	−
	MF	+	−	+	+	+	+	+	+	+	−
fernandeziana	MT	+	+	+	+	−	+	+	+	+	+
	MF	+	+	+	+	−	+	+	+	+	+
	Chile	+	+	+	+	−	+	+	+	+	−
margaritifera	MT	−	+	+	+	−	+	+	−	+	−
skottsbergii	MF	+	−	+	+	+	+	+	+	−	−
alata	Bol	+	+	+	+	−	+	+	+	+	−
coquimbensis	Chile	+	−	+	+	−	+	+	−	+	−
galiodes	Bol	+	+	+	+	−	+	+	−	−	−
olens	Bol	+	+	+	+	−	+	+	−	+	+

[a] F'one = flavone; Fl'ol = flavonol; C-Glys, Mono = one C-sugar, Di = two C-sugars; Sulf = sulfate(s); Oxygen = oxygenation at 4' or 3',4'-positions; Aca = acacetin; Dio = diosmetin; 7-OME = methylation at 7-position.
[b] MT = Masatierra; MF = Masafuera; Bol = Bolivia.

species produce flavonoids (of any type) with both 4'-mono- and 3',4'-dioxygenation. Likewise, all species produce flavone di-C-glycosides. Flavone mono C-glycosides are also produced by all species except that *P. berteroana* from Masatierra appears not to have them. For island species, flavonol production is limited to *P. fernandeziana*, but they were found in three of the mainland species studied. *Peperomia margaritifera* exhibited the simplest profile of island species with only five compounds (identical in two populations). It was the only island species found to have kaempferol, occurring as the 3,7-diglucoside. Three of the mainland species also exhibited this compound.

The O-methylation at position 4' to form the acacetin- or diosmetin-substitution patterns occurs in all island species except *P. margaritifera* and in one and three species, respectively, from the mainland. The 7-O-methylation is much more restricted in occurrence having been encountered only in *P. fernandeziana* from both islands, but not from mainland material; it was also seen in one mainland species, *P. olens*. It may be of interest to note that both luteolin 7- methyl ether and quercetin 7-methyl ether were identified from island collections of *P. fernandeziana*, which suggests that the flavonoid 7-O-methyltransferase may function with both luteolin and quercetin as substrates. Only the quercetin methyl ether was detected in *P. olens*. Since only one collection of each mainland species was studied there is no way to comment on variability of these taxa.

Flavonoid sulfate derivatives were observed in only two species examined in this study, *P. berteroana* from both islands, and *P. skottsbergii*. Sulfates were not detected in mainland material, but again, no information on variation is available. There appears to be selectivity in either species as to whether direct sulfation of a phenolic hydroxyl or a sugar hydroxyl occurs. The distribution of sulfate derivatives again raises the question of non-concordance between flavonoid data and the phylogenetic possibilities indicated by morphological data. Valdebenito *et al.* 1992*a*) presented graphic indications of evolutionary relationships within the island species which showed *P. berteroana* on one line and *P. margaritifera* and *P. skottsbergii* on the other. The presence of sulfate derivatives in *P. berteroana* and *P. skottsbergii* requires either that the capacity to produce sulfates arose independently in these two species, or that an ancestral taxon had the capacity to produce sulfates, with the capacity having been retained by these two species, and that *P. margaritifera* has lost the capacity. As the authors point out, there is nothing to suggest that one hypothesis is better than the other. Problems of this sort can only be resolved through detailed studies of the character in question which in this case would involve macromolecular and genetic studies of the sulfation enzymes and their controlling genes.

An interesting situation exists with regard to the evolutionary origin of *P. fernandeziana*. A phenogram of 16 species of *Peperomia* (Valdebenito *et al.*, 1992*a*) showed a close grouping of the three endemic species, but *P. fernandeziana* emerged from the treatment in a cluster at some phenetic distance from its island 'relatives'. The data were taken to support the view that this species represents a separate introduction to the Juan Fernandez Islands. Several features of *P. fernandeziana* are concordant with this view: (1) it possesses a unique array of flavonols; (2) it lacks sulfated derivatives; and (3) it possesses a uniquely complex array of C-glycosylflavones. The authors wisely point out that a more thorough study of the genus is necessary to test these ideas.

Gunnera, the sole genus of Gunneraceae, consists of about 40 species distributed in tropical and South Africa, Malesia, Tasmania, the Antarctic Islands, South America and the Hawaiian Islands (Mabberley, 1987). Three of these, *G. bracteata* Steud., *G. masafuerae* Skottsb. and *G. peltata* Phil., are endemic to the Juan Fernandez Islands. A detailed analysis of the flavonoids of the endemic species along with five species from mainland South America and one from Mexico has been described by Pacheco *et al.* (1993). The flavonoid profiles described consist of an assortment of glycosides of kaempferol, quercetin and isorhamnetin, in various combinations, an unidentified flavone glycoside and two unknowns. The unknowns were essentially ubiquitous and will not be discussed further. The differences between the endemic species lie in the nature of the quercetin glycosides present in each. They are otherwise essentially identical (overlooking a small amount of interpopulational variation).

The flavonoid data have been condensed somewhat and appear in Table 11.3. Flavonoid data show a strong similarity between the island species and *G. tinctoria* (eight samples from Chile and one from Argentina). Each of the island species and *G. tinctoria* exhibited a characteristic array of quercetin glycosides, but were otherwise essentially identical in possessing both isorhamnetin (or a glycoside in the case of *G. tinctoria*) and the unidentified flavone glycoside, but lacking kaempferol derivatives. Other species tested, two from Peru, two from Bolivia and one from Mexico, have kaempferol, lack the flavone glycoside (possibly present in *G. bolivari*) and lack isorhamnetin (except *G. bolivari*). Morphological data suggest that *G. tinctoria* is the closest relative of the

Table 11.3. *Occurrence of flavonoids in* Gunnera *from the Juan Fernandez Islands, mainland South America and Mexico*

Gunnera species	Origin[a]	Q-Glys[b]					K-Glys	IR	Flavone
		A	B	C	D	E			
bracteata	MT	+	–	–	–	–	–	+	+
masafuerae	MF	–	+	–	–	–	–	+	+
peltata	MT	–	–	+	–	–	–	+	+
tinctoria	Ch, Ar	–	–	–	+	–	–	+	+
bolivari	Peru	+	–	–	–	–	+	+	?
boliviana	Bol	–	–	–	–	+	+	–	–
margaretae	Bol	–	–	–	–	+	+	–	–
peruviana	Peru	+	–	–	–	–	+	–	–
mexicana	Mex	+	–	–	–	–	+	–	–

[a]MT = Masatierra; MF = Masafuera; Ch = Chile; Ar = Argentina; Bol = Bolivia; Mex = Mexico
[b]Q = quercetin; K = kaempferol; IR = isorhamnetin; Glys = glycoside; A–E are sets of glycosides

island species. In this instance, at least, flavonoid data are in complete agreement. The four species share the same chromosome number as well ($2n = 34$).

The likely path of evolution within this group is thought to have involved a propagule from the mainland (*G. tinctoria* as presumed ancestor) becoming established on Masatierra to give rise to *G. peltata* and *G. bracteata*, followed by colonization of Masafuera by a propagule from the older island, with subsequent speciation to afford *G. masafuerae*. Interesting changes in flavonoid glycosylation patterns have occurred during this series of events. *Gunnera bracteata* and *G. peltata* have lost the capacity to make quercetin 3-O-digalactoside, while *G. bracteata* has additionally lost the capacity to make quercetin 3-O-xylosylglucoside. Since *G. masafuerae* continues to make both of these compounds it seems reasonable to suggest that some common ancestor, which presumably already had become established on the older island, had the capacity to make all of these compounds. It would have been a propagule from this ancestral line that would have given rise to the new species, *G. masafuerae*, on the younger island. Sometime during its residence on Masafuera, which would have been within 1-2 million years, this species gained the capacity to make quercetin 3,7-diglucoside. Since this is the only taxon in which this compound occurs (in the taxa involved), there appears to have been an alteration of a glucosyltransferase such that it can now recognize the quercetin 7-hydroxyl group instead of a sugar hydroxyl of an existing quercetin 3-O-monoglycoside. Perhaps there has been a duplication of a glycosylation gene with the original continuing to control formation of linear diglycosides leaving the duplicated form free to take on new tasks. Once again, it must be pointed out that we do not have enough information concerning the nature of glycosylation enzymes to offer a satisfactory resolution to this quandry. Detailed study of such a system would repay the efforts.

A recent study (Bittner *et al.*, 1994) examined *G. peltata* and *G. masafuerae*, and two species from the Chilean mainland, *G. tinctoria* and *G. magellanica*, for terpene derivatives. Compounds identified were oleanolic acid, ursolic acid, lupeol and its acetate, erythrodiol, a dihydroxyursene, vomifoliol, loliolide, β-sitosterol, daucosterol, uvaol and ionone. Pinoresinol was also identified.

The Juan Fernandez Islands are home to Lactoridaceae, an endemic family that consists of the single species *Lactoris fernandeziana* Phil. Lammers, Stuessy & Silva O. (1986) examined the systematic and evolutionary relationships of the family. The flavonoid chemistry of leaf material from *Lactoris* was examined by Crawford, Stuessy & Silva O. (1986) who identified kaempferol and isorhamnetin as the 3-O-glucosides, 3-O-diglucosides and 3-O-diglycosides involving arabinose and galactose (order of attachment not determined). Owing to the simplicity of the flavonoid profiles of 'primitive' families it is difficult to draw any firm conclusions from these observations. The absence of quercetin as an accumulated flavonoid appears to distinguish Lactoridaceae from other members of Laurales, although this is the order into which Lactoridaceae seems to fit most comfortably. Wood anatomy, however, suggests placement in Piperaceae (Carlquist, 1990).

The widespread legume genus *Sophora* is also represented on the Juan Fernandez Islands by *S. masafuerana* (Phil.) Skottsb. and *S. fernandeziana* (Phil.) Skottsb. These species along with *S. microphylla* Ait. and *S. macrocarpa* J.E. Sm. from the Chilean mainland, *S. linearifolia* Griseb. from Argentina and a collection of *S. microphylla* Ait. from New Zealand were examined for their alkaloids (Hoeneisen *et al.*, 1993). The alkaloid profiles of species from the islands differ from those of species from the other sites. The authors concluded that the islands have been isolated from the mainland for a sufficient period of time to allow evolution of different structural features. Loss of overall capacity to make alkaloids has not occurred. It is useful to note that the alkaloid profiles of

S. microphylla from New Zealand and from the Chilean mainland were not identical.

New Zealand and associated islands

Allan (1982) comments that endemism in New Zealand reaches 80% of the flora and includes some 40 endemic genera. Chemical data are also available for endemic species of some of New Zealand's associated islands including Norfolk and Lord Howe Islands. Some information on the specific locations of these islands is in order. The Chatham Islands lie at 44°00′ S and 176°30′ W which translates into a distance of about 800 km east of Christchurch. The Chatham flora includes some 40 endemic species (Allan, 1982). The Kermadec Islands lie at 30°00′ S and 178°30′ W placing them roughly 1000 km northeast of Auckland. According to Allan (1982), the flora of the Kermadec Islands consists of 120 species, 15 of which are endemics. Stewart Island lies about 50 km off the south coast of the South Island of New Zealand. It became isolated from the South Island as a result of seismic activity. Despite its proximity to a major land mass it has some dozen endemic species (Allan, 1982). The plant communities on Stewart Island have been thoroughly discussed by Wilson (1987). Lord Howe Island (Australia) lies at 31°28′ S and 159°09′ E at a distance of about 600 km from the Australian mainland. Norfolk Island (Australia) lies at 29°05′ S and 167°59′ E at a distance of about 1450 km from the mainland.

Several groups of workers have undertaken studies of New Zealand (broad sense) plants for flavonoids. From time to time other types of compounds have also emerged. A case in point involves the work of Massias, Carbonnier & Molho (1981) who described their isolation of three C-glycosylflavones, eight xanthones, a C-glycosylxanthone and gentiopicrin (an iridoid) from the endemic *Gentiana corymbifera* T. Kirk. Included among the xanthones was the newly reported 1,2,8-trimethoxy-4,5-dihydroxyxanthone (Fig. 11.1 [25]). This collection of compounds, C-glycosylflavones, xanthones and an iridoid derivative, represents a typical gentianaceous array.

One is always on the alert for differences between 'mainland' and island species that may indicate that some evolutionarily interesting changes have occurred. Two examples of this, from quite unrelated families, follow. The first involves the genus *Coprosma* (Rubiaceae), which consists of some 90 species that are widely distributed in eastern Malesia, Australia and the islands of the Pacific (Mabberley, 1987). Wilson (1984), working with nearly 50 species from the New Zealand region, found that most had flavonoid profiles based on glycosides of kaempferol and quercetin while some had isorhamnetin and myricetin glycosides as well. A species from the Kermadec Islands, *C. petiolata* Hook. f., had only isorhamnetin glycosides, while a species from Norfolk Island, *C. baueri* Endl., had only quercetin glycosides. The significance of these apparent simplifications, whether they were the result of the founder effect or whether they indicate selection for specific flavonoids on the respective islands, was not determined.

The second example comes from the work of Purdie (1984), who examined 31 (of 39) species of the papilionaceous genus *Carmichaelia*. There was very little variance among the New Zealand taxa which had profiles comprising glycosides of kaempferol, quercetin, isorhamnetin and isoflavones as well as C-glycosylflavones, although not all taxa had all compounds. *Carmichaelia exsul* F. Muell., a Lord Howe Island endemic, lacked isorhamnetin, C-glycosylflavones, flavonol 7-O-glycosides and flavonol 3,7-di-O-glycosides. With only kaempferol and quercetin 3-O-glycosides and isoflavones, *C. exsul* exhibited the simplest profile seen in the survey.

The New Zealand white pine, *Dacrycarpus dacrydioides* (A. Rich.) de Laub., a segregate from *Podocarpus*, was subjected to a detailed flavonoid examination by Markham & Whitehouse (1984). They identified vitexin, isovitexin, orientin and isoörientin 2″-rhamnosides, 5,7,3′,4′,5′-pentahydroxyflavone (tricetin) 3′,5′-di-O-glucoside, and a series of quercetin and myricetin derivatives, some of which were O-methylated and existed as the unusual 7-O-rhamnoside- 3′-O-xylosides. Monoglycosides involving both of those positions were also among the pigment profile.

As an example of the flavonoid chemistry of a *Podocarpus* species, we can compare the pigments of *P. nivalis* Hook., whose flavonoid profile was determined by Markham, Webby & Vilain (1984). The major compound, 7-O-methyl-(2*R*:3*R*)-dihydroquercetin 5-O-β-D-glucoside, was accompanied by the corresponding dihydrokaempferol analogue and the respective flavonol derivatives. Of interest, especially in view of the glycosylations noted for *Dacrycarpus* above, was the finding of luteolin 3′-O-xyloside and luteolin 7-O-glucoside-3′-O-xyloside. The biosynthetic relationship between these two taxa would seem to be a fairly close one considering the rarity of this pattern of sugar attachment.

Phyllocladus (the celery pines) consists of four (Mabberley, 1987) or six species (Foo, Hrstich & Vilain, 1985) that occur from Borneo to New Guinea, New Zealand (three endemics) and Tasmania. Markham, Vilain

& Molloy (1985) reported that the three New Zealand endemics, *P. alpinus* Hook. f., *P. glaucus* Carr. and *P. trichomanoides* D. Don, the Tasmanian endemic *P. aspleniifolius* (Labill.) Hook., and *P. hypophyllus* Hook. from the Philippines and New Guinea, are characterized by a 'remarkably homogeneous' array of luteolin 7-O- and 3'-O-glycosides with lower levels of flavonol O-glycosides. In the same year, Foo, Hrstich & Vilain (1985) reported the occurrence of an unusual flavan-3-ol derivative that appears to be characteristic of the genus. The compound, called phylloflavan, was identified as *ent*-epicatechin-3-δ-(3,4-dihydroxyphenyl)-β-hydroxypentanoate (Fig. 11.1 [26]). It was accompanied by catechin and epicatechin. Other flavan-3-ol derivatives, including some that appear to have arisen via phenol coupling of the flavonoid moiety with a dihydroxycinnamoyl unit, have been reported from *P. trichomanoides* by Foo (1987, 1989).

One of the most thoroughly studied genera worldwide (of comparable size) is *Fuchsia* (Onagraceae) with information available for 74 of the 100 known species (Averett *et al.*, 1986). Of interest to us in this chapter are the flavonoid profiles of sect. *Skinnera* which consists of four species endemic to New Zealand, *F. colensoi* Hook. f., *F. excorticata* (J. R. & G. Forst.) L., *F. perscandens* Ckn. & Allan and *F. procumbens* R. Cunn. & A. Cunn., and one species from Tahiti, *F. cyrtandroides* R. J. Moore. Two of the New Zealand species, *F. colensoi* (likely to be of hybrid origin between *F. perscandens* and *F. excorticata*) and *F. excorticata*, have pigment profiles consisting of kaempferol, quercetin and flavone glycosides along with an eriodictyol derivative. The other two, *F. perscandens* and *F. procumbens*, lack flavonol glycosides altogether, but have flavone glycosides and the eriodictyol derivative. The species from Tahiti has only quercetin 3-O-glycosides, which is the simplest profile in the section. Cladistic analysis of the flavonoid data suggests that section *Skinnera* is monophyletic with the Tahitian species, *F. cyrtandroides*, as a sister clade to the species comprising the New Zealand group. Derivation of the Tahitian species does not necessarily involve one of the New Zealand species. A final note on *Fuchsia* concerns a study of anthocyanins in the genus, in which it was shown that *F. excorticata* exhibited a pigment profile consisting of cyanidin, pelargonidin, peonidin, delphinidin and malvidin, all of which occurred as both 3-O-glucosides and 3,5-di-O-glucosides. This profile was among the most complex seen in the genus (Crowden, Wright & Harborne, 1977).

Recent work from our laboratory has focused on flavonoids of several endemic taxa of New Zealand including *Brachyglottis* (Senecioneae), and

Cassinia and *Raoulia* (both Inuleae). In the case of *Brachyglottis*, five species from New Zealand proper and one from the Chatham Islands were examined for flavonoid aglycones present on leaf surfaces as well as for vacuolar (glycosidic) flavonoids. Common glycosides of kaempferol and quercetin occurred in all six species, with *B. cassinioides* (Hook. f.) B. Nordenstam having the largest number of compounds (four) (Reid & Bohm, 1994). Flavonoid aglycones were obtained from four species (Reid & Bohm, 1994): *B. cassinioides*, *B. compacta* (Kirk) B. Nordenstam, *B. huntii* (F. Muell.) B. Nordenstam and *B. monroi* (Hook. f.) B. Nordenstam. Kaempferol 3-methyl ether was obtained from all four of these species, naringenin from *B. huntii* and *B. monroi*, eriodictyol from *B. huntii* and 2',4',6',4-tetrahydroxychalcone from *B. monroi*. *Brachyglottis huntii*, which is endemic on the Chatham Islands, is distinguished from the other taxa studied by having eriodictyol but otherwise its flavonoid chemistry falls within the group of compounds seen in the genus. *Brachyglottis greyi* (Hook. f.) B. Nordenstam and *B. repanda* J. R. & G. Forst. did not appear to have leaf surface flavonoids, but their vacuolar components, quercetin 3-O-glucoside and 3-O-rutinoside, were otherwise common in the genus.

Leaf surface flavonoid aglycones and vacuolar flavonoid glycosides were identified from four New Zealand species of *Cassinia*: *C. amoena* Cheeseman, *C. fulvida* Hooker f., *C. leptophylla* (Forst. f.) R. Brown and *C. vauvilliersii* (Hombron & Jacquinot) Hook. f. (Reid & Bohm, 1994). Exudate compounds identified included chalcones, dihydrochalcones, flavanones and quercetin methyl ethers. Several of these compounds lacked B-ring oxygenation in common with other members of the Gnaphalieae (tribe Inuleae). Vacuolar components identified were kaempferol and quercetin 3-O- and 7-O-glycosides, eriodictyol 7-O-methyl ether and galangin (3,5,7-trihydroxyflavone). It is interesting to note that populations of *C. vauvilliersii* growing on different substrates, weathered igneous substrates from Tongariro National Park as opposed to sedimentary-derived soil from Mount Holdsworth, gave different flavonoid profiles. The suggestion that these pigment profiles may be fixed in each of these species is supported by studies of plants grown in the Otari Native Plant Museum which gave profiles identical to the 'native' ones.

Seven species (nine taxa) of *Raoulia* Hook. f. ex Raoul, the vegetable sheep, were examined for vacuolar flavonoids (Reid & Bohm, unpublished; no glandular flavonoids were detected). The pigment profile of the group comprised kaempferol, quercetin, isorhamnetin and myricetin

3-O-glucosides, kaempferol and quercetin 3-O-rutinosides and luteolin. No intra- or interspecific qualitative variation was found.

Other lipophilic flavonoids, some quite unusual, were obtained from the leaf gum of *Traversia baccharoides* Hook. f. (Asteraceae: Senecioneae) by Kulanthaivel & Benn (1986). More or less normal compounds isolated were (2S)-eriodictyol 7-methyl ether, quercetin 7-methyl ether (rhamnetin) and quercetin 7,3'-dimethyl ether (rhamnazin). The unusual compounds described by these workers were the 3-, 3'- and 4'-isobutyrates of quercetin. Also described in that paper was a derivative of truxillic acid, formed by dimerization of two cinnamic acid units to form a cyclobutyl system.

In their study of flavonoids of the Cyperaceae, Williams & Harborne (1977) included representatives of three genera from New Zealand: *Scirpus frondosa* Banks & Soland ex Boeck., which gave cyanidin and a C-glycosylflavone, *Carex secta* Boott, which had a C-glycosylflavone and tricin 5-O-glucoside, and an *Uncinia* species that afforded a C-glycosylflavone and tricin. *Lepidosperma tortuosum* F. Muell. from Tasmania gave quercetin. The authors observed that all taxa were very similar to other members of their respective genera.

The following section deals with a variety of terpenoids that have been isolated from New Zealand plant species. The compounds range in complexity from iridoids to cucurbitacins. Jensen & Neilsen (1980) studied several species of the cornaceous genus *Griselinia* from which they obtained the iridoid glucoside 'griselinoside' (Fig. 11.1 [27]). The New Zealand endemic species *G. lucida* Forst. f. and *G. littoralis* Raoul, as well as the Chilean *G. ruscifolia* (Clos) Taub., yielded griselinoside whereas two Chilean species, *G. alata* Ball and *G. jodinifolia* (Griseb.) Taub., appeared not to have iridoids.

In a second paper, concerned with antibiotic substances from New Zealand plants (for the first paper, see Muir, Cole & Walker, 1982, below), McCallion *et al.* (1982) described the isolation of a sesquiterpene dialdehyde (Fig. 11.1 [28]) from *Pseudowintera colorata* (Raoul) Dandy (Winteraceae). This compound, which possesses anti-*Candida* activity, was not found in *P. axillaris* (J. R. & G. Forst.) Dandy or *P. traversii* (Buchan.) Dandy. It was found, however, in *Drimys winteri* J. R. & G. Forst. (Winteraceae).

Dacrydium, a member of Podocarpaceae, consists of some two dozen or so species ranging from South East Asia to New Zealand. *Dacrydium cupressinum* Sol. ex Lambert, the 'rimu' or red pine, is an important timber tree of New Zealand. Berry, Perry & Weavers (1985) reported as major foliar sesquiterpenes the following compounds: α-longipinene, longifolene, longibornyl acetate, caryophyllene and its oxide, humulene, α- and β-selinenes, β- and δ-elemenes, aromadendrene and 9βH-caryophyllene. In an accompanying paper, Perry & Weavers (1985) identified a suite of diterpenes from the foliage of the endemic *D. intermedium* T. Kirk: rimuene (Fig. 11.1 [29]), *ent*-rosadiene, phyllocladene, *ent*-kaurene, *ent*-beyerene, sclarene and *ent*-sclarene. A third paper (Hayman, Berry & Weavers, 1986) described equally complex arrays of mono- and sesquiterpenes in the foliage of *D. biforme* (Hook.) Pilger, which occurs on both the South and North Islands of New Zealand and on Stewart Island. The same paper documented the existence of juvenile–adult dimorphism in the terpenoid composition of *D. biforme*.

The most structurally complex compounds in this section are the triterpenoids isocucurbitacin-D (Fig. 11.1 [30]) and its 3-epi isomer which Kupchan, Meshulan & Sneden (1978) isolated from *Phormium tenax* Forst. & Forst. f. (Agavaceae). This plant, which occurs on New Zealand, Lord Howe Island and Norfolk Island, was used by the Maoris as a source of fibre.

Coriariaceae, consisting of the single genus *Coriaria*, is noteworthy for its unusual distribution range which involves Mexico south to Chile, the western Mediterranean, an area from the Himalayas east to Japan, New Guinea, New Zealand and islands in the South Pacific (Mabberley 1987). The New Zealand *Coriaria*, *C. arborea* Linds, was examined chemically by several workers in an effort to determine the nature of the convulsive toxin present in the plant. Early work (Easterfield & Aston, 1901) and more recent efforts by Craven (1963) established the molecular structure of the toxic component 'tutin'. Examination of Fig. 11.1 [31] shows that tutin is a close relative of picrotoxin. Further discussion of the toxic properties of *Coriaria* species, including the European *C. myrtifolia* and the Japanese *C. japonica*, can be found in *New Zealand Medicinal Plants* (Brooker, Cambie & Cooper, 1987).

The flavonoids of *Coriaria* were investigated by Bohm & Ornduff (1981), who reported an overall profile consisting of kaempferol and quercetin glycosides and a naringenin glycoside from 12 species, including five from New Zealand: *C. angustissima* Hook. f., *C. arborea* Lindsay, *C. lurida* T. Kirk, *C. ruscifolia* L. and *C. sarmentosa* Forst., and one from Taiwan, *C. intermedia* Matsum. There were no significant differences between the pigment profiles of these island taxa and species from China, Japan, Mexico or Tibet. Quercetin 3-O-rutinoside was a major component in the flavonoid profile which observation affords us the opportunity to point out that this may be one of the few groups of plants that possess both rutin and tutin.

The similarity of the toxic components of *Coriaria*, with its broad distribution, prompts a few remarks about a plant of similar widespread occurrence, which was also included in the Brooker, Cambie & Cooper (1987) study of medicinal plants of New Zealand. The root of bracken fern, *Pteridium esculentum* (Forst. f.) Cockayne [*P. aquilinum* (L.) Kuhn var. *esculentum*], was an important food for the Maoris despite the toxic properties of the plant. As Brooker, Cambie & Cooper (1987) point out, it would appear that many, if not all, of the toxic components of bracken have been retained despite its isolation in New Zealand through time. It is not known when bracken first appeared in New Zealand, and, as in other spore-dispersed organisms, it is not possible to rule out the possibility that 'fresh' spores may arrive at any time.

The genus *Corynocarpus* Foster & Foster f., the sole member of the Corynocarpaceae, consists of four species variously distributed in New Guinea, northeastern Australia, New Caledonia, Vanuatu and New Zealand s.l. (Mabberley, 1987). *Corynocarpus laevigatus* J. R. & G. Forst., which occurs commonly on the North Island of New Zealand, and to a lesser extent on the South Island, Kermadec Islands and Chatham Island, has been subjected to detailed chemical examination. Interest in the 'karaka tree' stemmed from its intensely bitter and toxic bark and seeds. Several compounds involving 3-nitropropionic acid and glucose were identified by Moyer *et al.* (1979). 'Karakin' is 1,2,6-tri-(3-nitropropanoyl)-β-D-glucoside (Fig. 11.1 [32]) while 'corynocarpin' is the 1,4,6-triester. Other members of this family are the 1,6-diester ('cibarin'), the 2,6-diester ('coronarian') and the 2,3,6-triester ('corollin'). More recently, Majak & Benn (1994) reported isolation of 11 compounds, including mono, di, tri and tetra esters, from the fruit of this species. Several of these esters are also known from members of the legume genera *Astragalus*, *Coronilla* and *Indigofera* (see Majak & Benn, 1994, for citations).

In a survey for antibiotic compounds in New Zealand plants, Muir, Cole & Walker (1982) discovered that the anti-dermatophytic agent from *Schefflera digitata* J. R. & G. Forst. (Araliaceae), was the acetylenic derivative falcarindiol whose structure is (9Z)-heptadeca-1,9-diene-4,6-diyn-3,8-diol (Fig. 11.1 [33]). Four of the New Zealand species of *Chionochloa* (Poaceae) (22 species are known 21 of which grow in New Zealand), *C. flavescens* Zotov, *C. pallens* Zotov, *C. rigida* (Raoul) Zotov (also occurs on the Kermadec Islands) and *C. rubra* Zotov, were examined for their epicuticular wax composition (Cowlishaw, Bickerstaffe & Connor, 1983). These investigators reported fatty acids, alcohols and aldehydes in the range of C_{24}–C_{32}, wax esters in the range of C_{36}–C_{52} and alkanes in the range C_{29}–C_{33}. The last in this collection of miscellaneous natural products involves a study of the absolute configuration of tetraphyllin-B, a cyanoglucoside from the New Zealand passion fruit, *Tetrapathaea tetrandra* Cheeseman (Passifloraceae). Based on the efforts of Gainsford, Russell & Reay (1984), the structure is as shown in Fig. 11.1 [34]. Tetraphyllin-B, and the isomeric A, were originally isolated from *T. tetrandra* by Russell & Reay (1971).

Laurelia is a small genus in the Monimiaceae (Atherspermataceae R. Br.) that consists of two species (Mabberley, 1987), one from New Zealand and one from western South America. *Laurelia novae-zelandiae* A. Cunn. was shown by Adjaye *et al.* (1984) to accumulate the oxoaporphine alkaloids, liriodenine and lauterine, the aporphine alkaloid, pukateine, the benzylisoquinoline derivative, romneine, and a fifth alkaloid, corydine. Pinoresinol dimethyl ether and lirioresinol B dimethyl ether, the first lignans from the genus, were also reported in that paper. *Laurelia* was among a group of genera examined for alkaloids in an earlier study (Urzua & Cassels, 1978).

Alkaloids were also involved in a study of *Discaria toumatou* Raoul by Panichanum, Bick & Blunt (1982). This genus belongs to Rhamnaceae and consists of some 15 species, with only *D. toumatou* known from New Zealand. Species from South America and Australia are known for their cyclopeptide alkaloids but the New Zealand species has R(-)-N-methylcoclaurine, which is one of the benzylisoquinoline alkaloids reported from a Chilean species of *Discaria* (Torres, Delle Monache & Marini-Bettolo, 1979).

Flavonoids of the Lord Howe Island endemic *Carmichaelia exsul* were discussed in the section on New Zealand above. Cucurbitacins from *Phormium* were included in the section on terpenoids above. Here we look briefly at the work of Brophy, Goldsack & Goldsack (1993) who were interested in the essential oils of indigenous Myrtaceae of Lord Howe Island. The taxa investigated and the major oil components reported were: *Leptospermum polygalifolium* Salisb. subsp. *howense* J. Thompson, α, β and γ-eudesmols; *Melaleuca howeana* Cheel, 1,8-cieole and gluobulol; *Cleistocalyx fullageri* (F. Muell.) Merrill & Perry, α-pinene; *Metrosideros sclerocarpa* Dawson ex P. S. Green, *allo*-aromadendrene; and *M. nervulosa* C. Moore & F. Muell, sesquiterpene alcohols and *n*-hexadecanol.

Tasmania

Tasmania, the island state of Australia, lies off the southeast coast of the continent separated from the mainland by the Bass Strait. Several studies of Tasmanian plants, endemic or otherwise, deserve mention here. Bate-Smith (1977) reported ellagic acid, kaempferol and quercetin from *Anodopetalum biglandulosum* (Hook.) Hook f., a Tasmanian endemic, in his study of polyphenolic compounds in the Cunoniaceae.

The genus *Richea* (Epacridaceae) is represented in Tasmania by 10 endemic species (one species occurs on the Australian mainland). Two of the endemic species, the widespread and abundant *R. scoparia* Hook. f. and *R. angustifolia* B. L. Burtt, which is much more restricted in its distribution, were subjected to a detailed analysis of morphological and chemical variation by Menadue & Crowden (1983). A complex array of compounds yielded glycosides of kaempferol, quercetin, myricetin and two aurones. Although the data failed to provide clearcut discontinuities between the two taxa, there was a decided grouping of populations according to the nature of the substrates upon which the plants were growing. For example, the aurones were most prominent in plants growing on igneous (dolerite) or sedimentary rocks, but did not occur in plants growing on metamorphic rocks. Relationships between plant chemistry and edaphic factors are not unique to this system; different flavonoid patterns for two populations of *Cassinia vauvilliersii*, a New Zealand endemic (see above), were reported by Reid & Bohm (1994).

Biflavonoids were reported from the primitive monocot genus *Isophysis* by Williams, Harborne & Tomas-Barberan (1987). *Isophysis tasmanica* (Hook. f.) Moore (Iridaceae), endemic to Tasmania, afforded amentoflavone, dihydroamentoflavone, hinokiflavone and podocarpusflavone A.

Corolla colour variation (*inter alia*) in *Epacris impressa* Labill. (Epacridaceae) was studied throughout its range in southeastern Australia and Tasmania (Stace & Fripp, 1977). Populations were characterized by plants with white, pink or red corollas (cyanidin glycosides were involved). Populations in Tasmania were either white-flowered or pink; red-flowered populations apparently were restricted to Victoria on the Australian mainland.

Villarsia exaltata (Soland. ex Sims) G. Don. and *Liparophyllum gunnii* J. D. Hooker, both Menyanthaceae and from Tasmania, were included in a study of the flavonoids of the family by Bohm, Nicholls & Ornduff (1986). For discussion of this family see the discussion of the Caribbean islands below.

Acradenia Kippist (Rutaceae) was thought to consist of the single, Tasmanian endemic species *A. franklinieae* Kippist until a second species, *A. euodiiformis* (F. Muell.) Hartley was described. The new species, transferred from *Bosistoa*, occurs in northern New South Wales and adjacent Queensland. Earlier chemical studies (Baldwin *et al.*, 1961) showed the presence of C-prenylated acetophenones, a furoquinoline alkaloid, a limonoid and triterpenes in these species. A recent chemical examination of both species (Quader *et al.*, 1991) confirms that acetophenones, e.g., Fig. 11.1 [35], are major metabolites of the genus.

New Caledonia

New Caledonia (France) lies between 20–22° S and 164–167° E at a distance of about 1400 km in a northeasterly direction from Brisbane, Australia. The flora of New Caledonia is so unusual that Takhtajan (1986) accords the island status of a floral subkingdom in its own right. The Neocaledonian Subkingdom includes seven endemic families, over 130 endemic genera and over 90% endemic species.

Bate-Smith (1977) included a number of New Caledonian taxa in his study of the chemistry and taxonomy of the Cunoniaceae. He reported: ellagic acid, kaempferol, quercetin and cyanidin from hydrolysed extracts of *Cunonia pulchella* Brongn. & Gris; kaempferol and quercetin from *C. purpurea* Brongn. & Gris; kaempferol, quercetin and cyanidin from *Geissois primosa* Brongn. & Gris; ellagic acid, kaempferol, quercetin and cyanidin from *Weinmannia serrata* Brongn. & Gris; and ellagic acid, kaempferol, quercetin and myricetin from *Paucheria sebertii* Guill.

Leaf flavonoids of members of the Winteraceae, an archaic family well represented in New Caledonia, were examined by Williams & Harvey (1982). These authors accumulated data for representatives of the following genera: *Belliolum, Bubbia, Drimys, Exospermum, Pseudowintera, Tasmannia* and *Zygogynum. Degeneria vitiensis* I. W. Bailey & A. C. Smith (Degeneriaceae) and *Illicium munipurense* Watt ex King (Illiciaceae) were included in the study as well. *Belliolum* occurs on New Caledonia and the Solomon Islands, *Bubbia* in Australia, New Guinea and New Caledonia, *Drimys* in Central and South America, *Exospermum* only on New Caledonia, *Pseudowintera* in New Zealand, and *Zygogynum* only on New Caledonia. The high incidence of luteolin 7,3′-dimethyl ether (77% of Winteraceae species) isolates the family from any other. This flavone was found in neither *Degeneria* nor *Illicium*. Kaempferol and quercetin were met in all genera, while simple flavones (apigenin and luteolin) were seen only in species of *Drimys*,

Pseudowintera and *Tasmannia*, although not necessarily in all species of each. The authors pointed out that the presence of procyanidin in 60% of Winteraceae species and dihydroquercetin in 40% reflects the primitive woody nature of the family (*Belliolum*, *Bubbia*, *Drimys*, *Exospermum* and *Zygogynum* are vesselless).

Zieridium ignambiensis (Guillaum.) Harley (Rutaceae), one of three species of a genus endemic to New Caledonia, was shown by Elavumoottil *et al.* (1991) to accumulate two flavonols, 3,5-dimethoxy-6,7-methylenedioxy-3′,4′-methylenedioxyflavone (known) and 3,7-dihydroxy-5,6-dimethoxy-3′,4′-methylenedioxyflavone (new) (Fig. 11.1 [36]). Newly reported from this taxon was the furoquinoline alkaloid evolitrine. Three other alkaloidal compounds were also identified: 5-methoxy-N,N-dimethyltryptamine, 6-methoxy-2-methylcarboline and 6-methoxy-2-methyltetrahydrocarboline N-oxide. Recently, Lichius *et al.* (1994) described four highly methoxylated flavonols from *Z. pseudoobtusifolium* (Guillaum.) Guillaum., several of which showed anti-mitotic and cytotoxic properties. One of these compounds, 5,3′-dihydroxy-3,6,7,8,4′-pentamethoxyflavone, was also found to be a constituent of the Malaysian species *Acronychia porteri* Hook. f., also a member of Rutaceae. No systematic significance was attributed to the appearance of this compound in these taxa.

Cambie, Cox & Sidwell (1984) included representatives of five genera of Podocarpaceae from New Caledonia in their study of phenolic diterpenoids of the family. *Dacrycarpus viellardii* yielded three such compounds, *Decussocarpus comptonii* (Buchholz) de Laub two and *Falcatifolium taxoides* (Brongn. & Gris) de Laub only one. *Podocarpus ustus* Brongn. & Gris and *Prumnopitys ferruginoides* (Compt.) de Laub afforded sitosterol but no phenolic diterpenoids. Also included in this study were *Podocarpus affinis* Seem. from Fiji, *P. falcatus* R. Br. ex Mirb. from South Africa and *P. koordersii* Pilger from Java. These three taxa also yielded phenolic diterpenoids. 'Totarol' (Fig. 11.1 [37]) is typical of the type.

The bled resins of three Pacific kauri, *Agathis lanceolata* Lindley ex Warbury from New Caledonia, *A. vitiensis* (Seeman) Benth. & Hook. from Fiji and *A. macrophylla* (Lindley) Masters from Santa Cruz on the Solomon Islands were studied by Smith, Marty & Peters (1981). The acids, identified as their methyl esters, were *cis*- and *trans*-communic acids, sandarocopimaric acid, abietic acid (Fig. 11.1 [38]), dehydroabietic acid, neoabietic acid, agathalic acid and agathic acid. Comparisons with Australian and New Zealand species showed different percentages of individual components but the overall chemistries were sufficiently similar to suggest close relationships. For example, comparison of *A. lanceolata*, the New Caledonian species, with *A. atropurpurea*, the 'black kauri' of Queensland, showed a high degree of similarity, particularly with regard to levels of agathic acid, 41% in *A. lanceolata* compared to 56% in the Queensland specimen.

Extensive studies of New Caledonian plants for alkaloids have yielded a rich array of compounds. Thus, *Boronella* aff. *verticillata* Baill. (Bévalot, Vaquette & Cabalion, 1980) and *B. pancheri* Baill. (Muyard *et al.* 1994) (Rutaceae) have been shown to accumulate an array of quinolione, furoquinoline, dihydrofuroquinoline and pyranoquinoline alkaloids. *Dutaillyea*, an endemic genus with about ten species, also a member of the Rutaceae, yielded a similarly rich assortment of characteristic alkaloids (Baudouin *et al.*, 1981; Muyard *et al.*, 1992). A *bis*-indole alkaloid, plumocraline, was reported from *Alstonia plumosa* var. *communis* Boiteau forma *glabra* Boiteau, a New Caledonian member of the Apocynaceae, by Massiot *et al.* (1981). This particular paper was number 72 in a series on the chemistry of plants of New Caledonia, which indicates a fair sampling of island endemics over the years.

New Guinea, Fiji and Samoa

The New Guinea flora consists of 172 families, 844 genera and 2310 species (Hartley *et al.*, 1973). These workers tested 2250 species in the field for alkaloids and reported positive tests for 235 species. In a study of Atherospermataceae (Monimiaceae) Urzua & Cassels (1978) reported aporphine, oxoaporphine and biscodarine alkaloids from *Dryadodaphne novoguineensis*. Genera from other Southern Hemisphere sites [southern Chile, eastern Australia, New Caledonia, New Zealand (see *Laurelia* above) and Tasmania] were included in the study.

In addition to species of Cunoniaceae from New Caledonia, Bate-Smith (1977) reported detecting kaempferol and quercetin with sporadic occurrence of ellagic acid and cyanidin from the following taxa, which are given along with their islands of origin: *Aistopetalum viticoides* Schulte, *Gillbeea papuana* Schultr., *Opocunonia nymanii* K. Schum., *Spiraeopsis brassei* Perry and *Weinmannia* species from Papua New Guinea; *Spiraeanthemum samoense* A. Gray from Samoa; and *S. vitiense* A. Gray from Fiji.

In a recent paper, Cambie, Lal & Pausler (1992) described the isolation and identification of several compounds from *Degeneria vitiensis*, which is endemic to Fiji. Among the compounds described are β-sitosterol, myristicin, myristicin diol, 2,6-dimethoxybenzoic acid and the phenolic ester curculoside (Fig. 11.1 [39]). Earlier work on this plant (Pojer, Ritchie & Taylor, 1968) had shown the presence of 2-hydroxy-6-methoxybenzoic acid, β-sitosterol and waxes, but negligible alkaloids.

The Atlantic Ocean

The Faeroe Islands

The Faeroe Islands (Denmark) are located about halfway between Iceland and the northern tip of Scotland. The only information that came to light involving the chemistry of a Faeroe Island plant, not an endemic, was that of Williams & Harborne (1977) who included *Eriophorum vaginatum* L. in their study of flavonoids and geography of the Cyperaceae. They reported the presence of kaempferol, quercetin and cyanidin in hydrolysed extracts from this species.

The Canary Islands and Madeira

The Canary Islands consist of seven major islands that lie just off the northwest coast of Africa at *c.* 28°30′ N, 15°10′ W. Madeira lies further to the northwest at 32°45′ N, 17°00′ W. According to Kunkel (1980), the Canary Islands enjoy 46.6% species endemism. The Canary Islands have a complex volcanic geology (Schmincke, 1976, in Kunkel, 1976; Coello *et al.*, 1992). Chemical studies from several laboratories have treated a number of the endemic species in some detail yielding information on flavonoids, coumarins, lignans and diterpenes.

Senecio kleinia Sch. Bip. (*Kleinia nerifolia* Haw.), one of the island endemics, was included in a a detailed phytogeographic study of the *S. radicans* complex by Glennie *et al.* (1971). *Senecio kleinia* was one of three taxa (the other two are subspecies of *S. longiflorus*, see section on Madagascar) whose flavonoid profile consisted solely of quercetin 3-O-glucoside. In contrast, other members of the complex, except the non-succulent *S. angulatus* which lacks flavonols, have more complex profiles involving other flavonols, a flavone and a C-glycosylflavone. Other than pointing out the correlation between flavonoid profile and geographic origin of the species examined, the authors did not comment on the evolutionary significance of the data.

In a survey of some 20 endemic species of *Sonchus*, Bramwell & Dakshini (1971) reported luteolin 7-O-glucoside from 19 species and the coumarins cichoriin (6,7-dihydroxycoumarin 6-O-glucoside) from 15 species, esculin (6,7-dihydroxycoumarin 7-O-glucoside) from five species and scopoletin (6-methoxy-7-hydroxycoumarin) from all 20. A more recent study of phenolic compounds of *Sonchus* showed a richer array of compounds than reported earlier (Mansour, Saleh & Boulos, 1983). Thus, *S. canariensis* (Sch. Bip.) Boulos subsp. *orotavensis* Boulos, *S. brachylobus* Webb & Berth var. *canariae* (Pitard) Boulos and *Taeckholmia canariensis* Boulos exhibited a series of luteolin 7-O-mono- and 7-O-diglycosides, while the latter two taxa added apigenin 7-O-glucoside to the array. These flavonoid profiles were in general agreement with information from some 30 other species of *Sonchus*, *Embergeria*, *Babcockia* and *Taeckholmia*.

Allagopappus viscosissimus Cass., an endemic member of the Asteraceae (Inuleae), was shown by Gonzalez *et al.* (1992) to accumulate kaempferol and quercetin 3-methyl ethers, naringenin and eriodictyol as well as several germacranolides (cyclodecane-based sesquiterpene lactones). The last flavonoid example from the Canary Islands' flora involves the report of several glycosides of isoetin (5,7,2',4',5'-pentahydroxyflavone) isolated from the endemic *Heywoodiella oligocephala* Svent. & Bramwell (Harborne, 1991). *Heywoodiella* is closely related to the widespread genus *Hypochaeris*.

The most extensive studies of flavonoids in endemic species from the Canary Islands involve members of the genus *Sideritis* (Lamiaceae). Early papers (Gonzalez *et al.*, 1978, 1979) described isolation of kaempferol 3,7,4'-trimethyl ether from *S. bolleana* Bornm., 5,4'-dihydroxy-6,7- dimethoxyflavone (cirsimaritin) and 5,4'-dihydroxy-6,7,8-trimethoxyflavone (xanthomicrol) from *S. dasygnaphala* (Webb) Clos., and a series of O-methylated flavones and flavonols from *S. gomerae* Bolle. More recent work (Fernandez, Fraga & Hernandez, 1988) detailed isolation of a similar array of flavones and flavonols along with some flavanones: 5-hydroxy-6,7,3',4'-tetramethoxyflavanone from *S. cystosiphon* Svent.; 5-hydroxy-7,4'-dimethoxyflavanone and 5-hydroxy-7,3',4'-trimethoxyflavanone from *S. infernalis* Bolle; 5,4'-dihydroxy-6,7-dimethoxyflavanone from *S. sventenii* (Kunk) Mend-Heu; and 5-hydroxy-6,7,4'-trimethoxyflavanone from *S. gomerae*. A recent comprehensive study of flavonoid aglycones and glycosides involved 25 species (26 taxa) of *Sideritis* from six of the Canary Islands plus Madeira (Gil *et al.*, 1993). Both sets of flavonoids assort themselves in an interesting manner along sectional lines. Derivatives of isoscutellarein (8-hydroxyapigenin) and hypolaetin (8- hydroxykaempferol) occur in ten of the 15 species of sect. *Marrubiastrum*, not at all in the three species of sect. *Empedocleopsis* and in only one (*S. lotsyi*) of six species from sect. *Creticae*. With some differences in the size of complement, both 5,7-dioxygenated flavones and flavone 7-O-*p*-coumaroylglucosides occur in most members of all three sections. A major distinction between sectional chemistries was observed with aglycone distribution. Only scattered appearances of surface aglycones were observed in sect. *Marrubiastrum* and *Creticae* with only a few compounds present in any case. The compounds most frequently found were luteolin and chrysoeriol. Glycosides of these two flavonoids are common components of the polar fraction of species in these sections. In the three species that comprise section *Empedocleopsis*, however, 14 compounds were identified from non-polar fractions with very little variation except for some apparent quantitative differences among species. It is interesting to note that neither free luteolin nor chrysoeriol was present in these three species although chrysoeriol 7-O-glycosides do occur in all three.

Gonzalez *et al.* (1979) showed that *Sideritis canariensis* accumulated the coumarin 'siderin' which they identified as 4-methyl-6,7-dimethoxy-coumarin. This paper also described identification of the lignan (+)-sesamin from the same species. Lignans were also reported from the endemic *Bupleurum salicifolium* Soland. (Apiaceae) by Estévez-Reyes, Estévez-Braun & Gonzalez (1993). The three compounds identified were the butenolides, chasnarolide, methyl chasnarolide and benchequiol (Fig. 11.1 [40]).

It is important to refer the reader to the extensive work done on the diterpenoids of *Sideritis canariensis* by Gonzalez *et al.* (1978, 1979). Add to this the recent work by Fraga *et al.* (1991), who reported eight diter-penes from *S. canariensis* var. *pannosa*. A typical representative of this group is 'ribenol', 3-α-hydroxy-(-) 13-epimanoyl oxide (Fig. 11.1 [41]).

The last example in this section involves the endemic *Persea indica* L., which has been known for its toxic properties for some time. Gonzalez-Coloma *et al.* (1990) isolated two terpenoids and identified them as 'ryanodol' (Fig. 11.1 [42]) and 'cinnceylanol' (Fig. 11.1 [43]). Compound [43] is also known from the Sri Lankan species *Cinnamonum zelanicum* Nees. (Lauraceae) from which it was isolated by Isogai *et al.* (1976, 1977). These complex pentacyclic polyhydroxy-diterpenes are closely related to a compound known from *Ryania speciosa* (Flacourtiaceae), a shrub native to Trinidad and the Amazon basin, a preparation of which, called ryania powder, has been used to control agricultural pests.

Caribbean Islands

Menyanthaceae consist of five genera, *Fauria*, *Menyanthes*, *Liparophyllum*, *Nymphoides* and *Villarsia*. The first three are monotypic, with the bulk of the species, 40 in all (Mabberley 1987), about equally distributed in the latter two genera. A detailed flavonoid study of the family has been published (Bohm, Nicholls & Ornduff, 1986). Of interest to us here are *Nymphoides grayana* (Griseb.) Kuntze obtained from Andros Island in the Bahamas and *N. indica* (L.) Kuntze from Jamaica. These were found to be among the richest producers of flavonol aglycones of the family, sharing this distinction with *N. geminata* (obtained from Australia) and *N. fallax* (Mexico). Both produced compounds based on kaempferol, quercetin, quercetin 7-methyl ether and quercetin 3,7-dimethyl ether. *Nymphoides indica* also had isorhamnetin derivatives. All species had moderately complex arrays of glycosides. Two island representatives from the Pacific were examined in this study as well, *Villarsia exaltata* (Soland. ex Sims) G. Don from Tasmania and *Liparophyllum gunnii* J. D. Hooker, which occurs in Tasmania and on Stewart Island. The *Villarsia* sample showed essentially the same flavonoid chemistry as other members of the genus (from Australia and South Africa). Material of *Liparophyllum* (from Tasmania) gave the simplest flavonoid profile seen in the study, which consisted of only a few quercetin glycosides.

In a study of Cyperaceae, Williams & Harborne (1977) included two Caribbean species, *Calyptrocarya glomerulata* (Brongn.) Urb. and *Diplacrum longifolium* (Griseb.) C. B. Clarke. The former gave luteolin 5-methyl ether, while the latter gave C-glycosylflavones and cyanidin. These profiles were, in general, similar to other members of the genera *Calyptrocarya* and *Diplacrum*.

Weinmannia hirta Sur., which ranges from Cuba to Martinique, was included in Bate-Smith's (1977) flavonoid survey of the Cunoniaceae. He reported the presence of ellagic acid and quercetin in a hydrolysed extract of this species. This profile was similar to many other members of the family including species from northern South America, e.g., *W. fagaroides* H. B. Kunth and *W. glabra* L. f.

Four endemic Cuban species of *Vernonia* Schreb. were screened for their sesquiterpene lactones by Budesinsky, Perez Souto & Holub (1994): *V. acunnae* Alain., *V. angusticeps* Ekm., *V. menthaefolia* (Papp. ex Spreng.) Less. and *V. moaensis* Alain. All taxa yielded arrays of common lactones. Genkwanin (apigenin 7-methyl ether) and velutin (luteolin

7,3'-dimethyl ether), both widely distributed in the Asteraceae, were identified as components of *V. menthaefolia*.

South Atlantic Islands

Studies of island systems in the South Atlantic Ocean involve the
Falkland Islands, the Tristan da Cunha Island group (including Gough
Island) and St Helena. In discussions of these several island groups,
relationships with taxa from mainland South America, naturally, play
an important role. More surprisingly, however, are relationships between
species on the Tristan da Cunha group and the Juan Fernandez Islands in
one instance, and with New Zealand in another.

The Tristan da Cunha group consists of three main islands: Tristan
(0.5 million years old), Inaccessible (2.9 million years old) and
Nightingale (18 million years old), and two islets, Middle and
Stoltenhoff. The group lies at about 37° S 12° W in the middle of the
South Atlantic Ocean and is associated with the Mid-Atlantic Ridge.
Active volcanism occurred on Tristan as recently as 1961 (Preece,
Bennett & Carter, 1986). Gough Island lies about 350 km south-south-
east of Tristan at 40° S 10° W.

We will begin by looking at the Juan Fernandez–Tristan da Cunha
Islands connection. As part of an intensive study of evolutionary relationships between endemic plants of the Juan Fernandez Islands and
mainland South America, discussed in some detail above, Valdebenito,
Stuessy & Crawford (1990*a*, *b*) and Valdebenito *et al.* (1992*a*, *b*) examined the genus *Peperomia* (Piperaceae). An unexpected result of this study
was the finding that *P. tristanensis*, endemic to Inaccessible Island, is
closely related to *P. berteroana* of the Juan Fernandez Islands, and that
they are most reasonably treated as subspecies of *P. berteroana*: *P. berteroana* Miq. subsp. *berteroana* of the Juan Fernandez Islands and *P.
berteroana* Miq. subsp. *tristanensis* (Christoph.) Valdebenito of
Inaccessible Island (Valdebenito, Stuessy & Crawford, 1990*b*). The flavonoid profile of *P. berteroana* from Masatierra is different from that of
P. berteroana from Masafuera. Significantly, the flavonoid profile of
P. berteroana subsp. *tristanensis* is identical to that of plants from
Masafuera. The flavonoid data are summarized in Table 11.4. The
observation led these workers to conclude that the presence of
Peperomia on Inaccessible Island was due to long distance dispersal by
birds from the Juan Fernandez Islands with the propagules having originated from plants on Masafuera. Movement of propagules in the reverse
direction was rejected because of the presence of four well-distributed

Table 11.4. *Flavonoids of* Peperomia berteroana *subspecies*

Taxon	Flavonoids[a]				
	1–7	8	9	10	11–14
P. b. subsp. *berteroana* (MT)	+	+	+	+	−
P. b. subsp. *berteroana* (MF)	+	−	−	−	+
P. b. subsp. *tristanensis*	+	−	−	−	+

a) 1 = Acacetin; 2 = Acacetin 7-sulfate; 3 = Diosmetin 7-Glc; 4 = Diosmetin 7-Glc-Rhm; 5 = Luteolin 7-Glc; 6 = Luteolin 7-Ara-Ara; 7 = Luteolin 7-sulfate; 8 = Apigenin 7-Glu-sulfate; 9 = Luteolin 7-Glc-Rhm; 10 = Diosmetin 7-Glc; 11 = Acacetin 7-Glc-disulfate; 12 = diosmetin C-Glc-Ara; 13 = Diosmetin 7-Glc-Ara; 14 = Diosmetin 7-sulfate; Ara = arabinose; Glc = glucose; Rhm = rhamnose; MT = Masatierra; MF = Masafuera

species of *Peperomia* on the Juan Fernandez Islands in contrast to the rare status of *Peperomia* on Inaccessible Island.

Our second example from the South Atlantic involves *Empetrum rubrum* L. (Empetraceae) which occurs in the Chilean Andes, Patagonia and Tierra del Fuego, and on the Juan Fernandez Islands, the Falklands and Gough Island (Tristan da Cunha group). A study of the flavonoid chemistry of the entire family has been described by Moore, Harborne & Williams (1970), although only a few of the compounds observed on chromatograms were identified. Relevant to our interests here are their observations of *E. rubrum* from Gough Island and the Falklands. The Gough Island specimens showed a depauperate flavonoid profile relative to the mainland samples although the spot pattern observed fell within the limits observed for *E. rubrum*. Plants from the Falklands exhibited a flavonoid profile much closer in appearance to profiles observed with plants from mainland South America. The similarities were not, however, with plants sampled from northeast Tierra del Fuego which represented the geographically closest populations to the Falkland Islands (*c*. 400 km). Rather, the most similar flavonoid profiles were from samples from central and southern Tierra del Fuegan forests. This is in accordance with the results of floristic analysis conducted in the areas by Skottsberg (1913) and Moore (1968). No comments were made concerning *E. rubrum* on the Juan Fernandez Islands.

We go somewhat further afield in our next example, which involves Gough Island and continental South America along with New Zealand,

and represents a pair of long distance dispersal events. The sophoras of New Zealand, southern Chile and Gough Island attracted the attention of Sykes & Godley (1968) who demonstrated, using common garden studies, that they could reasonably be considered as races of *Sophora microphylla* Ait. Plants from southern Chile and Gough Island lack a juvenile form as do some plants of New Zealand, especially those from the Chatham Islands. Flavonoid studies of individual plants from New Zealand, the Chatham Islands, southern Chile and Gough Island (Markham & Godley, 1972) showed a higher level of similarity between the Gough Island plants and those from mainland Chile than between plants from any other two sites. Comparison with plants from New Zealand and the Chatham Islands suggested that the closest relationship with Chilean plants was with certain populations from New Zealand, not from the Chatham Islands. However, due to variation between populations it was not possible to determine which New Zealand population provided the propagules.

Our last example from the South Atlantic Ocean involves the report of a positive test for anthochlors (aurones and/or chalcones) in the St Helena endemic *Petrobium arboreum* (Asteraceae, Heliantheae, Coreopsidinae) by Crawford & Stuessy (1981). St Helena is located at 15°58′S and 5°43′W, which places it about 1200 km west of the coast of Africa (Angola).

The Indian Ocean

Kerguelen Islands

The Kerguelen Islands (France) lie in the southern Indian Ocean at 49°30′S 69°30′E. The flora of the islands is small but not without interesting features. Hooker (1847, 1864) described two new genera, *Lyallia* Hooker from the Kerguelen Islands (*L. kerguelensis*) and *Hectorella* J. D. Hooker (*H. caespitosa*) from New Zealand, and placed them in Portulacaceae. They have experienced some taxonomic uncertainty since then, but evidence involving leaf, floral, pollen, seed and fruit structural features and habit led Nyananyo & Heywood (1987) to suggest that these two species are best treated as members of the same genus, *Lyallia*, and that Portulacaceae is the appropriate resting place for them. This requires the new combination *L. caespitosa* (J. D. Hooker) Nyananyo & Heywood. Both taxa were included in Nyananyo's (1986) study of leaf flavonoids of Portulacaceae. Whereas members of the 13 other genera tested (24 species) had either flavones, C-glycosylflavones or flavonols, neither of the species of *Lyallia* appeared to have any flavonoids at all.

A study by Hennion, Fiasson & Gluchoff-Fiasson (1994) focused on the relationships among three species of *Ranunculus*, *R. biternatus* and *R. pseudotrullifolius* which occur in both the Magellanic region of South America and on the Kerguelen Islands, and *R. moseleyi*, which is endemic to the island. The disjunct species are thought to have inhabited the islands for approximately 8000 years. Karyological, morphological and chemical studies, which showed the existence of quercetin glycosides acylated with *p*-coumaric, caffeic and ferulic acids in common, suggest a much more recent origin of *R. moseleyi* from *R. pseudotrullifolius*.

Madagascar

Madagascar (the Malagasy Republic) lies off the east coast of Africa roughly between 10° and 25° S, 43° and 50° E. It has been isolated from continental Africa for about 100 000 000 years and as a result has a significant level of endemism in both its flora and its fauna. I am aware of only two studies involving flavonoids of plant species endemic to Madagascar. The first was a lengthy work devoted to chemistry and geography of the *Senecio radicans* complex (Glennie *et al.*, 1971). These workers found that *Senecio longiflorus* Sch. Bip. var. *madagascariensis* Rowl., *S. kleinia* Sch. Bip. (*Kleinia nerifolia* Haw.) from the Canary Islands and *S. longiflorus* Sch. Bip. var. *violacea* Berg. from Kenya shared the simplest flavonoid profile observed, which consisted solely of quercetin 3-O-glucoside. Other profiles consist of, in various combinations, quercetin 3-O-rutinoside (most abundant), kaempferol 3-O-glucoside and 3-O-rutinoside, quercetin 3-methyl ether, apigenin 7-O-glucoside, and 6,8-di-C-rhamnosylapigenin. The identical profiles of the two *S. longiflorus* varieties argues for a close relationship between them.

A recent paper described the identification of 35 lipophilic phenolic compounds from the leaf surface of eight species of endemic *Helichrysum* (Asteraceae) (Randriaminahy *et al.*, 1992). The compounds identified belonged to several structural types: chalcones with and without B-ring substitution; chalcones with C-prenyl groups, some of which had undergone cyclization to form pyran derivatives; a dihydrochalcone-type compound with complex A-ring substitution; flavanones with and without B-ring substitution; flavones with 6- and 6,8-oxygenation; flavonols that lacked B-ring substitution; and others with high levels of O-methylation. Some simple phenolic compounds were also identified, i.e., benzophenones and prenylated phloroglucinol derivatives. The phenolic chemistry of these species is consistent with what has been described from many other members of this very large genus of worldwide distribution. The authors concluded by stating: 'The *Helichrysum* species analyzed in this study are very similar to African species of this genus with regard to their lipophilic phenolic constituents' and that the profiles of compounds appear to have been '. . . more or less unchanged during the origin of endemic taxa in Madagascar'.

   Cabucala (Apocynaceae) is a genus of 18 species restricted in their distribution to Madagascar and the Comoro Islands, which lie approximately 450 km off the northwest coast of Madagascar. *Caucala caudata* Mfg. was shown by Massiot *et al.* (1982) to accumulate a new *bis*-indole alkaloid, which they named 'cabufiline'. The compound is similar in overall structure to other *bis*-indole alkaloids from the Apocynaceae.

Mauritius

Mauritius lies east of Madagascar at 57°30′E and 20°15′S. Amongst Bate-Smith's (1977) study of flavonoids of the Cunoniaceae were three species of *Weinmannia*: *W. laevis* Lamk., *W. macrostachys* DC. and *W. tinctoria* Sm. from Mauritius. He identified ellagic acid, kaempferol, quercetin and cyanidin from these samples. Other island species tested in this study came from Java, Papua New Guinea, New Caledonia, New Zealand and Tasmania. He analysed specimens from Australia, South America and South Africa as well. All had much the same array of phenolic compounds.

Miscellaneous

After completion of the manuscript my attention was drawn to a paper by Giannasi (1978) on flavonoid chemistry of the Ulmaceae s.l. Taxa from several islands were studied in that work and should be included in the present treatment. Flavonols (kaempferol, quercetin and/or myricetin in some combination) were identified from *Ampelocera cubensis* Griseb. (Cuba), *Aphananthe philippinensis* Planch (Phillipines) and *Phyllostylon brasiliensis* Capan (Cuba). Several C-glycosylflavones (based on apigenin, luteolin and/or chrysoeriol) were identified as constituents of *Celtis trinerva* Lam. (Jamaica), *Gironniera celtidifolia* Gaudich., (Fiji), *G. hirta* Ridley (Papua New Guinea), *Parasponia andersonii* Planch. (Fiji), *P. rugosa* Blume (Papua New Guinea) and in species of *Trema* from the Caroline Islands, the Bahamas, Cuba and Okinawa. If differences between island samples and continental samples existed they were not commented upon in the paper.

Function of island plant chemicals

Plants have evolved many strategies for defence against attack by viruses, bacteria, fungi, large and small herbivores, and by other plants with whom they compete for nutrients and space. Some of the defences are physical, such as barbs or spines, others are mutualistic, as in the associations between ants and acacias. Still others involve production of chemicals that are repulsive in one way or another to attackers. The subject has been thoroughly reviewed (Levin, 1976a; Rosenthal & Janzen, 1979; Rice, 1984; Cooper-Driver, Swain & Conn, 1985; Putnam & Tang, 1986; Harborne, 1988; Rizvi & Rizvi, 1992). The role of phenolic compounds (Levin, 1971) and alkaloids (Levin, 1976b), in particular, have been discussed in detail, but other secondary metabolites have been implicated in plant defence as well, e.g., cyanogenic glycosides (Ganders, 1990), glucosinolates (Louda & Rodman, 1983), cuticular waxes (Espelie & Hermann, 1988), polyacetylenes (Marchant & Towers, 1987) and terpenoid derivatives (Lincoln & Langenhein, 1979).

Although the putative selective advantages conferred on plants by their secondary metabolites seem reasonable, hard evidence in many cases is difficult to obtain due to the complexity of the systems involved (Muller, 1953). Ideally, it would be desirable to place a plant species in a totally new, but natural, environment and observe how it responds. Will new defensive strategies develop in response to newly encountered predators, will older strategies be discarded in the absence of old challenges, and if these changes do occur, what is the time scale involved? Oceanic islands, especially those that enjoy a marked degree of isolation from continental land masses, provide us with just such experimental opportunities.

This issue was addressed directly by Carlquist (1970), who described the flora of the Hawaiian Islands as being 'exceptionally poor in poisonous plants'. Carlquist based his view on a description of poisonous plants in Hawaii by Arnold (1968), in which it was pointed out that most of the dangerous species are recently introduced ones. This is misleading insofar as Arnold was a practising physician treating patients who had eaten poisonous plants. It seems reasonable to suggest that these individuals had encountered introduced species in gardens, parks and waste places in or near Honolulu. They would have been much less

likely to encounter endemic Hawaiian species, most of which had been
extirpated in developed areas in any event.

Implicit in Carlquist's conclusion seems to be the belief that a plant has
to be toxic to be effective in fending off attackers. This may be the case in
plant interactions with micro-organisms but in the case of herbivores it
seems an extreme position. Death has little educational value for the
attacking herbivore. Would other members of the pest's group learn
from their unfortunate cohort's fate? It would seem unlikely. Simply
put, in order for a chemical to be effective as a deterrent to herbivores,
it need only taste bad. If a herbivore encounters a bitter, or otherwise
obnoxious, compound while feeding, it is unlikely to sample that plant
again. Many, if not most, secondary chemicals belonging to the compound types listed above, in fact, have an unpleasant taste.

If evolution in the absence of former predators or pathogens involves
loss of deterrent chemicals then we would expect to see a reduction in the
frequency of occurrence of these compounds. Is this what one sees? Even
a cursory reading of the information above shows that the majority of
island species, endemic or not, continue to accumulate a rich array of
secondary metabolites including ones that are quite toxic, e.g., tutin in
Coriaria, the glucose nitropropyl esters in *Corynocarpus* and many alkaloid-bearing species in Hawaii. In some instances, however, a reduction in
the capacity to make a given compound or family of compounds is seen.
Noteworthy in this group is the case involving cyanogenesis on the
Galapagos Islands where a reduction in the number of cyanogenic plants
on the islands was interpreted in terms of reduced selection pressure by
predators (Adsersen, Adsersen & Brimer, 1988). Examples of reduced
flavonoid profiles include: *Hesperomannia* and *Remya* (both Asteraceae
sampled in Hawaii) (Bohm, unpublished); two of three species of
Dendroseris (Asteraceae) on the Juan Fernandez Islands (Pacheco *et
al.*, 1991*b*); members of the *Senecio radicans* complex (Asteraceae) from
the Canary Islands and Madagascar, along with a related species from
Kenya (Glennie *et al.*, 1971); *Fuchsia crytandroides* (Onagraceae) from
Tahiti (Williams & Warnock-Jones, 1986); *Carmichaelia exsul* (Fabaceae)
from Lord Howe Island (Purdie, 1984); and *Liparophyllum gunnii*
(Menyanthaceae) from Tasmania (Bohm, Nicholls & Ornduff, 1986).
The ultimate in reduction would, of course, be the elimination of the
compound(s) in question. The capacity to make cyanogenic compounds
has obviously been eliminated in some Galapagos species, while flavonoids appear to have been eliminated altogether from the two species that
comprise *Lyallia* (Nyananyo, 1986). The overall flavonoid synthetic

capacity in other systems appears to have been, if not lost, at least subdued.

There is scant information available to account mechanistically for these phenomena except in the broadest sense. The genetic control of cyanogenesis is known for clover wherein one locus controls the formation of the cyanogenic glycoside and one controls the glucohydrolyase necessary for hydrolytic cleavage of the cyanohydrin-sugar linkage. Absence of either enzyme (system) results in an acyanogenic form. It would be of interest to determine at what stage in the biosynthesis of cyanogenic glycosides in the Galapagos endemics the pathway is blocked. Similarly, it would be instructive to determine at what stage of the flavonoid biosynthetic pathway the blockage exists in *Lyallia*. Also a puzzle is the mechanism by which a flavonoid profile undergoes simplification. Are enzymes actually lost, or are they simply turned off? Much remains to be learned about the control of these systems.

The trend toward simplified chemical profiles in island plants does not appear to be a general one, however. Arguing against loss of putatively defensive chemicals is the fact that in several well documented cases a richer array of compounds exists in the endemic species than in their continental relatives/ancestors. For example, species of *Gunnera* (Gunneraceae) from the Juan Fernandez Islands possess flavonoid profiles that are either similar to or more complex than closely related species in Mexico and South America (Pacheco *et al.*, 1993). In the case of *Peperomia* from the Juan Fernandez Islands, a richer array of flavonoids was found in the endemic species than in mainland species (Valdebenito *et al.*, 1992*a*). Other instances of island chemical profiles being at least as complex as related continental systems would include: flavonoids of Hawaiian *Bidens* (Ganders, Bohm & McCormick, 1990); flavonoids of Hawaiian *Vaccinium* (Bohm & Koupai-Abyazani, 1994); flavonoids of Cunoniaceae worldwide (Bate-Smith, 1977); flavonoids of *Nymphoides indica* in Jamaica (Bohm, Nicholls & Ornduff, 1986); anthocyanins of *Fuchsia excoricata* from New Zealand (Crowden, Wright & Harborne, 1977); diterpenes of Canary Islands *Sideritis canariensis* (Fraga *et al.*, 1991); phenolic compounds and alkaloids of Hawaiian *Pelea* (Higa & Scheuer, 1974*a, b*); sesquiterpenes of *Cyperus rotundus* in the Pacific Basin (Komai, Tang & Nishimoto, 1991); and diterpenes of *Dacrydium intermedium* of New Zealand (Perry & Weavers, 1985).

It is clear that arrays of secondary metabolites in island plants may indeed undergo simplification, remain essentially unchanged or become more elaborate, relative to mainland species. In the case of HCN production in Galapagos plants, where a conspicuous loss has occurred, feeding deterrency may be the major factor at work (Adsersen, Adsersen & Brimer, 1988). It may be, however, that some of these secondary metabolites perform several functions, only one of which involves deterrency. Certain flavonoids, for example, can serve as ultraviolet shields, others can function in the attraction of pollination vectors (Harborne, 1988), some play a role in controlling pollen tube germination and growth (Taylor, 1994), while others act as messengers between roots and nodulating bacteria (Maxwell *et al.*, 1989). If flavonoids are no longer required for their deterrent properties because former herbivores are no longer a threat, the capacity to make the compounds may still be maintained because of their other properties.

A more general explanation may also be offered for the conservation of island species' capacities to make diverse secondary metabolites. Might it simply be a feature of plants in general that they are equipped with the means, in this context chemical, to protect themselves against the attendant problems of moving into a new neighbourhood?

Acknowledgements

I wish to thank Professor Tod F. Stuessy for discussions of the plants of the Juan Fernandez Islands, my collaborators in our Hawaiian Island studies, Dr Alan Reid for providing information on the flora and geography of New Zealand, Dr Ken Marr for providing some of the Hawaiian plant material, and Professors F. Hennion, J. L. Fiasson and K. Gluchoff-Fiasson for permission to use their work on *Ranunculus* from the Iles Kerguelen prior to publication. Work from our laboratory cited herein was supported by operating and equipment grants from the Natural Sciences and Engineering Research Council of Canada.

Addendum

After completion of the manuscript for this chapter a review of the chemistry of medicinal plants of New Caledonia (Sévenet *et al.*, 1991) was brought to my attention. The work involved natural product studies during the period 1967–91 done at CNRS (Gif sur Yvette) in association with scientists working at ORSTOM (Nouméa). The table on pp. 296–306 summarizes the chemical information in that review. Families, genera and species are listed alphabetically for convenience. Type of compound only is given. The plant names shown are those given by Sévenet *et al.* (1991). In a few cases I have included, in parentheses, alternative family names as given by Mabberley (1987).

The work described shows an understandably heavy bias toward alkaloids, which is not surprising since these compounds most often exhibit physiological properties that have led people to use the plants in question for medicinal purposes. It is also not surprising that the list of plants studied involves so many species from Apocynaceae and Rutaceae. These families are not only well represented in the New Caledonian flora but are extremely well known as alkaloid producers. Other classes of natural products encountered in the work include iridoids, quassinoids, triterpenes and a smattering of phenylpropanoid derivatives.

Although some chemotaxonomic observations were made by Sévenet *et al.* (1991), the occurrence of compounds was not discussed in evolutionary terms. However, we can infer from the richness and structural complexity of the compounds described that island life does not seem to have altered the chemical synthetic capacities of New Caledonian plants. This is in agreement with the general conclusion drawn above.

Sévenet, T., Pusset, J., Bourret, D. & Potier, P. (1991). *Études sur les Plantes Médicinales de Nouvelle-Calédonie*. Gif sur Yvette: CNRS Publication.
Mabberley, D. J. (1987). *The Plant Book*. Cambridge: Cambridge University Press.

Literature cited

Adjaye, A. E., Dobberstein, R. H., Venton D. L. & Fong, H. H. S. (1984). Phytochemical investigations of *Laurelia novae-zelandiae*. *Journal of Natural Products*, **47**, 553-4.
Adsersen, A., Adsersen, H. & Brimer, L. (1988). Cyanogenic constituents in plants from the Galapagos Islands. *Biochemical Systematics and Ecology*, **16**, 65–77.
Adsersen, A., Brimer, L., Olsen, C. E. & Jaroszewski, J. W. (1993). Cyanogenesis of *Passiflora colinvauxii*, a species endemic to the Galapagos Islands. *Phytochemistry*, **33**, 365–7.
Allan, H. H. (1982). *Flora of New Zealand*, vol. I. Wellington: P. D. Hasselberg, Government Printer.
Arnold, H. L. (1968). *Poisonous plants of Hawaii*. Rutland, Vermont:Charles E. Tuttle & Co.
Arslanian, R. L., Mondragon, B., Stermitz, F. R. & Marr, K. L. (1990). Acyl histamines and a rare protopine type alkaloid from leaves of *Zanthoxylum dipetalum*. *Biochemical Systematics and Ecology*, **18**, 345–7.

Averett, J. E., Hahn, W. J., Berry, P. E. & Raven, P. H. (1986). Flavonoids and flavonoid evolution in *Fuchsia* (Onagraceae). *American Journal of Botany*, **73**, 1525–34.

Baldwin, M. E., Bick, I. R. C., Komzak, A. A. & Price, J. R. (1961). Ketones from *Acradenia franklinii*. *Tetrahedron*, **16**, 206–11.

Bate-Smith E. C. (1977). Chemistry and taxonomy of the Cunoniaceae. *Biochemical Systematics and Ecology*, **5**, 95-105.

Bartlett, M. F., Korzun, B., Sklar, R., Smith, A. F. & Taylor, W. I. (1963). The alkaloids of *Hunteria eburnea* Pichon. II. The quaternary bases. *Journal of Organic Chemistry*, **28**, 1445–9.

Baudouin, G., Tillequin, F., Koch, M., Pusset, J. & Sévenet, T. (1981). Plantes de Nouvelle-Caleconie. LXXIII. Alcaloides de *Dutaillyea oreophila* et de *D. drupacea*. *Journal of Natural Products*, **44**, 546–50.

Berry, K. M., Perry N. B. & Weavers, R. T. (1985). Foliage sesquiterpenes of *Dacrydium cupressinum*: identification, variation and biosynthesis. *Phytochemistry*, **24**, 2893–8.

Bévalot, F., Vaquette, J. & Cabalion, M. P. (1980). Alcaloide et coumarine de *Boronella* aff. *verticillata* Baill. *Plantes Medicine et Phytotherapie* **14**, 218–20.

Bittner, M., Silva O., M., Rozas, Z., Jakupovic J. & Stuessy, T. (1994). Chemical study of the genus *Gunnera* in Chile. II. Secondary metabolites of two continental species and two species from the Juan Fernandez Islands. *Boletin de la Sociedad Quimica de Chile*, **39**, 79–83.

Bohm, B. A. (1987). Intraspecific flavonoid variation. *The Botanical Review*, **53**, 197–279.

Bohm, B. A., Crins, W. J. & Wells, T. C. (1994). Exudate flavonoids of *Holocarpha*. *Biochemical Systematics* and *Ecology*, **22**, 859.

Bohm, B. A. & Fong, C. (1990). Non-polar flavonoids of *Wilkesia* and *Argyroxiphium*. *Phytochemistry*, **29**, 1175–7.

Bohm, B. A. & Koupai-Abyazani, M. R. (1994). Flavonoids and condensed tannins from the leaves of Hawaiian *Vaccinium reticulatum* and *V. calycinum*. *Pacific Science*, **48**, 458–63.

Bohm, B. A., Nicholls, K. W. & Ornduff, R. (1986). Flavonoids of the Menyanthaceae: Intra- and interfamilial relationships. *American Journal of Botany*, **73**, 204–13.

Bohm, B. A. & Ornduff, R. (1981). Leaf flavonoids and ordinal affinities of Coriariaceae. *Systematic Botany*, **6**, 15–26.

Bramwell, D. & Dakshini, K. M. M. (1971). Luteolin 7-glucoside and hydroxycoumarins in Canary Island *Sonchus* species. *Phytochemistry*, **10**, 2245–6.

Brooker, S. G., Cambie, R. C. & Cooper, R. C. (1987). *New Zealand Medicinal Plants*, revised edn. Auckland, New Zealand: Heinemann Publishers.

Brophy, J. J., Goldsack, R. J. & Goldsack, G. (1993). The essential oils of the indigenous Myrtaceae of Lord Howe Island. *Flavour and Fragrance Journal*, **8**, 153–9.

Budesinsky, M., Perez Souto, N. & Holub, M. (1994). Sesquiterpene lactones of some species of *Vernonia* Schreb. *Collection of Czechoslovak Chemical Communications*, **59**, 913–28.

Bushnell, O. A., Fukuda, M. & Makinodan, T. (1950). The antibacterial properties of some plants found in Hawaii. *Pacific Science*, **4**, 167–83.

Cambie, R. C., Cox, R. E. & Sidwell, D. (1984). Phenolic diterpenoids of *Podocarpus ferrugineus* and other podocarps. *Phytochemistry*, **23**, 333–6.

Cambie, R. C., Lal, A. R. & Pausler, M. G. (1992). Constituents of *Degeneria vitiensis* heartwood. *Biochemical Systematics and Ecology*, **20**, 265.

Cannon, J. R., Hughes, G. K., Ritchie, E. & Taylor, W. C. (1953). The chemical constituents of Australian *Zanthoxylum* species. *Australian Journal of Chemistry*, **6**, 86–9.

Carlquist, S. (1970). *Hawaii. A Natural history. Geology, Climate, Native Flora and Fauna above the Shoreline*. New York: Natural History Press.

Carlquist, S. (1990). Wood anatomy and relationships of Lactoridaceae. *American Journal of Botany*, **77**, 1498–505.

Carr, G. D. (1985) Monograph of the Hawaiian Madiinae (Asteraceae): *Argyroxiphium*, *Dubautia* and *Wilkesia*. *Allertonia*, **4**, 1–123.

Carr, G. D. (1987). Beggar's ticks and tarweeds: masters of adaptive radiation. *Trends in Ecology and Evolution*, **2**, 192–5.

Carr, G. D. & Kyhos, D. W. (1981). Adaptive radiation in the Hawaiian silversword alliance (Compositae–Madiinae). I. Cytogenetics of spontaneous hybrids. *Evolution*, **35**, 543–56.

Coello, J., Cantagrel, J. M., Hernan, F., Fúston, J. M., Ibarrola, E., Ancochea, E., Casquet, C., Jamond, C., Díaz de Téran, J. R. & Cendrero, A. (1992). Evolution of the eastern volcanic ridge of the Canary Islands based on new K-Ar data. *Journal of Volcanology & Geothermal Research*, **53**, 251–74.

Cooper-Driver, G. A., Swain, T. & Conn, E. E., (ed.) (1985). *Chemically Mediated Interactions between Plants and Other Organisms*. New York: Plenum Press.

Cowlishaw, M. G., Bickerstaffe, R. & Connor, H. E. (1983). Intraspecific variation in the epicuticular wax composition of four species of *Chionochloa*. *Biochemical Systematics and Ecology*. **11**, 247–59.

Craven, B. M. (1963). Molecular structure of tutin. *Nature*, **197**, 1193–4.

Crawford, D. J. & Stuessy, T. F. (1981). The taxonomic significance of anthochlors in the subtribe Coreopsidinae (Compositae: Heliantheae). *American Journal of Botany*, **68**, 107–17.

Crawford, D. J., Stuessy, T. F. & Silva O., M. (1986). Leaf flavonoid chemistry and the relationships of the Lactoridaceae. *Plant Systematics and Evolution*, **153**, 133–9.

Crins, W. J. & Bohm, B. A. (1989). Flavonoid diversity in relation to the systematics and evolution of the tarweeds. *Annals of the Missouri Botanical Garden*, **77**, 73–83.

Crins, W. J., Bohm, D. A. & Carr, G. D. (1988). Flavonoids as indicators of hybridization in a mixed population of lava-colonizing Hawaiian tarweeds (Asteraceae: Heliantheae: Madiinae). *Systematic Botany*, **13**, 567–71.

Crowden, R. K., Wright, J. & Harborne, J. B. (1977). Anthocyanins of *Fuchsia* (Onagraceae). *Phytochemistry*, **16**, 400–2.

Easterfield, T. H. & Aston, B. C. (1901). Tutu. Part I. Tutin and coriamyrtin. *Journal of the Chemical Society. Transactions*, **79**, 120–6.

Elavumoottil, L., Garnier, J., Mahuteau, J. & Plat, M. (1991). Two methylenedioxyflavones from *Zieridium ignambiensis*. *Phytochemistry*, **30**, 3488–9.

Espelie, K. E. & Hermann, H. R. (1988). Congruent cuticular hydrocarbons: biochemical convergence of a social wasp, an ant and a host plant. *Biochemical Systematics and Ecology*, **16**, 505–8.

Estévez-Reyes, R., Estévez-Braun, A. & Gonzalez, A. G. (1993). New lignan butenolides from *Bupleurum salicifolium*. *Journal of Natural Products*, **56**, 1177–81.

288 *Bruce A. Bohm*

Fernandez, C., Fraga, B. M. & Hernandez, M. G. (1988). Flavonoid aglycones from some Canary Islands species of *Sideritis*. *Journal of Natural Products*, **51**, 591–3.

Fish, F., Gray, A. I., Waigh, R. D. & Waterman, P. G. (1976). Dipetalolactone: a novel pyranocoumarin from the root bark of *Zanthoxylum dipetalum*. *Phytochemistry*, **15**, 313–16.

Fish, F., Gray, A. I. & Waterman, P. G. (1975). Alkaloids, courmarins, triterpenes and a flavanone from the root of *Zanthoxylum dipetalum*. *Phytochemistry*, **14**, 2073–6.

Floyd, M. E. (1979). Phenolic chemotaxonomy of the genus *Pelea* A. Gray (Rutaceae). *Pacific Science*, **33**, 153–60.

Folkers, K. & Koniuszy, F. (1939). *Erythrina* alkaloids. III. Isolation and characterization of a new alkaloid, erythramine. *Journal of the American Chemical Society*, **61**, 1232–5.

Foo, L. Y. (1987). Phenylpropanoid derivatives of catechin, epichatechin and phylloflavan from *Phyllocladus trichomanoides*. *Phytochemistry*, **26**, 2825–30.

Foo, L. Y. (1989). Flavanocoumarins and flavanophenylpropanoids from *Phyllocladus trichomanoides*. *Phytochemistry*, **28**, 2477–81.

Foo, L. Y., Hrstich, L. & Vilain, C. (1985). Phylloflavan, a characteristic constituent of *Phyllocladus* species. *Phytochemistry*, **24**, 1495–8.

Fraga, B. M., Guillermo, R., Hernandez, M. G., Mestres, T. & Artega, J. M. (1991). Diterpenes from *Sideritis canariensis*. *Phytochemistry*, **30**, 3361–4.

Gainsford, G. J., Russell, G. B. & Reay, P. F. (1984). Absolute configuration of tetraphyllin B, a cyanoglucoside from *Tetrapathaea tetrandra*. *Phytochemistry*, **23**, 2527–9.

Ganders, F. R. (1990). Altitudinal clines for cyanogenesis in introduced populations of white clover near Vancouver, Canada. *Heredity*, **64**, 387–90.

Ganders, F. R., Bohm, B. A. & McCormick, S. P. (1990). Flavonoid variation in Hawaiian *Bidens*. *Systematic Botany*, **15**, 231–9.

Ganders, F. R. & Nagata, K. M. (1983). New taxa and new combinations in Hawaiian *Bidens* (Asteraceae). *Lyonia*, **2**, 1–16.

Gardner, R. C. (1976a). Systematics of *Lipochaeta* DC. (Compositae: Heliantheae) of the Hawaiian Islands. Ph.D. dissertation, Ohio State University.

Gardner, R. C. (1976b). Evolution and adaptive radiation in *Lipochaeta* (Compositae: Heliantheae) of the Hawaiian Islands. *Systematic Botany*, **1**, 383–91.

Giannasi, D. E. (1978). Generic relationships in the Ulmaceae based on flavonoid chemistry. *Taxon*, **27**, 331–44.

Gil, M. I., Ferreres, F., Marrero, A., Tomas-Lorente, F. & Tomas-Barberan, F. A. (1993). Distribution of flavonoid aglycones and glycosides in *Sideritis* species from the Canary Islands and Madeira. *Phytochemistry*, **34**, 227–32.

Gonzalez, A. G., Barrera, J. B., Mendez, J. T., Martinez, J. E. & Sanchez, M. L. (1992). Germacranolides from *Allagopappus viscosissimus*. *Phytochemistry*, **31**, 330–1.

Gonzalez, A. F., Fraga, B. M., Hernandez, M. G., Larruga, F., Luis, J. G. & Ravelo, A. G. (1978). Flavones from some Canary Islands species of *Sideritis*. *Lloydia*, **41**, 279–80.

Gonzalez, A. G., Fraga, D. M., Hernandez, M. G., Luis, J. G. & Larruga, F. (1979). Comparative phytochemistry of the genus *Sideritis* from the Canary Islands. *Biochemical Systematics and Ecology*, **7**, 115–20.

Gonzalez-Coloma, A., Hernandez, M. G., Perales, A. & Fraga, B. M. (1990). Chemical ecology of Canarian laurel forest: toxic diterpenes from *Persea indica* (Lauraceae). *Journal of Chemical Ecology*, **16**, 2723–33.

Glennie, C. W., Harborne, J. B., Rowley, G. D. & Marchant, C. J. (1971). Correlations between flavonoid chemistry and plant geography in the *Senecio radicans* complex. *Phytochemistry*, **10**, 2413–17.

Gorman, M., Neuss, N., Djerassi, C., Kutney, J. P. & Scheuer, P. J. (1957). Alkaloid studies. XIX. Alkaloids of some Hawaiian *Rauwolfia* species. The occurrence of sandwicine and its interconversion with ajmaline and ajmalidine. *Tetrahedron*, **1**, 328–37.

Govindachari, T. R. & Thyagarajan, B. S. (1956). Alkaloids of *Toddalia aculeata*: identity of toddaline with chelerythrine. *Journal of the Chemical Society* 769–71.

Harborne, J. B. (1967). *Comparative biochemistry of the flavonoids*. New York: Academic Press.

Harborne, J. B. (1988). *Introduction to Ecological Biochemistry*, 3rd edn. New York: Academic Press.

Harborne, J. B. (1991). Revised structures for three isoetin glycosides; yellow floral pigments in *Heywoodiella oligocephala*. *Phytochemistry*, **30**, 1677–8.

Hartley, T. G., Dunstone, E. A., Fitzgerald, J. S., Jones, S. R. & Lambertson, J. A. (1973). A survey of New Guinea plants for alkaloids. *Lloydia*, **36**, 217–319.

Hartley, T. G. & Stone, B. C. (1989). Reduction of *Pelea* with new combinations in *Melicope* (Rutaceae). *Taxon*, **38**, 119–23.

Hashimoto, T. & Yamada, Y. (1994). Alkaloid biogenesis: molecular aspects. *Annual Reviews of Plant Physiology and Plant Molecular Biology*, **45**, 257–85.

Hayman, A. R., Perry, N. B. & Weavers, R. T. (1986). Juvenile-adult chemical dimorphism in foliage of *Dacrydium biforme*. *Phytochemistry*, **25**, 649–53.

Hennion, F., Fiasson, J. L. & Gluchoff-Fiasson, K. (1994). Morphological and phytochemical relationships between *Ranunculus* species from Iles Kerguelen. *Biochemical Systematics and Ecology*, **22**, 533–42.

Higa, T. & Scheuer, P. J. (1974*a*). Alkaloids from *Pelea barbigera*. *Phytochemistry*, **13**, 1269–72.

Higa, T. & Scheuer, P. J. (1974*b*). Hawaiian plant studies. XVI. Coumarins and flavones from *Pelea barbigera* (Gray) Hillebrand (Rutaceae). *Journal of the Chemical Society, Perkin Transactions 1*, 1350–2.

Hoeneisen, M., Silva O., M., Wink, M., Crawford, D. J. & Stuessy, T. (1993). Alkaloids of *Sophora* of Juan Fernandez Islands and related species. *Boletin de la Sociedad Quimica de Chile*, **38**, 167–71.

Holm, L. (1971). The role of weeds in human affairs. *Weed Science*, **19**, 485–90.

Hooker, J. D. (1847). *Flora Antarctica*, part 2. London: Reeve & Co.

Hooker, J. D. (1864). *Handbook of the New Zealand Flora*, part 1. London: Reeve & Co.

Hudgins, W. R. & Scheuer, P. J. (1964). Chemotaxonomic value of some fingerprint gas chromatograms of oils of the genus *Pelea*. *Naturwissenschaften*, **51**, 511.

Isogai, A., Murakoshi, S., Suzuki, A. & Tamura, S. (1977). Chemistry and biological activities of cinnceylanine and cinnceylanol, new insecticidal substances from *Cinnamonum zeylanicum* Nees. *Agricultural and Biological Chemistry*, **41**, 1779–84.

Isogai, A., Suzuki, A., Tamura, S., Murakoshi, S., Ohasi, Y. & Sasada, Y. (1976). Structures of cinnceylanine and cinnceylanol, polyhydroxy pentacyclic diterpenes from *Cinnamonum zeylanicum* Nees. *Agricultural and Biological Chemistry*, **40**, 2305–6.

Jensen, S. R. & Nielsen, B. J. (1980). Iridoid glucosides in *Griselinia, Aralidium* and *Toricellia*. *Phytochemistry*, **19**, 2685–8.

Jordan, W. & Scheuer, P. J. (1965). Hawaiian plant studies. XIV. Alkaloids of *Ochrosia sandwicensis* A. Gray. *Tetrahedron*, **21**, 3731–40.

Komai, K., Iwamura, J. & Ueki, K. (1977). Isolation, identification and physiological activities of sesquiterpenes in purple nutsedge. *Weed Research* (Japan) **22**, 14.

Komai, K. & Tang, C.-S. (1989). A chemotype of *Cyperus rotundus* in Hawaii. *Phytochemistry*, **28**, 1883–6.

Komai, K., Tang, C.S. & Nishimoto, R. K. (1991). Chemotypes of *Cyperus rotundus* in Pacific Rim and Basin: distribution and inhibitory activities of their essential oils. *Journal of Chemical Ecology*, **17**, 1–8.

Komai, K. & Ueki, K. (1981). Geographical variation of essential oils in tubers of purple nutsedge. In *Proceedings of the 8th Asian Pacific Weed Science Society Conference*, pp. 387–9. Bangalore: India.

Kulanthaivel, P. & Benn, M. H. (1986). A new truxillate and some flavonoid esters from the leaf gum of *Traversia baccharoides* Hook. f. *Canadian Journal of Chemistry*, **64**, 514–18.

Kunkel, G. (ed.) (1976). *Biogeography and Ecology in the Canary Islands*. Monographiae Biologicae 30. The Hague: Dr W. Junk BV Publishers.

Kunkel, G. (1980). *Die Kanarischen Inseln und Ihre Pflanzenwelt*. Stuttgart: Gustav Fischer.

Kupchan, S. M., Meshulan, H. & Sneden, A. T. (1978). New cucurbitacins from *Phormium tenax* and *Maran oreganus*. *Phytochemistry*, **17**, 767–9.

Lammers, T. J., Stuessy, T. F. & Silva O., M. (1986). Systematic relationships of the Lactoridaceae, an endemic family of the Juan Fernandez Islands, Chile. *Plant Systematics and Evolution*, **152**, 243–66.

Levin, D. A. (1971). Plant phenolics: an ecological perspective. *American Naturalist*, **105**, 157–81.

Levin, D. A. (1976a). The chemical defenses of plants to pathogens and herbivores. *Annual Reviews of Ecology and Systematics*, **7**, 121–59.

Levin, D. A. (1976b). Alkaloid-bearing plants: An ecogeographic perspective. *American Naturalist*, **110**, 261–84.

Lichius, J. J., Thoison, O., Montagnac, A., Pais, M., Cosson, J. P. & Hadi, A. H. A. (1994). Antimitotic and cytotoxic flavonols from *Zieridium pseudoobtusifolium* and *Acronychia porteri*. *Journal of Natural Products*, **57**, 1012–16.

Lincoln, D. E. & Langenheim, J. H. (1979). Variation of *Satureja douglasii* monoterpenoids in relation to light intensity and herbivory. *Biochemical Systematics and Ecology*, **7**, 289–98.

Louda, S. M. & Rodman, J. E. (1983). Concentration of glucosinolates in relation to habitat and insect herbivory for the native crucifer *Cardamine cordifolia*. *Biochemical Systematics and Ecology*, **11**, 199–207.

Mabberley, D. J. (1987). *The Plant Book*. Cambridge: Cambridge University Press.

Macias, F. A. & Fischer, N. H. (1992). Melampolides from *Lecocarpus pinnatifidens*. *Phytochemistry*, **31**, 2747–54.

Macias, F. A., Molinillo, J. M. G. & Fischer, N. H. (1993). Melampolides and *cis,cis*-germacranolides from *Lecocarpus lecocarpoides*. *Phytochemistry*, **32**, 127–31.

Majak, W. & Benn, M. (1994). Additional esters of 3-nitropropanoic acid and glucose from fruit of the New Zealand karaka tree, *Corynocarpus laevigatus*. *Phytochemistry*, **35**, 901–3.

Mansour, R. M. A., Saleh, N. A. M. & Boulos, L. (1983). A chemosystematic study of the phenolics of *Sonchus*. *Phytochemistry*, **22**, 489–92.

Marchant, Y. Y., Ganders, F. R., Wat, C.K. & Towers, G. H. N. (1984). Polyacetylenes in Hawaiian *Bidens* (Asteraceae). *Biochemical Systematics and Ecology*, **12**, 167–78.

Marchant, Y. Y. & Towers, G. H. N. (1987). Phylloplane fungi of Hawaiian plants and their photosensitivitiy to polyacetylenes from *Bidens* species. *Biochemical Systematics and Ecology*, **15**, 9–14.

Marchant, Y., Turjman, M., Flynn, T., Balza, F., Mitchell, J. C. & Towers, G. H. N. (1985). Identification of psoralen, 8-methoxypsoralen, isopimpinellin and 5,7-dimethoxycoumarin in *Pelea anisata* H. Mann, a plant used in Hawaiian leis and a source of photodermatitis. *Contact Dermatitis*, **12**, 196–9.

Markham, K. R. & Godley, E. J. (1972). Chemotaxonomic studies in *Sophora*. 1. An evaluation of *Sophora microphylla* Ait. *New Zealand Journal of Botany*, **10**, 627–40.

Markham, K. R., Vilain, C. & Molloy, B. P. J. (1985). Uniformity and distinctness of *Phyllocladus* as evidenced by flavonoid accumulation. *Phytochemistry*, **24**, 2607–9.

Markham, K. R., Webby, R. F. & Vilain, C. (1984). 7-O-Methyl-(2R:3R)-dihydroquercetin 5-O-β-D-glucoside and other flavonoids from *Podocarpus nivalis*. *Phytochemistry*, **23**, 2049–52.

Markham, K. R. & Whitehouse, L. A. (1994). Unique flavonoid glycosides from the New Zealand white pine, *Dacrycarpus dacrydioides*. *Phytochemistry*, **23**, 1931–6.

Marr, K. L. & Tang, C-S. (1992). Volatile insecticidal compounds and chemical variability of Hawaiian *Zanthoxylum* (Rutaceae) species. *Biochemical Systematics and Ecology*, **20**, 209–17.

Massias, M., Carbonnier, J. & Molho, D. (1981). Xanthones and C-glycosylflavones from *Gentiana corymbifera*. *Phytochemistry*, **20**, 1577–8.

Massiot, G., Vercauteren, J., Jacquier, N. J., Leux, J. & Le Men-Olivier, L. (1981). Plantes de Nouvelle-Calédonie. 72. Plumocraline, alkaloide *bis*-indolique noveau d'*Alstonia plumosa* var. *communis* Boiteau forma *glabra* Boiteau. *Comptes Rendue Seances Academie Scientifique*, ser. 2, **292**, 191–4.

Massiot, G., Vercauteren, J., Richard, B., Jacquier, M. J. & Le Men-Olivier, L. (1982). Cabufiline, a new *bis*-indole alkaloid from *Cabucala caudata* Mfg. (Apocynaceae). *Comptes Rendue Seances Academie Scientifique*, ser. 2, **294**, 579–82.

Maxwell, C. A., Hartwig, U. A., Joseph C. M. & Phillips, D. (1989). A chalcone and two related flavonoids released from alfalfa roots induce *nod* genes of *Rhizobium meliloti*. *Plant Physiology*, **91**, 842–7.

McCallion, R. F., Cole, A. L. J., Walker, J. R. L., Blunt, J. W. & Munro, M. H. G. (1982). Antibiotic substances from New Zealand plants. II. Polypodial, an anti-*Candida* agent from *Pseudowintera colorata*. *Planta Medica*, **44**, 134–8.

Menadue, Y. & Crowden, R. K. (1983). Morphological and chemical variation in populations of *Richea scoparia* and *R. angustifolia* (Epacridaceae). *Australian Journal of Botany*, **31**, 73–84.

Miles, D. H., Chittawong, V., Hedin P. A. & Kokpol, U. (1993). Potential agrochemicals from leaves of *Wedelia biflora*. *Phytochemistry*, **32**, 1427–9.

Moore, D. M. (1968). The vascular flora of the Falkland Islands. *British Antarctic Survey Bulletin*, no. **60**, 1–202.

Moore, D. M., Harborne, J. B. & Williams, C. A. (1970). Chemotaxonomy, variation and geographical distribution of the Empetraceae. *Botanical Journal of the Linnean Society*, **63**, 277–93.

Moyer, B. G., Pfeffer, P. E., Valentine, K. M. & Gustine, D. L. (1979). 3-Nitropropanoyl-D-glucopyranoses of *Corynocarpus laevigatus* J.R. & G. Forst. *Phytochemistry*, **18**, 111–13.

Muir, A. D., Cole, A. L. J. & Walker, J. R. L. (1982). Antibiotic compounds from New Zealand plants. I. Falcarindiol, an anti-dermatophyte agent from *Schefflera digitata*. *Planta Medica*, **44**, 129–33.

Muller, C. H. (1953). The association of desert annuals with shrubs. *American Journal of Botany*, **40**, 842–7.

Muyard, F., Bévalot, F., Laude, B. & Vaquette, J. (1992). Alkaloids from stem bark of *Dutaillyea baudouinii*. *Phytochemistry*, **31**, 1087–9.

Muyard, F., Bévalot, F., Regnier, A. & Vaquette, J. (1994). Plants from New Caledonia: constituents of *Boronella pancheri*. *Biochemical Systematics and Ecology*, **22**, 434.

Nyananyo, B. L. (1986). A survey of leaf flavonoids in the Portulacaceae. *Biochemical Systematics and Ecology*, **14**, 633–5.

Nyananyo, B. L. & Heywood, V. H. (1987). A new combination in *Lyallia* (Portulacaceae). *Taxon*, **36**, 640–1.

Pacheco, P., Crawford, D. J., Stuessy T. F. & Silva O., M. (1985). Flavonoid evolution in *Robinsonia* (Compositae) of the Juan Fernandez Islands. *American Journal of Botany*, **72**, 989–98.

Pacheco, P., Stuessy, T. F. & Crawford, D. J. (1991*a*). Natural interspecific hybridization in *Gunnera* (Gunneraceae) of the Juan Fernandez Islands. *Pacific Science*, **45**, 389–99.

Pacheco, P., Crawford, D. J., Stuessy, T. F. & Silva O., M. (1991b). Flavonoid evolution in *Dendroseris* (Compositae, Lactuceae) from the Juan Fernandez Islands, Chile. *American Journal of Botany*, **78**, 534–43.

Pacheco, P., Crawford, D. J., Stuessy, T. F. & Silva O. M. (1993). Flavonoid chemistry and evolution of *Gunnera* (Gunneraceae) in the Juan Fernandez Islands, Chile. *Gayana Botany*, **50**, 17–30.

Panichanun, S., Bick, I. R. C. & Blunt, J. W. (1982). Alkaloid occurrence in *Discaria toumatou*. *Journal of Natural Products*, **45**, 777–8.

Patama, T. T. & Widén, C. J. (1991). Phloroglucinol derivatives from *Dryopteris fusco-atra* and *D. hawaiiensis*. *Phytochemistry*, **30**, 3305–10.

Patterson, R. (1984). Flavonoid uniformity in diploid species of Hawaiian *Scaevola* (Goodeniaceae). *Systematic Botany*, **9**, 263–5.

Perry, N. B. & Weavers, R. T. (1985). Foliage diterpenes of *Dacrydium intermedium*: identification, variation and biosynthesis. *Phytochemistry*, **24**, 2899–2904.

Pojer, P. M., Ritchie, E. & Taylor, W. C. (1968). Some constituents of *Degeneria vitiensis* Bailey & Smith. *Australian Journal of Chemistry*, **21**, 1379–80.

Preece, R. C., Bennett, K. D. & Carter, J. R. (1986). The Quaternary palaeobotany of Inaccessible Island (Tristan da Cunha group). *Journal of Biogeography*, **13**, 1–33.

Proksch, P., Budzikiewicz, H., Tanowitz, B. D. & Smith, D. M. (1984). Flavonoids from the external leaf resins of four *Hemizonia* species (Asteraceae). *Phytochemistry*, **23**, 679–80.

Purdie, A. W. (1984). Some flavonoid components of *Carmichalelia* (Papilionaceae) – a chemotaxonomic survey. *New Zealand Journal of Botany*, **22**, 7–14.

Putnam, A. R. & Tang, C. S. (ed.) (1986). *The science of allelopathy*. New York: John Wiley & Sons.

Quader, A., Armstrong, J. A., Gray, A. I., Hartley T. G. & Waterman, P. G. (1991). Chemosystematics of *Acradenia* and general significance of acetophenones in the Rutaceae. *Biochemical Systematics and Ecology*, **19**, 171–6.

Randriaminahy, M., Proksch, P., Witte, L. & Wray, V. (1992). Lipophyllic phenolic constituents from *Helichrysum* species endemic to Madagascar. *Zeitschrift für Naturforschung*, **47c**, 10–16.

Reid, A. R. & Bohm, B. A. (1994). Vacuolar and exudate flavonoids of New Zealand *Cassinia* (Asteraceae: Gnaphalieae). *Biochemical Systematics and Ecology*, **22**, 501.

Rice, E. (1984). *Allelopathy*. New York: Academic Press.

Rizvi, S. J. H. & Rizvi, V. (ed.) (1992). *Allelopathy – Basic and Applied Aspects*. London: Chapman & Hall.

Rosenthal, G. A. & Janzen, D. H. (ed.) (1979). *Herbivores. Their Interaction with Secondary Plant Metabolites*. New York: Academic Press.

Russell, G. B. & Reay, P. F. (1971). The structures of tetraphyllin A and B, two new cyanoglucosides from *Tetrapathaea tetrandra*. *Phytochemistry*, **10**, 1373–7.

Scheuer, P. J. (1955). The constituents of mokihana (*Pelea anisata* Mann). *Chemistry & Industry (London)*, 1257–8.

Scheuer, P. J., Horigan, L. P. & Hudgins, W. R. (1962). A survey for alkaloids in Hawaiian plants. III. *Pacific Science*, **16**, 63–9.

Scheuer, P. J. & Hudgins, W. R. (1964). Hawaiian plant studies. XII. Major constituents of the essential oil of *Pelea christophersonii*. *Perfumery and Essential Oil Record*, **55**, 723–4.

Scheuer, P. J. & Metzger, J. T. H. (1961). Hawaiian plant studies. VI. The structure of holeinine. *Journal of Organic Chemistry*, **26**, 3069–71.

Scheuer, P. J. & Pattabhiraman, T. R. (1965). Hawaiian plant studies. XIII. Isolation of a canthinone from a member of the family Amaranthaceae. *Lloydia*, **28**, 95–100.

Schmid, M. (1982). Endemisme et speciation en Nouvelle Caledonie. *Compte Rendue Societe Biogeographique*, **58**, 52–60.

Schminke, H. U. (1976). The geology of the Canary Islands. In *Biogeography and Ecology in the Canary Islands*, ed. G. Kunkel, pp. 67–184. Monographie Biologicae 30. The Hague: Dr W. Junk BV Publishers.

Skottsberg, C. J. F. (1913). A botanical survey of the Falkland Islands. *Kungliga Svenska Vetenskaysakademien Handlungen*, **50**, 1–129.

Smith, R. M., Marty R. A. & Peters, C. F. (1981). The diterpene acids in the bled resins of three Pacific kauri, *Agathis vitiensis*, *A. lanceolata* and *A. macrophylla*. *Phytochemistry*, **20**, 2205–7.

Stace, H. M. & Fripp, Y. J. (1977). Raciation in *Epacris impressa*. I. Corolla colour and corolla length. *Australian Journal of Botany*, **25**, 299–314.

Stuessy, T. F., Foland, K. A., Sutter, J. F., Sanders, R. W. & Silva O., M. (1984). Botanical and geological significance of potassium–argon dates from the Juan Fernandez Islands. *Science*, **225**, 49–51.

Stuessy, T. F., Marticorena, C., Rodríguez R., R., Crawford, D. J. & Silva O., M. (1992). Endemism in the vascular flora of the Juan Fernandez Islands. *Aliso*, **13**, 297–307.

Swanholm, C. E., St John, H. & Scheuer, P. J. (1959). A survey for alkaloids in Hawaiian plants. I. *Pacific Science*, **13**, 295–305.

Swanholm, C. E., St John, H. & Scheuer, P. J. (1960). A survey for alkaloids in Hawaiian plants. II. *Pacific Science*, **14**, 68–74.

Sykes, W. R. & Godley, E. J. (1968). Transoceanic dispersal in *Sophora* and other genera. *Nature*, **218**, 495–6.

Takhtajan, A. (1986). *Floristic regions of the world.* Berkeley, California: University of California Press.

Tanowitz, B., Leeder, G., Ross, P. N. & Proksch, P. (1987). Foliar flavonoid exudates and sectional taxonomy of *Hemizonia. Biochemical Systematics and Ecology*, **15**, 535–40.

Taylor, L. P. (1994). Flavonols and functional pollen. *Polyphenols Actualities*, **10**, 15–17.

Torres, R., Delle Monache, F. & Marini-Bettolo, G. B. (1979). Alkaloids from the genus *Discaria. Journal of Natural Products*, **42**, 430–5.

Urzua, A. & Cassels, B. K. (1978). Alkaloid chemosystematics, chemotaxonomy and biogenesis in the Atherospermataceae. *Lloydia*, **41**, 98–113.

Valdebenito, H. A., Stuessy, T. F. & Crawford, D. J. (1990a). Synonymy in *Peperomia berteroana* (Piperaceae) results in biological disjunction between Pacific and Atlantic Oceans. *Brittonia*, **42**, 121–4.

Valdebenito, H. A., Stuessy, T. F. & Crawford, D. J. (1990b). A new biogeographic connection between islands in the Atlantic and Pacific Oceans. *Nature*, **347**, 549–50.

Valdebenito, H. A., Stuessy, T. F., Crawford, D. J. & Silva O., M. (1992a). Evolution of *Peperomia* (Piperaceae) in the Juan Fernandez Islands, Chile. *Plant Systematics and Evolution*, **182**, 107–19.

Valdebenito, H. A., Stuessy, T. F., Crawford, D. J. & Silva O., M. (1992b). Evolution of *Erigeron* (Compositae) in the Juan Fernandez Islands, Chile. *Systematic Botany*, **17**, 470–80.

Wagner, H., Geyer, B., Kiso, Y., Hikino, H. & Rao, G. S. (1986). Coumestans as the main active principles of liver drugs *Eclipta alba* and *Wedelia calendulacea. Planta Medica*, **52**, 370–8.

Wagner, W. L., Herbst, D. R. & Sohmer, S. H. (1990). *Manual of the Flowering Plants of Hawaii.* Honolulu: University of Hawaii Press; Bishop Museum Press.

Werny, F. & Scheuer, P. J. (1963). Hawaiian plant studies. IX. Alkaloids of *Platydesma campanulata. Tetrahedron*, **19**, 1293–305.

Williams, C. A. & Harborne, J. B. (1977). Flavonoid chemistry and plant geography in the Cyperaceae. *Biochemical Systematics and Ecology*, **5**, 45–51.

Williams, C. A. & Harvey, W. J. (1982). Leaf flavonoids in the Winteraceae. *Phytochemistry*, **21**, 329–37.

Williams, C. A. & Warnock-Jones, P. J. (1986). Leaf flavonoids and other phenolic glycosides and the taxonomy and phylogeny of *Fuchsia* sect. *Skinnera* (Onagraceae). *Phytochemistry*, **25**, 2547–9.

Williams, C. A., Harborne, J. B. & Tomas-Barberan, F. A. (1987). Biflavonoids in the primitive monocots *Isophysis tasmanica* and *Xerophyta plicata*. *Phytochemistry*, **26**, 2553–5.

Wilson, H. D. (1987). Plant communities of Stewart Island (New Zealand). *New Zealand Journal of Botany* (supplement), 1–124.

Wilson, R. D. (1984). Chemotaxonomic studies in the Rubiaceae. 2. Leaf flavonoids of New Zealand coprosmas. *New Zealand Journal of Botany*, **22**, 195–200.

Witter, M. S. & Carr, G. D. (1988). Adaptive radiation and genetic differentiation in the Hawaiian silversword alliance (Compositae: Madiinae). *Evolution*, **42**, 1278–87.

Table for Addendum. *Reports of natural products from New Caledonian plants*

Taxon	Compound	Reference
Gymnosperms		
Araucariaceae		
Agathis lanceolata Lindley & Warb.	Diterpenes	Manh *et al.*, 1983. J. Nat. Prod., **46**, 262, and citations therein
Cupressaceae		
Libocedrus yateensis Guillaumin	Lignans	Erdtman & Harmatha, 1979. Phytochem., **18**, 1495
Taxaceae (syn. Austrotaxaceae (EF)[a]		
Austrotaxus spicata Compton (EG)	Taxane derivatives	Ettouvati *et al.*, 1988. Bull. Soc. Chim. Fr., **4**, 749
Monocotyledons		
Sévenet *et al.* (1991) state that species of *Cryptostylis*, *Liparis* and *Phaius* (all Orchidaceae) are alkaloid-rich, but give no details. No other monocots had been studied to the end of 1991.		
Dicotyledons		
Alangiaceae		
Alangia bussyanum (Baill.) Harms	Alkaloids	Plat, unpublished results
Annonaceae		
Desmos tiebaghiensis	Alkaloids	Leboeuf *et al.*, 1982. J. Nat. Prod., **45**, 617
Polyalthia nitidissima Benth.	Alkaloids	Jossang *et al.*, 1983. Pl. Med., **49**, 20
Xylopia pancheri Baill. (ES)	Alkaloids	Nieto *et al.*, 1976. Pl. Med., **30**, 48
X. vieillardii Baill.	Alkaloids	Jossang *et al.*, 1991. J. Nat. Prod., **54**, 466
Apocynaceae		
Alstonia boulindaensis Boiteau	Alkaloids	Lewin *et al.*, 1975. Phytochem., **14**, 2067
		Lewin *et al.*, 1975. C. R. Acad. Sci. C, **280**, 987
		Lewin *et al.*, 1978. J. Ind. Chem. Soc., **60**, 1096
A. coriacea Pancher & S. Moore	Alkaloids	Cherif *et al.*, 1987. Heterocycles, **26**, 3055
		Cherif *et al.*, 1989. Phytochem., **28**, 667

Table for Addendum. *Continued*

Taxon	Compound	References
A. deplanchei van Heurck & Muell. Arg.	Alkaloids	Das *et al.*, 1974. Tetrahedron letters, **49**, 4299
		Das *et al.*, 1977. J. Org. Chem., **42**, 2785
A. lanceolata van Heurck & Muell. Arg.	Alkaloids	Besselievre *et al.*, 1980. Tetrahedron letters, **21**, 63
		Vercauteren *et al.*, 1981. Phytochem., **20**, 1411
		Vercauteren *et al.*, 1981. Tetrahedron letters, **22**, 2871
A. lanceolifera S. Moore	Alkaloids	Petitfrere-Auvray *et al.*, 1981. Phytochem., **20**, 1987
		Ravao *et al.*, 1982. Phytochem., **21**, 2160
A. legouixieae van Heurck & Muell. Arg.	Alkaloids	Lewin *et al.*, 1981. Ann. Pharm. Fr., **39**, 273
A. lenormandii van Heurck & Muell. Arg.	Alkaloids	Legseir *et al.*, 1986. Phytochem., **25**, 1735
A. odontophora Boiteau	Alkaloids	Vercauteren *et al.*, 1979. Phytochem., **18**, 1729
A. plumosa Boiteau var. *communis* Boiteau	Alkaloids	Massiot *et al.*, 1981. C. R. Acad.Sc. II, **292**, 191
		Massiot *et al.*, 1981. C. R. Acad. Sci. II, **294**, 2973
		Jacquier *et al.*, 1982. Phytochem., **21**, 2973
A. quaternata van Heurck & Muell. Arg.	Alkaloids	Mamatas-Kalamaras *et al.*, 1975. Phytochem., **14**, 1849
A. sphaerocapitata Boiteau	Alkaloids	Caron *et al.*, 1984. Phytochem., **23**, 2355
A. undulata Guillaumin	Alkaloids	Vercauteren *et al.*, 1989. C. R. Acad. Sc. II, **309**, 33
		Nuzillard *et al.*, 1983. C. R. Acad. Sc. II, **296**, 977
		Nuzillard *et al.*, 1983. C. R. Acad. Sc. II, **309**, 195
		Guillaume *et al.*, 1984. Phytochem., **23**, 2407
		Pinchon *et al.*, 1990. Phytochem., **29**, 3341
A. vieillardii van Heurck & Muell. Arg.	Alkaloids	Cosson, 1975. Ph.D. dissertation, Univ. Paris-Sud, Orsay
A. vitiensis Seeman	Alkaloids	Mamatas-Kalamaras *et al.*, 1975. Phytochem., **14**, 1637
Ervatamia lifuana Boiteau	Alkaloids	Bruneton *et al.*, 1980. Pl. Med., **39**, 180
Melodinus aemeus Baill.	Alkaloids	Baasou *et al.*, 1978. Phytochem., **17**, 1449

Table for Addendum. *Continued*

Taxon	Compound	References
M. balansae Baill.	Alkaloids	Mehri *et al.*, 1972. Ann. Pharm. Fr., **30**, 643
		Mehri *et al.*, 1972. Bull. Soc. Chim. Fr., **8**, 3291
		Damak *et al.*, 1976. Tetrahedron Letters, **3**, 167
M. celastroides Baill.	Alkaloids	Rabaron *et al.*, 1973. Phytochem., **12**, 2537
		Rabaron *et al.*, 1973. Pl. Méd. Phytothér., **7**, 319.
		Rabaron *et al.*, 1978. Phytochem., **17**, 1452
		Baassou *et al.*, 1981. Ann. Pharm. Fr., **32**, 167
		Baassou *et al.*, 1983. Tetrahedron Letters, **24**, 761
		Baassou *et al.*, 1987. Ann. Pharm. Fr., **45**, 49
		Mehri *et al.*, 1991. J. Nat. Prod., **54**, 372
M. guillauminii Boiteau	Alkaloids	Zeches *et al.*, 1984. Phytochem., **23**, 171
M. insulaepinorum Boiteau	Alkaloids	Batchily *et al.*, 1985. Ann. Pharm. Fr., **43**, 359
M. phylliraeoides Labill.	Alkaloids	Mehri *et al.*, 1984. Ann. Pharm. Fr., **42**, 145
M. polyadenus Baill.	Alkaloids	Rabaron *et al.*, 1981. Ann. Pharm. Fr., **39**, 369
M. reticulatus Boiteau	Alkaloids	Mehri *et al.*, 1983. Pl. Med., **48**, 72
M. scandens Forst.	Alkaloids	Mehri *et al.*, 1971. Ann. Pharm. Fr., **29**, 291
		Plat *et al.*, 1970. Tetrahedron Letters, **39**, 3395
		Daudon *et al.*, 1975. J. Org. Chem., **40**, 2838
		Daudon *et al.*, 1976. J. Org. Chem., **41**, 3275
		Mehri & Plat, 1974. Pl. Méd. Phytothér., **8**, 143
		Chazelet *et al.*, 1987. Ann. Pharm. Fr., **45**, 355
M. tiebaghiensis Boiteau	Alkaloids	Mehri *et al.*, 1878. Phytochem., **17**, 1451
Neisosperma brevituba (Boiteau) Boiteau	Alkaloids	Bruneton, 1973. Ph.D. dissertation, Univ. Paris-Sud, Orsay

Table for Addendum. *Continued*

Taxon	Compound	References
N. kilneri (F. v. Muell.) Fosberg & Sachet	Alkaloids	Batchily *et al.*, 1985. Ann. Pharm. Fr., **43**, 513
N. lifuana (Guillaum.) Boiteau	Alkaloids	Peube-Locou *et al.*, 1971. C. R. Acad. Sc. C, **273**, 905
		Peube-Locou *et al.*, 1972. Ann. Pharm. Fr., **30**, 775
		Peube-Locou *et al.*, 1973. Phytochem., **12**, 199
		Koch *et al.*, 1973. Bull. Soc. Chim. Fr., **9–10**, 2868
N. miana (Baill. ex White) Boiteau	Alkaloids	Buzas *et al.*, 1958. C. R. Acad. Sc., **247**, 1390
		Preaux *et al.*, 1974. Pl. Méd. Phytothér., **8**, 250
N. oppositifolia (Lam.) Fosberg & Sachet	Alkaloids	Peube-Locou *et al.*, 1972. Ann. Pharm. Fr., **30**, 821
		Peube-Locou *et al.*, 1972. Phytochem., **11**, 2109
N. sevenetii (Boiteau) Boiteau	Alkaloids	Akhter *et al.*, 1978. Tetrahedron Letters, **43**, 4137
		Bruneton, 1973. Ph.D. dissertation, Univ. Paris-Sud, Orsay
N. thiollierei (Montrouzier) Boiteau	Alkaloids	Colin *et al.*, unpublished results
Ochrosia balansae (Guill.) Guill. var. *excelsior* Boiteau	Alkaloids	Bruneton & Cave, 1972. Phytochem. **11**, 846
		Bruneton & Cave, 1972. Ann. Pharm. Fr., **30**, 629
		Bruneton *et al.*, 1971. C. R. Acad. Sc. C, **273**, 442
O. elliptica Labill.	Alkaloids	Buzas *et al.*, 1958. C. R. Acad. Sc., **247**, 1390
O. moorei (F. v. Muell.) F. v. Muell.	Alkaloids	Doy & Moore, 1962. Aust. J. Chem., **15**, 548
		Ahond *et al.*, 1981. J. Nat. Prod., **44**, 193
O. mulsanti Montr.	Alkaloids	Kan-Fan *et al.*, 1970. Phytochem., **9**, 1351
		Bruneton *et al.*, 1972. Phytochem., **11**, 3073
O. silvatica Däniker	Alkaloids	Cosson & Schmid, 1970. Phytochem., **9**, 1353
		Kunesch *et al.*, 1977. C. R. Acad. Sc. C, **285**, 89
		Potier & Janot, 1973. C. R. Acad. Sc. C, **276**, 1727

Table for Addendum. *Continued*

Taxon	Compound	References
Pagiantha cerifera Mgf.	Alkaloids	Harmouche *et al.*, 1976. Ann. Pharm. Fr., **34**, 31
		Ros *et al.*, 1978. Z. Naturforsch., **33c**, 290
		Bert *et al.*, 1985. Heterocycles, **23**, 2505
		Bert *et al.*, 1985. Heterocycles, **24**, 1567
		Bert *et al.*, 1989. Fitoter., **60**, 141
		Baudouin *et al.*, 1986. J. Chem. Soc., Chem. Commun., 3
Rauvolfia balansae (Baill.) Boiteau	Alkaloids	Majumdar *et al.*, 1972. Tetrahedron Letters, **16**, 1563
R. semperflorens (Muell. Arg.) Schlechter	Alkaloids	Schlittler & Furlenmeir, 1953. Helv. Chim. Acta, **36**, 996
Atherospermataceae (syn. Monimiaceae)		
Nemuaron vieillardii Baillon (EG)	Alkaloids	Bick *et al.*, 1972. J. Chem. Soc., Chem Commun., 980
		Bick *et al.*, 1973. Aust. J. Chem., **26**, 455
Bignoniaceae		
Deplanchea speciosa Vieill.	Phenylpropanoids, Iridoids	Davioud *et al.*, 1989. Pl. Med., **57**, 87
Celastraceae		
Dicarpellum pronyensis (Guillaum.) A. C. Smith (EG)	Alkaloids	Adeoti *et al.*, 1978. Phytochem., **17**, 831
Peripterygia marginata Loes (EG)	Alkaloids	Leboeuf *et al.*, 1971. Pl. Méd. Phytothér., **5**, 126
		Hocquemiller *et al.*, 1974. C. R. Acad. Sc. C, **278**, 525
		Hocquemiller *et al.*, 1977. Tetrahedron, **33**, 645
Pleurostylia opposita	Alkaloids	Bruneton, unpublished results
Clusiaceae		
Calophyllum inophyllum L.	4-Phenylcoumarines	Gopalakrishnan *et al.*, 1980. Indian J. Pharmacol., **12**, 181
Elaeocarpaceae		
Elaeocarpus persicaefolius Brongn. & Gris	Alkaloids	Cosson, unpublished results

Table for Addendum. *Continued*

Taxon	Compound	References
Euphorbiaceae		
Macaranga vedeliana Muell. Arg.	Flavonol, Stilbene	Hnawia *et al.*, 1990. Phytochem., **29**, 2367; other work in preparation
Flacourtiaceae		
Homalium spp.	Alkaloids	Pais *et al.*, 1973. Tetrahedron, **29**, 1001
Flindersiaceae (= Rutaceae)		
Flindersia fournieri Pancher & Sébert	Alkaloids	Tillequin *et al.*, 1979. J. Nat. Prod., **42**, 92
		Tillequin & Koch, 1979. Phytochem., **18**, 1559, 2066
		Tillequin & Koch, 1980. Phytochem., **20**, 1282
		Tillequin *et al.*, 1974. Pl. Méd. Phytothér., **8**, 250
		Tillequin *et al.*, 1979. Ann. Pharm. Fr., **37**, 543
		Tillequin *et al.*, 1980. Pl. Med., **38**, 383
Goodeniaceae		
Scaevola montana Labill.	Iridoids	Skaltsounis *et al.*, 1989. Ann. Pharm. Fr., **47**, 249
S. racemigera Däniker	Iridoids	Skaltsounis *et al.*, 1985. Helv. Chim. Acta, **68**, 599
		Skaltsounis *et al.*, 1987. Heterocycles, **26**, 599
		Skaltsounis *et al.*, 1989. Pl. Med., **55**, 191
Hernandiaceae		
Hernandia cordigera Vieill.	Alkaloids, Lignans	Bruneton, 1980. C. R. Acad. Sc. C, **291**, 187
		Lavault *et al.*, 1980. Bull. Nat. Hist. Paris, **2**, 387
		Lavault *et al.*, 1981. Pl. Med., **42**, 50
		Richomme *et al.*, 1984. J. Nat. Prod., **47**, 879
H. peltata Meissner	Alkaloids, Lignans	Citations as for *H. cordigera*

Table for Addendum. *Continued*

Taxon	Compound	Reference
Icacinaceae		
Lasianthera austrocaledonica Baill.	Iridoids	Sévenet *et al.*, 1976. Phytochem., **15**, 576
Lauraceae		
Beilschmiedia oreophila Schltr.	Alkaloids	Tillequin *et al.*, 1985. Heterocycles, **23**, 1357
Cryptocarya longifolia Kosterm.	Alkaloids	Bick *et al.*, 1981. Aust. J. Chem., **34**, 195
C. odorata (Pancher & Sébert) Guillaum.	Essential oils, Alkaloids	Bick *et al.*, 1972. Bull. Soc. Chim. Fr., **12**, 4596
C. oubatchensis Schltr.	Alkaloids	Leboeuf *et al.*, 1989. Can. J. Chem., **67**, 947
C. phyllostemon Kosterm.	Alkaloids	Cave *et al.*, 1989. Aust. J. Chem., **42**, 2243
		Leboeuf *et al.*, 1989. Can. J. Chem., **67**, 947
C. velutinosa Kosterm.	Alkaloids	Leboeuf *et al.*, 1989. J. Nat. Prod., **52**, 516
Litsea lecardii Guillaum.	Alkaloids	Weber *et al.*, 1986. Pl. Med., **52**, 74
L. triflora Guillaum.	Alkaloids	Castedo *et al.*, 1980. An. Quim., **76c**, 171
Linaceae		
Hugonia penicillanthemum Baill. ex Pancher & Sébert	Alkaloids	Ikhiri *et al.*, 1987. J. Nat. Prod., **50**, 626
H. oreogena Schlecter	Alkaloids	Ikhiri *et al.*, 1987. J. Nat. Prod., **50**, 626
Meliaceae		
Dysoxylum roseum C. DC.	Diaryl epoxide, Triterpenes	Adesanya *et al.*, in preparation
Menispermaceae		
Pachygone vieillardii Diels	Alkaloids	Abouchacra, 1984. Ph.D. dissertation, Pharmacy, Paris-Sud, Chatenay
Mimosaceae		
Acacia spirobis Labill.	Alkaloids, Essential oils	Poupat & Sévenet, 1975. Phytochem., **14**, 1881
		Brophy *et al.*, 1987. Phytochem., **26**, 3971

Table for Addendum. *Continued*

Taxon	Compound	Reference
A. simplicifolia Druce	Alkaloids	Poupat *et al.*, 1976. Phytochem., **15**, 2019
Albizia spp.	Alkaloids	Work in progress
Tephrosia noctiflora Boj.	Peptides	Forgacs *et al.*, 1980. Phytochem., **19**, 1225
Monimiaceae		
Hedycarya baudouinii Baill. (ES)	Alkaloids	El Tohami, 1984. Ph.D. dissertation, Paris-Sud, Chatenay
H. parvifolia Perkins & Schltr.	Alkaloids	Lavault *et al.*, 1980. Bull. Nat. Hist., Paris 4th ser., **2**, 387
Oleaceae		
Osmanthus austrocaledonica (Vieill.) Knolb	Alkaloids, Iridoids	Benkrief *et al.*, in preparation
Phellinaceae (= Aquifoliaceae)		
Phelline billardieri Panch.	Alkaloids	Nhu *et al.*, 1970. C. R. Acad. Sc. C, **270**, 2154
		Seguineau & Langlois, 1980. Phytochem., **19**, 1279
		Langlois, 1981. Tetradedron Letters, **22**, 2263
		Debourges & Langlois, 1982. J. Nat. Prod., **45**, 163
P. brachyphylla Baill.	Alkaloids	Langlois *et al.*, 1969. C. R. Acad. Sc. C, **269**, 639
P. comosa Labill. varieties	Alkaloids	Langlois *et al.*, 1970. Bull. Soc. Chim. Fr., **10**, 3535
		Pusset *et al.*, 1989. Phytochem., **28**, 1298
		Langlois, 1990. Heterocycles, **30**, 659
P. aff. *lucida* Vieill. ex Baill.	Alkaloids	Langlois *et al.*, 1985. Heterocycles, **22**, 2453
		Langlois *et al.*, 1985. C. R. Acad. Sc., Paris II, **300**, 441
		Langlois *et al.*, 1985. C. R. Acad. Sc., Paris II, **301**, 519
		Langlois & Razafimbelo, 1988. J. Nat. Prod. **51**, 499

Table for Addendum. *Continued*

Taxon	Compound	Reference
Proteaceae		
Knightia deplanchei Vieill. ex Brongn. & Gris	Alkaloids	Kan-Fan & Lounasmaa, 1973. Acta Chem. Scand., **27**, 3, 1039
		Lounasmaa, 1975. Pl. Med., **27**, 83
		Lounasmaa *et al.*, 1975. J. Org. Chem., **40**, 25, 3694
		Lounasmaa & Johansson, 1974. Tetrahedron Letters, **29**, 2509
K. strobilina Labill.	Alkaloids	Bick *et al.*, 1981. Pl. Med., **41**, 379
Garnieria spathulaefolia Brongn. & Gris (EG)	Butyrolactone	Lounasmaa *et al.*, 1980. Phytochem., **19**, 949, 953
		Lounasmaa *et al.*, 1985. Heterocycles, **23**, 939
Rhamnaceae		
Emmenospermum pancherianum Baill.	Triterpenes	Baddeley *et al.*, 1980. Aust. J. Chem., **33**, 2071
Rhizophoraceae		
Crossostylis spp.	Alkaloids	Gnecco Medina *et al.*, 1983. J. Nat. Prod., **46**, 398
Rubiaceae		
Coelospermum billardieri Däniker	Alkaloids	Lopez *et al.*, 1988. J. Nat. Prod., **51**, 829
	Iridoids	Benkrief *et al.*, 1991. J. Nat. Prod., **54**, 532
Guettarda eximia Baill.	Alkaloids	Stöckigt *et al.*, 1977. J. Chem. Soc., Chem. Commun., 164
		Husson *et al.*, 1977. Tetrahedron Letters, **22**, 1889
		Kan-Fan & Husson, 1978. J. Chem. Soc., Chem. Commun., 618
G. heterosepala Guill.	Alkaloids	Brillanceau *et al.*, 1984. Tetrahedron Letters, **235**, 2767
		Brillanceau *et al.*, 1986. J. Nat. Prod., **49**, 1130
Lindenia austrocaledonica Brongn.	Alkaloids	Saad *et al.*, 1988. Tetrahedron Letters, **29**, 615

Table for Addendum. *Continued*

Taxon	Compound	Reference
Plectronia odorata Benth. & Hook	Iridoids, Alkaloids	Quournevis *et al.*, 1989. J. Nat. Prod., **52**, 306
Psychotria spp.	Alkaloids	Libot *et al.*, 1987. J. Nat. Prod., **50**, 468
Rutaceae		
Boronella aff. *verticillata* Baill. (EG)	Alkaloids, Coumarin	Bevalot *et al.*, 1980. Pl. Méd. Phytothér., **14**, 218
Comptonella spp. (EG)	Alkaloids	Baudouin *et al.*, 1981. J. Nat. Prod., **44**, 546
		Pusset *et al.*, 1991. Pl. Med., **57**, 153
Geijera balansae	Alkaloids	Ahond *et al.*, 1979. Phytochem. **18**, 1415
		Skaltsounis *et al.*, 1985. J. Nat. Prod., **48**, 772
Melicope lasioneura (Baill.) Guill.	Alkaloids	Tillequin *et al.*, 1982. J. Nat. Prod., **45**, 486
M. leptococca (Baill.) Guillaumin	Alkaloids	Skaltsounis *et al.*, 1983. J. Nat. Prod., **46**, 132
M. leratii Guill.	Alkaloids	Ahond *et al.*, 1978. Phytochem., **17**, 166
		Ahond *et al.*, 1978. Tetrahedron, **34**, 2385
Myrtopsis spp. (EG)	Alkaloids, Coumarines, Terpenes	Hifnawy *et al.*, 1976. Pl. Med., **29**, 346
		Hifnawy *et al.*, 1977. Pl. Med., **31**, 156
		Hifnawy *et al.*, 1977. Phytochem., **16**, 1035
Sarcomelicope argyrophylla Guill. (EG)	Alkaloids	Brum-Bousquet *et al.*, 1985. Pl. Med., **51**, 536
S. dogniensis Hartley	Alkaloids	Mitaku *et al.*, 1987. Heterocycles, **26**, 2057
		Mitaku *et al.*, 1989. Ann. Pharm. Fr., **47**, 149
		Brum-Bousquet *et al.*, 1988. Pl. Med., **54**, 470
S. glauca Hartley	Alkaloids	Mitaku *et al.*, 1986. J. Nat. Prod., **49**, 1091
S. leiocarpa (P. S. Green) Hartley	Alkaloids	Baudouin *et al.*, 1985. J. Nat. Prod., **48**, 260
S. pembaiensis Hartley	Alkaloids	Mitaku & Pusset, 1988. Pl. Méd. Phytothér., **22**, 83

Table for Addendum. *Continued*

Taxon	Compound	Reference
S. simplicifolia (Endl.) Hartley subsp. *neo-scotia*	Alkaloids	Couge *et al.*, 1980. Pl. Méd. Phytothér., **14**, 208
		Couge *et al.*, 1980. Pl. Med., **39**, 217
		Bert *et al.*, 1974. Phytochem., **13**, 301
		Tillequin *et al.*, 1980. J. Nat. Prod., **43**, 498
Zanthoxylum oreophilum	Alkaloids	Vaquette *et al.*, 1979. Pl. Med., **35**, 42
Z. sarasinii	Alkaloids	Simeray *et al.*, 1988. Pl. Med., **54**, 189
Simaroubaceae		
Soulamea fraxinifolia Brongn. & Gris	Alkaloids, Quassinoids	Charles *et al.*, 1986. J. Nat. Prod., **49**, 303
S. muelleri Brongn. & Gris	Quassinoids	Polonsky *et al.*, 1980. Tetrahedron, **36**, 2983
S. panceri Brongn. & Gris	Alkaloids, Quassinoids	Viala & Polonsky, 1970. C. R. Acad. Sc. C, **271**, 410
Verbenaceae		
Oxera morieri Vieill. (EG)	Iridoids, Alkaloids	Benkurd *et al.*, 1991. Pl. Med., **57**, 79
Winteraceae		
Zygogynum pauciflorum	Alkaloids	Ahond *et al.*, 1990. J. Nat. Prod., **53**, 875

a) ES = endemic species; EG = endemic genus; EF = endemic family

After Sévenet *et al.*, 1991

12

Chromosomal stasis during speciation in angiosperms of oceanic islands

TOD F. STUESSY AND DANIEL J. CRAWFORD

Abstract

Speciation in angiosperms is often accompanied by change in chromosome number via euploidy or aneuploidy. Evolution involving autopolyploidy and to a much greater extent allopolyploidy has been documented on numerous occasions. In fact, it is estimated that more than 60% of angiosperms exist at the polyploid level, having resulted from many different allopolyploid reticulate evolutionary events. Evolution via aneuploidy is also common, with both ascending and descending modes known (the latter more common, however). Stimuli for rapid chromosomal evolution include intertaxon hybridization and rapid environmental change. Floras of oceanic islands contain numerous endemic taxa which have evolved from continental ancestors and often have further speciated within the archipelago. Chromosomal surveys of the Hawaiian, Juan Fernandez, Bonin and Galapagos Islands reveal very little change in chromosome number during the evolution of endemic taxa, even though speciation is often accompanied by marked morphological divergence. More continental islands, such as the Queen Charlotte Islands, and older oceanic islands closer to mainland sources, for example the Canary Islands, likewise reveal patterns of chromosomal variation. Explanations for absence of change in chromosome number on oceanic islands include low levels of hybridization due to habitat exclusion of endemic taxa, short periods of geological time and selection against aneuploid cytotypes which might disrupt the adaptive complex of traits that led to successful establishment, colonization and radiation.

Speciation in angiosperms is often accompanied by change in chromosome number via euploidy or aneuploidy. Evolution involving autopolyploidy and to a much greater extent allopolyploidy has been documented on numerous occasions (Grant, 1981; Soltis & Soltis, 1993). In fact, it has been estimated (Grant, 1981) that more than 60% of angiosperms exist at the polyploid level, having resulted from many different reticulations during evolution. Evolution via aneuploidy is also common, with both ascending and descending modes documented (the latter more frequent; Goldblatt & Johnson, 1988; Kim, 1994). Stimuli for rapid chromosomal evolution such as polyploidy include (in combination) long-lived organisms usually possessing the means for vegetative propagation (or autogamy), primary speciation accompanied by chromosome repatterning and the common occurrence of natural interspecific hybridization (Grant, 1981).

Several studies have now revealed that certain aspects of speciation in angiosperms of oceanic islands are different in degree or kind from those seen in continental areas (Crawford, Whitkus & Stuessy, 1987b). The most conspicuous differences are small degrees of genetic change accompanied by marked morphological divergence. An excellent example would be *Dendroseris* of the Juan Fernandez Islands (Chapter 4) and the Hawaiian silversword alliance (Carr et al., 1989; see also Chapter 2). In *Dendroseris*, morphological divergence among the three taxonomic subgenera is substantial and indeed they were at one time regarded as distinct genera (Skottsberg, 1953a). Genetic divergence as measured via isozymes, and chloroplast and nuclear ribosomal DNA (Crawford, Stuessy & Silva O., 1987a; Crawford et al., 1992) show very little differentiation. By contrast, congeners in continental areas showing high similarity at isozyme loci and/or low divergence in DNA sequences are usually similar morphologically, the most notable examples are progenitor-derivative species pairs (Gottlieb, 1974; Crawford & Smith, 1982; Soltis, 1985; Sytsma & Gottlieb, 1986).

Another notable feature of the evolution of oceanic island angiosperms appears to be stasis of chromosome number during speciation. In continental situations, species showing high levels of genetic cohesiveness (similarity in molecular attributes and cross-compatibility) often differ in chromosome structure and/or aneuploid number; *Clarkia* (Lewis, 1973), *Stephanomeria* (Gottlieb, 1973) and *Coreopsis* (Smith, 1974) are three of many available examples. On islands, chromosomal stasis has been mentioned on several occasions (Sanders, Stuessy & Rodríguez R., 1983; Kim & Carr, 1990; Sun, Stuessy & Crawford, 1990; Stuessy &

Crawford, 1992) and the topic forms a principal focus of Chapter 1 of this book with reference to the Hawaiian Islands. In this chapter we extend consideration of this phenomenon from the Hawaiian Islands to include the Juan Fernandez Islands, Galapagos Islands and Bonin Islands, all isolated oceanic islands. We also present data for the Canary Islands, which consist of older oceanic islands that are located close to continental source areas and for which a recent chromosomal survey has been published (Ardévol Gonzales, Borgen & Pérez de Pax, 1993). For further interest, we also present data from the Queen Charlotte Islands, clearly continentally derived islands and for which a detailed cytological survey has also been completed (Taylor & Mulligan, 1968).

The purposes of this paper are to: (1) survey known levels of aneuploidy and euploidy within the endemic angiosperm floras of isolated oceanic archipelagos for which data are conveniently available (Hawaiian Islands, Juan Fernandez Islands, Bonin Islands and Galapagos Islands; all midoceanic island chains formed from hot spots); (2) compare levels of chromosomal alteration during speciation in an oceanic island system with islands of greater age and close to continental source areas (Canary Islands); (3) compare levels of chromosomal alteration in a continental island system (Queen Charlotte Islands); and (4) offer hypotheses to explain the absence of chromosomal variation during speciation in plants of oceanic archipelagos.

Surveys of chromosomal levels in floras of isolated oceanic archipelagos (Hawaii, Juan Fernandez, Bonin and Galapagos Islands)

Due to numerous cytological and biosystematic investigations by many workers over the past several decades, a great deal of information on chromosome numbers has accumulated for endemic angiosperm species of the Hawaiian Islands. Because Gerald Carr has provided an excellent review of these data in Chapter 1 of this book, only a brief summary will be given here for comparison with data from other archipelagos. Of the 956 species of angiosperms in the Hawaiian flora, 357 (37.3%) have been counted chromosomally at least once (Table 1.1, Chapter 1). Some 37 genera reveal at least some chromosomal variation, although this may be due in part to erroneous observations, especially in older literature. Discounting these problems, eight genera still appear to be cytologically variable (Table 1.2, Chapter 1), although some examples are much better documented than others. Two of the best cases are in *Dubautia*

(Compositae) and _Wikstroemia_ (Thymeleaceae). The former genus has been studied extensively by Carr and co-workers (e.g., Carr, 1985; Carr & Kyhos, 1981, 1986) and two chromosomal units exist, $n = 14$ and $n = 13$, the latter hypothesized as derived from the former. In _Wikstroemia_, three different levels occur with $n = 9$, 18 and 36 (Gupta & Gillett, 1969; Mayer, 1991), and all are believed to have derived from a single diploid colonizer (Wagner, Herbst & Sohmer, 1990). The obvious conclusion from these data is that very few examples of chromosomal variation among endemic species of the Hawaiian Islands exist. The islands are geologically youthful, 0.9–5.6 million years old, depending upon the specific island (Macdonald, Abbott & Peterson, 1983).

The results of chromosomal surveys of the Juan Fernandez angiospermous endemic flora (Sanders, Stuessy & Rodríguez R., 1983; Spooner _et al._, 1987; Sun, Stuessy & Crawford, 1990) are similar to those from the Hawaiian Islands. Of 104 endemic angiosperms, 38 have been counted chromosomally (all dicots) from 69 populations, yielding information for 37% of the endemic taxa (coincidentally the same percentage as known for the Hawaiian flora). No aneuploid or euploid chromosomal variation has been documented within this endemic flora and only one immigrant line may have originated via chromosomal change as it arrived on the islands: the endemic _Wahlenbergia_ species are all $n = 11$, a number otherwise unknown in the genus (Sun, Stuessy & Crawford, 1990). Within the largest genera, no variation has yet been encountered (Table 12.1). These islands are also geologically young, from 1.01–5.8 million years old (Stuessy _et al._, 1984).

The Bonin (Ogasawara) Islands, from three to five million years old (Asami, 1970), have some parallels with the Juan Fernandez archipelago, especially in the size of the flora. There are 110 endemic angiosperms, and 58 (53%) have been counted chromosomally (Ono, 1975; Ono & Masuda, 1981; Ono & Kobayashi, 1985). Of the 18 genera with two or more species (four is the maximum known in _Ilex_ and _Pittosporum_), data are available for nine of them (Table 12.2). Most of the genera show no chromosomal variation. Variation is encountered, however, in _Alpinia_, _Callicarpa_ and _Malaxis_.

Callicarpa is an interesting case, because all three endemic species are known as $n = 17$, but two of the taxa have additional chromosomal levels documented. _Callicarpa glabra_ Koidz. has been counted as both $n = 16$ and 17, and _C. subpubescens_ Hook. et Arn. as $n = 15$ and 17 (Ono & Masuda, 1981). _Callicarpa glabra_ and _C. nishimurae_ are less variable morphologically and restricted to habitats only on the Chichijima

Table 12.1. *Known chromosome counts among endemic species of the largest genera on the Juan Fernandez Islands*

Genus	Number of endemic species on islands	Number of species counted	Chromosome number(s)
Dendroseris (Compositae)	11	7	$n = 18$
Robinsonia (Compositae)	7	2	$n = 20$
Erigeron (Compositae)	6	6	$n = 27$
Wahlenbergia (Campanulaceae)	5	3	$n = 11$
Peperomia (Piperaceae)	4	4	$n = 22{-}24$[a]

[a]This variation reported for *Peperomia* is due to difficulties of obtaining clear counts
Data from Sanders, Stuessy & Rodríguez R. (1983), Spooner *et al.* (1986), and Sun, Stuessy & Crawford (1990)

Table 12.2. *Known chromosome counts among endemic species of the Bonin Islands*

Genus	Number of endemic species on islands	Number of species counted	Chromosome number(s)
Alpinia (Zingiberaceae)	3	2	$n = 18, 25$
Boninia (Rutaceae)	2	2	$n = 18$
Callicarpa (Verbenaceae)	3	3	$n = 15, 16, 17$
Crepidiastrum (Compositae)	3	3	$n = 5$
Hedyotis (Rubiaceae)	3	2	$n = 17$
Ilex (Aquifoliaceae)	4	2	$n = 17$
Malaxis (Orchidaceae)	2	2	$n = 18, 19$
Osteomeles (Rosaceae)	2	2	$n = 16$
Pittosporum (Pittosporaceae)	4	4	$n = 12$

Data from Ono & Kobayashi (1985)

Table 12.3. *Known chromosome counts among genera with two or more endemic species on the Galapagos Islands*

Genus	Number of endemic species on islands	Number of species counted	Chromosome number(s)	Reference
Acalypha (Euphorbiaceae)	4	3	$n = 10$	Seberg (1984)
Jaegeria (Compositae)	2	2	$n = 18$	Torres (1968)
Scalesia (Compositae)	14	14	$n = 14$	Eliasson (1974)

Islands; *C. subpubescens* is more variable and is found on both the Chichijima and Hahajima Island groups (Kawakubo, 1986). It appears, therefore, that infraspecific aneuploid variation exists within two of the three taxa. Even though the reproductive biology of *Callicarpa* is the subject of Chapter 7 of this book, this information does not deal with the issue of infraspecific aneuploidy; further investigations are obviously needed.

Less is known about *Alpinia* and *Malaxis*. Two of the species of *Alpinia* endemic to the Bonin Islands have been counted: *A. boninensis* Makino, $n = 18$, and the related *A. bilamellata* Makino, $n = 25$ (Ono & Masuda, 1981). The former is common on Chichijima and Hahajima whereas the latter occurs only on Chichijima, and is rare and endangered. Ono & Masuda (1981) stress that the $n = 25$ (as $2n = 50$) count is based on observations of only a few cells and that further detailed analyses are needed. The two endemic species of *Malaxis*, *M. boninensis* (Koidz.) Nackejima and *M. hakajimensis* Kobayashi, also differ cytologically ($n = 19$ and $n = 18$, respectively; Ono & Masuda, 1981) and may provide another example of infraspecific aneuploidy.

The Galapagos Islands, so stimulating to Darwin evolutionarily (Browne 1995), have suprisingly scant chromosomal data for the 221 endemic angiosperm species (Wiggins & Porter, 1971; Porter, 1983). Only 25 (11%) endemic species have been counted. Chromosome counts have been reported for only three genera with two or more endemic species (Table 12.3) and no variation in number is known. Obviously, it would be valuable to mount a chromosomal survey of the endemic angiosperm flora. The islands are young geologically, 0.1–4.2 million years old (Cox, 1983).

The above surveys of four oceanic archipelagos reveal that chromosomal variation, either aneuploidy or euploidy, is not common. In fact, it is conspicuously lacking in these endemic floras.

Chromosomal levels in floras of oceanic islands close to continental areas (Canary Islands)

The Canary Islands are a complex archipelago of ages varying from 0.1 to 20.5 million years old (Schmincke, 1976; Coello *et al.*, 1992). Previous perspectives have judged the western islands to be oceanic in origin and the two eastern islands, Fuerteventura and Lanzarote, as continental (e.g., Schmincke, 1976). Recent radiometric investigations on this eastern ridge have now ascertained the aerial parts of these islands also to be volcanic rather than continental (Coello *et al.*, 1992). Even with this new perspective, however, the easternmost islands are still extremely close to the African continent (*c.* 100 km) and hence more strongly affected by this source area than the western islands. The entire archipelago is much closer to the African continent than any of the isolated oceanic islands discussed above.

The number of endemic angiosperm species on the Canary Islands is 464, of which 244 (53%) have been counted chromosomally (Borgen, 1977). The picture is much more mixed with regard to genera with endemic species that demonstrate chromosomal variation. As for endemic genera, of the eight of 19 that have been recorded for the archipelago (Bramwell, 1976) and that have two or more species, six have been counted chromosomally (Table 12.4). No chromosomal variation is seen in these species. In selected larger genera that have endemic species (although the genera themselves are not endemic), some show consistency of chromosomal number and others reveal considerable variation (Table 12.5).

A good example to discuss further is *Aeonium* (Crassulaceae), recently monographed by Liu (1989). Abundant counts exist for this horticulturally desirable genus largely made by Uhl (1961) and Liu (1989), and most are $n = 18$ and 36 (discounting some additionally aberrant reports). Analysis of the distribution of diploid and tetraploid species on the islands of the Canary archipelago give the following results (from youngest to oldest islands): La Palma, eight $2n$; Hierro, five $2n$; Tenerife, ten $2n$, one $4n$; Gomera, nine $2n$; Gran Canaria, three $2n$, one $4n$, two both $2n$

Table 12.4. *Known chromosome counts among species of endemic genera on the Canary Islands with two or more species*

Genus	Number of endemic species on islands	Number of species counted	Chromosome number
Allagopappus (Compositae)	2	2	$n = 10$
Gonospermum (Compositae)	4	4	$n = 9$
Greenovia (Crassulaceae)	4	3	$n = 18$
Parolinia (Cruciferae)	4	4	$n = 11$
Schizogyne (Compositae)	2	2	$n = 9$
Spartocytisus (Leguminosae)	2	2	$n = 24$

Data from Ardévol González, Borgen & Pérez de Pax (1993)

and $4n$; Lanzarote, two $2n$. A weak trend exists for tetraploids to be distributed on some of the older islands, rather than on the younger islands. Studies by Liu (1989) and Mes & Hart (1996) reveal that polyploids are not just restricted to one evolutionary lineage within the genus. This suggests that polyploids could be accruing in endemic taxa during longer periods of geological time. Gran Canaria, for example, is approximately 14 million years old (Coello *et al.*, 1992).

Chromosomal levels in floras of islands of continental origin (Queen Charlotte Islands)

As a contrast with the chromosomal data from oceanic island archipelagos both isolated and near continental source areas, it is instructive also to compare levels of chromosomal variation in an island system that is truly continental. The best studied islands of this nature are the Queen Charlotte Islands off the western coast of Canada in British Columbia (at 53° N latitude). An outstanding flora was published in 1968 by Taylor & Mulligan which contained two volumes: the first dealing with the morphological and distributional information, and the second a chromosomal survey.

Due to their proximity to the Canadian mainland (less than 80 km at the closest point; 55 km to the nearest off-shore island), the Queen

Table 12.5. *Selected genera of the Canary Islands that have considerable numbers of endemic species and for which cytological data are available*

Genus	Number of endemic species on islands	Number of species counted	Chromosome number(s)
Aeonium (Crassulaceae)	28	28	$n = 18, 36$
Argyranthemum (Compositae)	17[a]	17	$n = 9$
Centaurea (Compositae)	10	7	$n = 14, 15, 16$
Convolvulus (Convolvulaceae)	10	4	$n = 12, 15$
Echium (Boraginaceae)	22	14	$n = 8$
Euphorbia (Euphorbiaceae)	11	6	$n = 10, 30$
Limonium (Plumbaginaceae)	13	12	$n = 6, 7, 12, 16$
Lotus (Leguminosae)	12	12	$n = 7, 14$
Monanthes (Crassulaceae)	13	9	$n = 18, 36$
Senecio (Compositae)	12	8	$n = 20, 30$
Sonchus (Compositae)	28	28	$n = 9$
Taeckholmia (Compositae)	3	3	$n = 9$
Tolpis (Compositae)	6	6	$n = 9$

[a] A total of 22 species in Macaronesia (Humphries, 1976; Bremer & Humphries, 1993)
From Ardévol Gonzáles, Borgen & Pérez de Pax (1993)

Charlotte Islands have only 11 endemic angiosperms. Eight (73%) of these have been counted chromosomally. The total number of angiosperm taxa (including infraspecific units) is 542, with 411 (76%) having been counted from at least one population. The overall chromosomal survey, therefore, is quite extensive.

Two presentations of data will be used to show the considerable degree of chromosomal variation within the angiosperms of the Queen Charlotte Islands. The first is a list of the four genera that have ten or more species on the islands, i.e., *Carex*, *Juncus*, *Potamogeton* and *Ranunculus* (Table 12.6). Substantial aneuploid (dysploid) chromosomal variation occurs

Table 12.6. *Chromosome numbers in genera with ten or more species on the Queen Charlotte Islands*

Genus	Number of species on islands (not endemic)	Number of species counted	Chromosome number(s) of these species on islands as well as elsewhere
Carex (Cyperaceae)	37	24	$n = 14, 25, 26, 27, 28, 30, 31, 33, 35, 37, 38, 39, 44$
Juncus (Juncaceae)	17	12	$n = 17, 19, 20, 22, c. 30, 40, c. 56, c. 50–60$
Potamogeton (Potamogetonaceae)	10	4	$n = 13$
Ranunculus (Ranunculaceae)	13	10	$n = 7, 8, 14, 16$

Data from Taylor & Mulligan (1968)

here within and between species. *Juncus* and *Carex* are especially known for extensive aneuploid and euploid variation in continental regions (e.g., Rothrock & Reznicek, 1996). The only exception is the aquatic *Potomogeton* with uniform $n = 13$ in all four species counted. Perhaps the more uniform aquatic environment helps maintain cytological uniformity. The second set of data show the genera of Compositae on the islands that have two or more species (Table 12.7). Almost all of these species have been counted and chromosomal variation is the rule rather than the exception, especially euploidy. The data in Table 12.7 may be misleading, however, because they contain counts for taxa found on the islands even when originally counted elsewhere (such as in the Canadian continent). It is hardly surprising that the Queen Charlotte Islands have a distinctly continental character chromosomally as they were part of the Canadian continent and were doubtless connected to the continent during lowered sea levels of the Pleistocene (Heusser, 1955, 1960, 1985; Brown, 1960; Brown & Nasmith, 1962). Furthermore, much of the islands were glaciated, which led to recolonization by continental populations after the glacial ice receded (Taylor & Mulligan, 1968). Nonetheless, these data do indicate that variation in chromosome num-

Table 12.7. *Chromosome variation in genera of Compositae with two or more species on the Queen Charlotte Islands*

Genus	Number of species on islands (not endemic)	Number of species counted	Chromosome numbers on islands
Arnica	2	2	$n = 19, 28$
Cirsium	3	2	$n = 17, 34$
Erigeron	2	2	$n = 9, 18$
Hieracium	3	3	$n = 9, 18$
Senecio	6	6	$n = 18, 20, 24,$ $40, 45$
Tanacetum	2	2	$n = 9, 27$

Data from Taylor & Mulligan (1968)

ber is quite common in this island flora, similar to that known for virtually any continental area, and quite distinct from the lack of variation seen on isolated oceanic islands.

Hypotheses for absence of chromosomal variation during speciation in angiosperms of isolated oceanic archipelagos

Workers have noted the lack of chromosomal variation among endemic plants of oceanic islands on several occasions, although not focusing on this specific point (e.g., Carr & Kyhos, 1981; Sanders, Stuessy & Rodríguez R., 1983; Stuessy *et al.*, 1984; Kim & Carr, 1990; Sun, Stuessy & Crawford, 1990). As Carr (Chapter 1) has also provided a good discussion of this issue, we add our own perspectives as a supplement.

With regard to the absence of euploidy on oceanic islands, we can refer to Grant (1981) and his list of factors (in combination) that promote polyploidy in vascular plants: (1) long-lived organisms usually possessing means of vegetative propagation (or autogamy); (2) primary speciation accompanied by chromosomal repatterning; and (3) common occurrence of natural interspecific hybridization. On the Juan Fernandez Islands, the distribution of life-forms in the native and endemic flora is as follows (modified from the Raunkier system presented by Skottsberg, 1953*b*): trees or rosette-trees (48%), shrubs (25%), perennial herbs (22%), and annuals (5%). Although 95% of the species of the flora are perennials, and hence potentially long-lived, we do not see particular means of vegetative propagation that could help preserve chromosomal variants, either euploid or aneuploid, mentioned by Grant as promoting polyploidy. Further, successful colonists of oceanic islands are often herbaceous perennials, capable of quick establishment and adaptation due to more rapid generation times.

Although on Juan Fernandez Islands there are no studies of chromosomal repatterning, there are excellent studies with regard to the Hawaiian silversword group by Carr and associates (e.g., Carr & Kyhos, 1981, 1986). In most cases, little chromosomal repatterning has occurred with high fertility observed. The same result has been documented in *Tetramolopium* (Lowrey, 1986; see summary in Chapter 1). It is often the case, therefore, that primary speciation on oceanic islands has not been accompanied by chromosomal repatterning. This is perhaps not too surprising, because this phenomenon occurs most frequently in annuals (Grant, 1981), which are typically not common on oceanic islands.

The occurrence of natural interspecific hybridization on oceanic islands varies depending upon the archipelago and perhaps also degree of human disturbance. Essentially no natural hybridization is known on the Juan Fernandez Islands, with the exception of *Gunnera bracteata* × *G. peltata* on Masatierra (Pacheco, Stuessy & Crawford, 1991) and the possible hybrid between *Eryngium bupleuroides* Hook. et Arn. and *E. inaccessum* Skottsb. (called *E. fernandezianum*; Skottsberg, 1921), still to be substantiated. On the Hawaiian Islands and Canary Islands, however, numerous hybrids are known, perhaps due to dramatic habitat disturbance over many centuries by human interventions which have hybridized habitats (Anderson, 1948). In our opinion, natural interspecific hybridization might be more common on very young geological islands with continuing volcanic activity that would itself cause disturbance of the habitats, disruption of populations, etc. On quieter islands, i.e., those more than two million years old that have ceased active vulcanism, hybridization should be less common. Taxa would be more ecologically and geographically isolated, thus greatly reducing the possibility of hybridization and hence allopolyploidy.

Another important point regarding low levels of polyploidy is that arriving immigrants already are at high levels of polyploidy (e.g., 66% on Juan Fernandez; 77% on Hawaii, Chapter 1), which may be one reason few additional polyploids have occurred (Kiehn & Lorance, 1996). There are clear exceptions, however, such as *Wikstroemia* on

Hawaii, in which the ancestor was probably diploid and polyploidy subsequently did occur on the islands (Chapter 1). Taking all of Grant's (1981) factors into consideration, they do not apply well to the endemic flora of oceanic islands and may help explain the absence of euploidy.

Kiehn & Lorance (1996) also stress that lack of polyploidy on islands could be simply reflections of successful immigrants coming from ancestral groups that themselves, for whatever reasons, lack chromosomal variation. Conversely, in the several instances in which chromosomal variation (both aneuploidy and euploidy) occurs in island endemics, their continental relatives are also known to be chromosomally variable. For example, the subtribe Madiinae (Heliantheae), the closest relatives of the Hawaiian silversword alliance, are known for extensive chromosomal restructuring and aneuploid variation (Carr, 1985; Kyhos, Carr & Baldwin, 1990). It is not particularly surprising, therefore, to find chromosome differences in the island endemics. Similarly, the subfamily Sempervivoideae (Crassulaceae) is known for extensive polyploidy (Uhl, 1961), making the occurrence of polyploidy in *Aeonium* on the Canary Islands not wholly unexpected. It has been suggested recently by Lowe & Abbott (1996) that the allopolyploid *Senecio teneriffae* originated on the Canary Islands. This endemic is part of a complex well known for the formation of polyploids (Harris & Ingram, 1992*a*, *b*; Abbott, Curnow & Irwin, 1995) and thus there may be some 'predisposal' to the formation of the insular polyploid.

Additional factors may also be important in the absence of chromosomal variation in endemic plants of oceanic islands. One factor mentioned earlier is time. The maximum age of an oceanic island formed as the result of passing over a hot spot is about six million years before it erodes and subsides under the sea. There simply may not be sufficient time for substantial chromosomal variation to accrue.

Another possible factor in observed low levels of chromosomal variation might be lower levels of genetic variation within populations, which lowers the probability of chromosomal alterations via genetically controlled breakage. This could also be due to absence of abnormal physiological conditions in cells (e.g., perhaps due to an absence of continental viral infections that increase the rate of chromosomal breakage). Low variation may be due to: founder effect, small population sizes, perennial habit, and self-compatibility of immigrants. These dimensions are not always the case, however. On the Juan Fernandez Islands some species are indeed quite uniform genetically, such as the rare *Lactoris fernandeziana* (Crawford *et al.*, 1994), but others are quite variable (e.g., in

Myrceugenia fernandeziana, Crawford *et al.*, unpublished). There is also experimental evidence that even after the stringent bottleneck represented by the founder effect, surprising quantitative genetic variation may be released in subsequent generations (e.g., Carson & Wisotzkey, 1989; Carson, 1990).

And finally, strong directional selection during adaptive radiation may eliminate deleterious variants (such as change in chromosome rearrangement and number; Sun, Stuessy & Crawford 1990); self- compatibility of immigrants can also help conserve adaptive gene complexes. This point has been well developed by Carr (Chapter 1) and he cites new data bearing on the adaptive nature of chromosomal number in the context of the cellular environment in which genes are activated (Kyhos & Carr, 1994).

While there seems no doubt that, in general, chromosome stasis is more pronounced in island taxa as compared to those of continents, it continues to be useful to compare particular endemics with their closest continental relatives to determine chromosomal change (e.g., as was done on Ullung Island; see Chapter 9). In addition to island conditions, phylogenetic history may also be an important factor in determining chromosomal change during the evolution of island endemics.

Acknowledgements

It is a pleasure to acknowledge: Susie Benkowski for help with the tabulation of chromosome numbers from floristic works; support from the National Science Foundation under grant no. DEB-9500499 for our continuing investigations on the vascular flora of the Juan Fernandez Islands; CONAF of Chile for permission to conduct field studies in the Robinson Crusoe Islands National Park which led to chromosomal studies from that archipelago summarized herein and which stimulated production of this more general review. The authors would especially like to thank Professor Gerald Carr, author of Chapter 1 in this book, who finished his chapter before ours and kindly made it available for consultation in the preparation of our own contribution.

Literature cited

Abbott, R. J., Curnow, D. J. & Irwin, J. A. (1995). Molecular systematics of *Senecio squalidus* L. and its close diploid relatives. In *Advances in Compositae Systematics*, ed. D. J. N. Hind & G. V. Pope, pp. 223–37. London: Royal Botanic Gardens, Kew.

Anderson, E. (1948). Hybridization of the habitat. *Evolution,* **2,** 1–9.

Ardévol Gonzáles, J. F., Borgen, L., & Pérez de Pax, P. L. (1993). Checklist of chromosome numbers counted in Canarian vascular plants. *Sommerfeltia,* **18,** 1–59.

Asami, S. (1970). Topography and geology in the Bonin Islands. In *The Nature in the Bonin Islands,* ed. T. Tuyama & S. Asami, pp. 91–108. Tokyo: Hirokawa Shoten. [In Japanese.]

Borgen, L. (1977). *Check-List of Chromosome Numbers Counted in Macaronesian Vascular Plants.* Oslo: published by author. Mimeo.

Bramwell, D. (1976). The endemic flora of the Canary Islands. In *Biogeography and Ecology in the Canary Islands,* ed. G. Kunkel, pp. 207–40. The Hague: W. Junk.

Bremer, K. & Humphries, C. J. (1993). Generic monograph of the Asteraceae–Anthemideae. *Bulletin Natural History Museum London (Botany),* **23,** 71–177.

Brown, A. S. (1960). Physiography of the Queen Charlotte Islands. *Canadian Geographic Journal,* **61,** 30–7.

Brown, A. S. & Nasmith, H. (1962). The glaciation of the Queen Charlotte Islands. *Canadian Field-Naturalist,* **76,** 209–19.

Browne, J. (1995). *Charles Darwin: Voyaging.* Princeton: Princeton University Press.

Carr, G. D. (1985). Monograph of the Hawaiian Madiinae (Asteraceae): *Argyroxiphium, Dubautia* and *Wilkesia. Allertonia,* **4,** 1–123.

Carr, G. D. & Kyhos, D. W. (1981). Adaptive radiation in the Hawaiian silversword alliance (Compositae–Madiinae). I. Cytogenetics of spontaneous hybrids. *Evolution,* **35,** 543–56.

Carr, G. D. & Kyhos, D. W. (1986). Adaptive radiation in the Hawaiian silversword alliance (Compositae–Madiinae). II. Cytogenetics of artificial and natural hybrids. *Evolution,* **40,** 959–76.

Carr, G. D., Robichaux, R. H., Witter, M. S. & Kyhos, D. W. (1989). Adaptive radiation of the Hawaiian silversword alliance (Compositae–Madiinae): a comparison with Hawaiian picture-winged *Drosophila.* In *Genetics, Speciation and the Founder Principle,* ed. W. Giddings, K. Y. Kaneshiro & W. W. Anderson, pp. 79–97. New York: Oxford University Press.

Carson, H. L. (1990). Increased genetic variance after a population bottleneck. *Trends in Ecology and Evolution,* **5,** 228–30.

Carson, H. L. & Wisotzkey, R. G. (1989). Increase in genetic variance following a population bottleneck. *American Naturalist,* **134,** 668–73.

Coello, J., Cantagrel, J. M., Hernan, F., Fúster, J. M., Ibarrola, E., Ancochea, E., Casquet, C., Jamond, C., Díaz de Téran, J. R., & Cendrero, A. (1992). Evolution of the eastern volcanic ridge of the Canary Islands based on new K-Ar data. *Journal Volcanology & Geothermal Research,* **53,** 251–74.

Cox, A. (1983). Ages of the Galapagos Islands. In *Patterns of Evolution in Galapagos Organisms,* ed. R. I. Bowman, M. Berson & A. E. Leviton, pp. 11–23. San Francisco: Pacific Division, American Association for the Advancement of Science.

Crawford, D. J. & Smith, E. B. (1982). Allozyme variation in *Coreopsis nuecensoides and C. nuecensis,* a progenitor-derivative species pair. *Evolution,* **36,** 379–86.

Crawford, D. J., Stuessy, T. F., Cosner, M. B., Haines, D. W., Silva O., M. & Baeza, M. (1992). Evolution of the genus *Dendroseris* (Asteraceae:

Lactuceae) on the Juan Fernandez Islands: evidence from chloroplast and ribosomal DNA. *Systematic Botany*, **17**, 676–82.

Crawford, D. J., Stuessy, T. F., Cosner, M. B., Haines, D. W., Wiens, D. & Peñailillo, P. (1994). *Lactoris fernandeziana* (Lactoridaceae) on the Juan Fernandez Islands: allozyme uniformity and field observations. *Conservation Biology*, **8**, 277–80.

Crawford, D. J., Stuessy, T. F. & Silva O., M. (1987a). Allozyme divergence and the evolution of *Dendroseris* (Compositae: Lactuceae) on the Juan Fernandez Islands. *Systematic Botany*, **12**, 435–43.

Crawford, D. J., Whitkus, R. & Stuessy, T. F. (1987b). Plant evolution and speciation on oceanic islands. In *Patterns of Differentiation in Higher Plants*, ed. K. Urbanska, pp. 183–99. London: Academic Press.

Eliasson, U. (1974). Studies in Galapagos plants. XIV. The genus *Scalesia* Arn. *Opera Botanica*, **36**, 1–117.

Goldblatt, P. & Johnson, D. E. (1988). Frequency of descending versus ascending aneuploidy and its phylogenetic implications. *American Journal of Botany*, **75** (6, part 2), 175–6. abstract.

Gottlieb, L. D. (1973). Genetic differentiation, sympatric speciation and the origin of a diploid species of *Stephanomeria*. *American Journal of Botany*, **60**, 545–53.

Gottlieb, L. D. (1974). Genetic confirmation of the origin of *Clarkia lingulata*. *Evolution*, **28**, 244–50.

Grant, V. (1981). *Plant Speciation*, 2nd edn. New York: Columbia University Press.

Gupta, S. & Gillett, G. W. (1969). Observations on Hawaiian species of *Wikstroemia* (Angiospermae: Thymelaeaceae). *Pacific Science*, **23**, 83–8.

Harris, S. A. & Ingram, R. (1992a). Molecular systematics of the genus *Senecio* L. I. Hybridization in a British polyploid complex. *Heredity*, **69**, 1–10.

Harris, S. A. & Ingram, R. (1992b). Molecular systematics of the genus *Senecio* L. II. The origin of *S. vulgaris* L. *Heredity*, **69**, 112–21.

Heusser, C. J. (1955). Pollen profiles from the Queen Charlotte Islands, British Columbia. *Canadian Journal of Botany*, **33**, 429–49.

Heusser, C. J. (1960). Late-Pleistocene environments of North Pacific North America. *American Geographical Society Special Publication* no. 35, 1–305.

Heusser, C. J. (1985). Quaternary pollen records from the Pacific Northwest: aleutians to the Oregon–California boundary. In *Pollen Records of Late-Quaternary North American Sediments*, ed. V. M. Bryant, Jr. & R. G. Holloway, pp. 141–66. Dallas: American Association of Stratigraphic Palynologists Foundation.

Humphries, C. J. (1976). A revision of the Macaronesian genus *Argyranthemum* Webb ex Schultz Bip. (Compositae–Anthemideae). *Bulletin Natural History Museum London (Botany)*, **5**, 147–240.

Kawakubo, N. (1986). Morphological variation of three endemic species of *Callicarpa* (Verbenaceae) in the Bonin (Ogasawara) Islands. *Plant Species Biology*, **1**, 59–68.

Kiehn, M. & Lorance, D. H. (1996). Chromosome counts on angiosperms cultivated at the National Tropical Botanical Garden, Kauai, Hawaii. *Pacific Science*, **50**, 317–23.

Kim, I. (1994). Aneuploidy in flowering plants: Asteraceae and Onagraceae. *Korean Journal of Plant Taxonomy*, **24**, 265–78.

Kim, I. & Carr, G. D. (1990). Cytogenetics and hybridization of *Portulaca* in Hawaii. *Systematic Botany*, **15**, 370–7.

Kyhos, D. W. & Carr, G. D. (1994). Chromosome stability and lability in plants. *Evolutionary Theory*, **10**, 227–48.

Kyhos, D. W., Carr, G. D. & Baldwin, B. G. (1990). Biodiversity and cytogenetics of the tarweeds (Asteraceae: Heliantheae–Madiinae). *Annals Missouri Botanical Garden*, **77**, 84–95.

Lewis, H. (1973). The origin of diploid neospecies in *Clarkia*. *American Naturalist*, **107**, 161–70.

Liu, H. Y. (1989). Systematics of *Aeonium* (Crassulaceae). *National Museum Natural Science (Taichung, Taiwan) Special Publication*, **3**, 1–102.

Lowe, A. J. & Abbott, R. J. (1996). Origins of the new allopolyploid species *Senecio cambrensis* (Asteraceae) and its relationship to the Canary islands endemic *Senecio teneriffae*. *American Journal of Botany*, **83**, 1365–72.

Lowrey, T. K. (1986). A biosystematic revision of Hawaiian *Tetramolopium* (Compositae: Astereae). *Allertonia*, **4**, 203–65.

Macdonald, G. A., Abbott, A. T. & Peterson, F. L. (1983). *Volcanoes in the Sea: The Geology of Hawaii*, 2nd edn. Honolulu: University of Hawaii Press.

Mayer, S. (1991). Artificial hybridization in Hawaiian *Wikstroemia* (Thymelaeaceae). *American Journal of Botany*, **78**, 122–30.

Mes, T. H. M. & Hart, H. T. (1996). The evolution of growth forms in the Macaronesian genus *Aeonium* (Crassulaceae) inferred from chloroplast DNA RFLPs and morphology. *Molecular Ecology*, **5**, 351–63.

Ono, M. (1975). Chromosome numbers of some endemic species of the Bonin Islands. I. *Botanical Magazine (Tokyo)*, **88**, 323–8.

Ono, M. & Kobayashi, S. (1985). Flowering plants endemic to the Bonin Islands. In *Endemic Plant Species and Vegetation of the Bonin (Ogasawara) Islands*, ed. M. Ono & K. Okutomi, pp. 1–96. Kamakura: Aboc-sha.

Ono, M. & Masuda, Y. (1981). Chromosome numbers of some endemic species of the Bonin Islands. II. *Ogasawara Research* no. 4, 1–24.

Pacheco, P., Stuessy, T. F. & Crawford, D. J. (1991). Natural interspecific hybridization in *Gunnera* (Gunneraceae) of the Juan Fernandez Islands, Chile. *Pacific Science*, **45**, 389–99.

Porter, D. M. (1983). Vascular plants of the Galapagos: origins and dispersal. In *Patterns of Evolution in Galapagos Organisms*, ed. R. I. Bowman, M. Berson & A. E. Leviton, pp. 33–96. San Francisco: Pacific Division, American Association for the Advancement of Science.

Rothrock, P. E. & Reznicek, A. A. (1996). Documented chromosome numbers 1996. 1. Chromosome numbers in *Carex* section *Ovales* (Cyperaceae) from eastern North America. *Sida*, **17**, 251–8.

Sanders, R. W., Stuessy, T. F. & Rodríguez R., R. (1983). Chromosome numbers from the flora of the Juan Fernandez Islands. *American Journal of Botany*, **70**, 799–810.

Schmincke, H. U. (1976). The geology of the Canary Islands. In *Biogeography and Ecology in the Canary Islands*, ed. G. Kunkel, pp. 67–184. The Hague: W. Junk.

Seberg, O. (1984). Taxonomy and phylogeny of the genus *Acalypha* (Euphorbiaceae) in the Galapagos archipelago. *Nordic Journal of Botany*, **4**, 159–90.

Skottsberg, C. (1921). The phanerogams of the Juan Fernandez Islands. In *The Natural History of Juan Fernandez and Easter Island*, vol. 2, *Botany*, ed. C. Skottsberg, pp. 95–240. Uppsala: Almqvist & Wiksells.

Skottsberg, C. (1953a). A supplement to the pteridophytes and phanerogams of Juan Fernandez and Easter Island. In *The Natural History of Juan Fernandez and Easter Island*, ed. C. Skottsberg, pp. 763–92. Uppsala: Almqvist & Wiksells.

Skottsberg, C. (1953b). The vegetation of the Juan Fernandez Islands. In *The Natural History of Juan Fernandez and Easter Island*, ed. C. Skottsberg, pp. 793–960. Uppsala: Almqvist & Wiksells.

Smith, E. B. (1974). *Coreopsis nuecensis* (Compositae) and a related new species from southern Texas. *Brittonia*, **26**, 161–71.

Soltis, D. E. (1985). Allozymic differentiation among *Heuchera americana, H. parviflora, H. pubescens* and *H. villosa* (Saxifragaceae). *Systematic Botany*, **10**, 193–8.

Soltis, D. E. & Soltis, P. S. (1993). Molecular data and the dynamic nature of polyploidy. *Critical Reviews Plant Sciences*, **12**, 243–73.

Spooner, D., Stuessy, T. F., Crawford, D. J. & Silva O., M. (1987). Chromosome numbers from the flora of the Juan Fernandez Islands. II. *Rhodora*, **89**, 351–6.

Stuessy, T. F. & Crawford, D. J. (1992). Chromosomal stasis during speciation of angiosperms in oceanic islands. *American Journal of Botany*, **79**, 163. Abstract.

Stuessy, T. F., Foland, K. A., Sutter, J. F., Sanders, R. W. & Silva O., M. (1984). Botanical and geological significance of potassium-argon dates from the Juan Fernandez Islands. *Science*, **225**, 49–51.

Stuessy, T. F., Sanders, R. W. & Silva O., M. (1984). Phytogeography and evolution of the flora of the Juan Fernandez Islands: a progress report. In *Biogeography of the Tropical Pacific*, ed. F. J. Radovsky, S. H. Sohmer & P. H. Raven, pp. 55–69. Honolulu: Association of Systematics Collections and Bishop Museum.

Sun, B-Y., Stuessy, T. F. & Crawford, D. J. (1990). Chromosome counts from the flora of the Juan Fernandez Islands, Chile. III. *Pacific Science*, **44**, 258–64.

Sytsma, K. J. & Gottlieb, L. D. (1986). Chloroplast DNA evolution and phylogenetic relationships in *Clarkia* sect. *Peripetasma* (Onagraceae). *Evolution*, **40**, 1248–61.

Taylor, R. L. & G. A. Mulligan. (1968). *Flora of the Queen Charlotte Islands*, part 1, *Systematics of the Vascular Plants*; part 2, *Cytological Aspects of the Vascular Plants*. Research Branch Monograph no. 4. Ottawa: Canada Department of Agriculture.

Torres, A. M. (1968). Revision of *Jaegeria* (Compositae–Heliantheae). *Brittonia*, **20**, 52–73.

Uhl, C. H. (1961). The chromosomes of the Sempervivoideae (Crassulaceae). *American Journal of Botany*, **48**, 114–23.

Wagner, W. L., Herbst, D. R. & Sohmer, S. H. (1990). *Manual of the Flowering Plants of Hawaii*. Honolulu: University of Hawaii Press; Bishop Museum Press.

Wiggins, I. W. & Porter, D. M. (1971). *Flora of the Galapagos Islands*. Stanford: Stanford University Press.

13

The current status of our knowledge and suggested research protocols in island archipelagos

TOD F. STUESSY AND MIKIO ONO

Abstract

Evolution of higher plants on oceanic islands is now understood to be different in certain respects from that in plant groups of continental regions. This volume re-emphasizes the long-held view that island plants often undergo rapid speciation under directional selection through adaptive radiation. Such dramatic morphological change is usually not accompanied by changes in chromosome number (nor structural rearrangements) nor in genetic composition as evidenced by isozyme and RAPD loci. Geographic isolation, especially between islands, is fundamental as a stimulus for speciation. Change in island size and habitat spectra over geological time causes rapid ecological shifts and concomitant extinctions. Modern sources of data and methods of analysis have provided us with some of these newer perspectives and suggest continued research protocols involving: (1) basic floristic inventories using consistent species concepts; (2) preliminary evolutionary hypotheses; (3) geological data and history (especially of vulcanism); (4) explicit phylogenetic hypotheses; (5) estimates of character evolution, including co-evolution and adaptive complexes; (6) postulates of modes of speciation; (7) rigorous biogeographic analyses; and (8) communication of conservation priorities to appropriate officials and agencies.

This volume illustrates the breadth of plant island biological studies from the morphological to the molecular and from the phylogenetic to the biogeographic. The chapters contained herein reflect the dynamic nature of studies of oceanic islands, which in large measure parallel existing ideas and approaches now prevalent in modern plant systematics and evolutionary biology (e.g., Hoch & Stephenson, 1995). The geographic focus here is definitely on the Pacific Basin, with contributions dealing with the western Pacific (Bonin Islands), South Pacific, Central Pacific (Hawaii) and eastern Pacific (Juan Fernandez). Information is also included within several of the chapters, however, on other archipelagos such as the Canary Islands, New Caledonia, etc. The chapters deal with differing taxa, too, such as whole floras (Bohm, Chapter 11; Carr, Chapter 1); Compositae (Baldwin, Chapter 2; Carr, Chapter 1; Crawford *et al.*, Chapter 4); Rhizophoraceae (Setagouchi, Ohba & Tobe, Chapter 10); and Verbenaceae (Kawakubo, Chapter 7). They also reveal use of different types of data, including vegetative and floral morphology (Kawakubo, Chapter 7; Setagouchi, Ohba & Tobe, Chapter 10), vegetative and reproductive anatomy, and ultrastructure (Setagouchi, Ohba & Tobe, Chapter 10); reproductive biology (Kawakubo, Chapter 7); cytology (Carr, Chapter 1; Stuessy & Crawford, Chapter 12); cytogenetics (Carr, Chapter 1); secondary products chemistry (Bohm, Chapter 11); and macromolecular data (Baldwin, Chapter 2; Crawford *et al.*, Chapter 4; Setagouchi, Ohba & Tobe, Chapter 10). Types of analyses include phenetics for morphology (Sun & Stuessy, Chapter 9), isozymes (Ito, Soejima & Ono, Chapter 6) and cladistics with morphological as well as macromolecular data for phylogeny reconstruction (Baldwin, Chapter 2; Crawford *et al.*, Chapter 4; Setagouchi, Ohba & Tobe, Chapter 10).

These investigations, and others from the scattered literature during the past decade, have shown that patterns and processes of evolution on islands are not always the same as those of continental regions. Islands are geologically ephemeral, rapidly ecologically changing, and geographically limited land masses in comparison to most continental regions. We have reconfirmed that there is often a strong ecological component to speciation, often reflected in adaptive radiation of island taxa (e.g., Carlquist, 1974), that promotes divergent morphological responses. But we now have seen that few chromosomal changes have accompanied their radiations (Carr, Chapter 1; Stuessy & Crawford, Chapter 12) and few genetic changes have occurred, as reflected at isozyme and RAPD loci (Crawford *et al.*, Chapter 4). We have also seen rates of speciation, as measured by rates of macromolecular divergence, accelerated over those of continental regions (Baldwin, Chapter 2; Crawford *et al.*, Chapter 4). We have also seen the importance that geographic isolation plays as a stimulus to speciation (Stuessy *et al.*, Chapter 3), both within but especially between islands. And furthermore, we now appreciate the significance that reduction of island land area and loss of habitat can play in causing extinction, obscuring past evolutionary events and complicating efforts to explain species diversity (e.g., Stuessy *et al.*, Chapter 5). A useful question regarding islands is whether rates of extinction of endemic taxa really are higher than in most continental areas, and if so, what should this say about establishing conservation priorities for political areas in which such islands occur?

Recent investigations, especially as revealed in the chapters of this book, now suggest a series of research protocols for evolutionary studies in plants of island archipelagos. This does not mean that there is only one way to study plants of oceanic islands, nor that one should recommend only one particular focus or research agenda. What it means is that with the arsenal of data and methods now available, it is desirable to utilize as many of them as possible for a fuller understanding of the biology of the archipelagos.

First, there must be a good inventory at the specific level of the plants of the particular islands under consideration. This is obviously fundamental or virtually none of the other questions about plant evolution can be addressed meaningfully. An excellent example of the importance of a modern inventory is the recent *Manual of the Flowering Plants of Hawaii* by Wagner, Herbst & Sohmer (1990). Prior to this inventory, the inconsistent use of concepts of species made it virtually impossible to consider evolutionary questions in a realistic fashion. For example, in *Cyrtandra* (Gesneriaceae), Hillebrand (1888) recognized 29 species whereas St John described 252 new species (in ten papers during 1987–88, see references in Wagner, Herbst & Sohmer, 1990) in addition to 131 species he had already recognized from just the island of Oahu (St John, 1966)! Having a consistent taxonomic treatment for the plants of the archipelago has now permitted advances in biogeographic analysis (Wagner & Funk, 1995).

Second, before working on details of evolutionary patterns and processes, it is desirable to have preliminary hypotheses of relationships based on morphology and basic distributions. This means having a phenetic (or preliminary cladistic or phyletic) framework of how island endemics relate to progenitors in continental source areas. A pertinent example would be Ullung Island, in which such basic phytogeographic origins are still in the process of being refined (Sun & Stuessy, Chapter 9). Another useful example is in *Erigeron* (Compositae) of the Juan Fernandez Islands in which phenetic studies were first completed to assess ties to mainland relatives followed by cladistic analyses among island endemics (Valdebenito, Stuessy & Crawford, 1992).

Third, the isolated nature of islands and their volcanic origins emphasizes the need for geological data to establish their time of origin and geomorphological evolution. Radiometric data (usually K-Ar) are essential for this understanding. Utility of such a time sequence has been employed in the biogeographic analyses in Hawaiian angiosperms (Wagner & Funk, 1995) and in determining evolutionary directionality and rates of evolutionary change on the Juan Fernandez Islands (Crawford *et al.*, Chapter 4). Submarine geology around islands and toward continental areas also needs to be determined, because these can often extend the suspected time of land surface area that was available as stepping stones for colonization (e.g., Christie *et al.*, 1992). An important point here, however, is that just the existence of such seamounts does not mean they were necessarily above the surface of the sea at any previous time period. If terrestrial fossils or evidence of terrestrial erosion and/or chemical weathering does provide such evidence, then it is important to establish that they remained above the sea long enough to provide a stepping stone to other islands borne later including those presently now seen. Because isolated oceanic islands have a life of usually not more than six million years, such sequences must be evaluated carefully. Having a detailed study of vulcanism describing the sequence of volcanic events is also most helpful, although this requires a more sophisticated level of geological expertise (such as revealed in Hawaii; Macdonald, Abbott & Peterson, 1983). Having such data can help suggest availability of land areas for colonization after the initial island formation. Fossil data are also important, if such exist, such as marine snails at higher elevations, to interpret changing sea levels and corresponding reduction and expansion of land areas (during Pleistocene). Erosional patterns need to be suggested also, as was done, for example, on the Juan Fernandez Islands (Sanders *et al.*, 1987), because this, too, can have a marked effect on island elevation, overall size and ecology. This is essential for successful interpretation of species diversity in archipelagos (i.e., equilibrium island biogeographic analyses, see Chapter 5). If stable depositional environments exist on islands in bogs or quiet ponds, such as on the Galapagos Islands, pollen spectra from cores can be extremely useful in showing changing vegetational patterns and ecological correlates (e.g., Colinvaux & Schofield, 1976).

Fourth, to be able to suggest modes of speciation or hypothesize on any other evolutionary phenomenon, it is essential to have a strong phylogeny that is itself a robust hypothesis of evolutionary relationships. The particular relationships that must be estimated are those among endemic taxa within individual islands, among taxa of different islands, and between island endemics and continental progenitors (often in a progenitor-derivative anagenetic context; e.g., Sun *et al.*, 1996; Sun & Stuessy, Chapter 9). Methods to assess those relationships are either explicit phenetics (especially distance measures such as minimum-spanning trees), explicit phyletics (various methods; Stuessy, 1987, 1990, 1997) or cladistics (such as parsimony algorithms used here in this volume). Morphology and DNA restriction site and sequence data are current types of information upon which these analyses are based.

Fifth, phylogenetic estimates, even minimal cladistic branching patterns, can be used to infer character state evolution by their optimization on dendrograms of whatever nature. This can lead to a better understanding of the correlations of evolutionarily significant character states during evolution of the group on the islands and in effect help seek character states that are compatible for the particular taxon. This information can be used to help refine new phylogenetic analyses (Lamboy, 1994). It can also be used to interpret adaptive change during phylogeny in consort with ecological and/or vegetational change data.

Sixth, modes of speciation should be inferred based on all available data. Full insights on speciation involve details of populational studies and population genetic information (e.g., Gottlieb, 1973), often not available to the systematist and simply not available for many island groups. Many initial hypotheses can be advanced and should be advanced (Sundberg & Stuessy, 1991), however, based on distributions, flowering times, phylogenetic relationships, etc., that can go a long way to focus the context of speciation within island groups. For example, on the Juan Fernandez Islands, simultaneous with studies at the populational level on different taxa (e.g., *Wahlenbergia*; Crawford *et al.*, 1990) are studies on patterns of phylogeny (Stuessy, Crawford & Marticorena, 1990), patterns of endemism (Stuessy *et al.*, 1992) and isolating mechanisms (Stuessy *et al.*, Chapter 3), that narrow concepts of speciation even more precisely. For detailed studies of modes of speciation it is helpful to have data on chromosomes and cytogenetics (e.g., Carr, Chapter 1; Baldwin, Chapter 2), genetic variation within and between populations (Crawford *et al.*, Chapter 4), and reproductive systems including breeding systems and pollinators (Kawakubo, Chapter 7). Seeking ecological correlations and adaptations during speciation is also essential (Sun & Stuessy, Chapter 9).

Seventh, rigorous biogeographic hypotheses must be formulated involving concepts of dispersal to the archipelago and vicariance/dispersal within them. Modes of dispersal need to be inferred based on morphology of diaspores and/or correlations with bird flight patterns (e.g., Cruden, 1966), which are no doubt of great importance for initial colonization of islands (e.g., Ono, 1991). The use of cladistic methods via area cladograms with reliable geological data for the Hawaiian Islands is used efficaciously in the book by Wagner & Funk (1995).

Eighth, we must bring all our gathered knowledge about the endemic flora of island archipelagos into concepts and priorities for conservation of the often fragile plants. For example, on the Juan Fernandez Islands, more than 70% of the endemic species of angiosperms are extinct, threatened, rare or occasional; all categories are at risk (Stuessy *et al.*, 1992). No one is better informed or better equipped to recommend research priorities for conservation biology in these island archipelagos than the systematist or evolutionary biologist working intimately with them. We can offer data on populational sizes, degrees of genetic variation, co-adaptations with other organisms which may also be at risk, obvious degradations of habitats, need for educational programmes for park personnel, etc. The challenge is not to plunge personally into the political arena, where unavoidably the real solutions lie (and the resources dwell), but to do a better job of informing politicians on issues so that they can take appropriate and timely action. Just speaking to each other at national and international scientific meetings does little to advance the conservation agenda on the islands to which we have devoted so much of our own personal time.

Oceanic islands are very special natural laboratories of plant evolution, unlike any other place on earth. They stimulate, they puzzle and they charm*. We would like to think that Alfred Russell Wallace (1880), and other early naturalists fascinated by islands, would be pleased with the progress made over the past century. At the very least, this volume is testimony to the dynamic spirit and positive accomplishments of present-day investigations.

---

*As fully attested by the senior editor of this book, who met his wife on board ship to the Juan Fernandez Islands!

Literature cited

Carlquist, S. (1974). *Island Biology.* New York: Columbia University Press.

Christie, D. M., Duncan, R. A., McBirney, A. R., Richards, M. A., White, W. M., Harpp, K. S. & C. G. Fox. (1992). Drowned islands downstream from the Galapagos hot spot imply extended speciation times. *Nature,* **355,** 246–8.

Colinvaux, P. A. & Schofield, E. K. (1976). Historical ecology in the Galapagos Islands. I. A Holocene pollen record from El Junco, Isla San Cristobal. *Journal of Ecology,* **64,** 989–1012.

Crawford, D. J., Stuessy, T. F., Lammers, T. G. & Silva O., M. (1990). Allozyme variation and evolutionary relationships among three species of *Wahlenbergia* (Campanulaceae) in the Juan Fernandez Islands, Chile. *Botanical Gazette,* **151,** 119–24.

Cruden, R. W. (1966). Birds as agents of long-distance dispersal for disjunct plant groups of the temperate western hemisphere. *Evolution,* **20,** 517–32.

Gottlieb, L. D. (1973). Genetic differentiation, sympatric speciation and the origin of a diploid species of *Stephanomeria. American Journal of Botany,* **60,** 545–53.

Hillebrand, W. (1888). *Flora of the Hawaiian Islands: A Description of Their Phanerogams and Vascular Cryptogams.* Heidelberg: Carl Winter.

Hoch, P. C. & Stephenson, A. G. (1995). *Experimental and Molecular Approaches to Plant Biosystematics.* St Louis: Missouri Botanical Garden.

Lamboy, W. F. (1994). The accuracy of the maximum parsimony method for phylogeny reconstruction with morphological characters. *Systematic Botany,* **19,** 489–505.

Macdonald, G. A., Abbott, A. T. & Peterson, F. L. (1983). *Volcanoes in the Sea: The Geology of Hawaii,* 2nd edn. Honolulu: University of Hawaii Press.

Ono, M. (1991). Flora of the Bonin Islands: endemism and dispersal mode. *Aliso,* **13,** 95–105.

Sanders, R. W., Stuessy, T. F., Marticorena, C. & Silva O., M. (1987). Phytogeography and evolution of *Dendroseris* and *Robinsonia,* tree-Compositae of the Juan Fernandez Islands, Chile. *Opera Botanica,* **92,** 195–215.

St John, H. (1966). Monograph of *Cyrtandra* (Gesneriaceae) on Oahu, Hawaiian Islands. *B. P. Bishop Museum Bulletin,* **229,** 1–465.

Stuessy, T. F. (1987). Explicit approaches for evolutionary classification. *Systematic Botany,* **12,** 251–62.

Stuessy, T. F. (1990). *Plant Taxonomy: The Systematic Evaluation of Comparative Data.* New York: Columbia University Press.

Stuessy, T. F. (1997). Classification: more than just branching patterns of evolution. *Aliso,* **15,** 113–24.

Stuessy, T. F., Crawford, D. J. & Marticorena, C. (1990). Patterns of phylogeny in the endemic vascular flora of the Juan Fernandez Islands, Chile. *Systematic Botany,* **15,** 338–46.

Stuessy, T. F., Marticorena, C., Rodríguez R., R., Crawford, D. J. & Silva O., M. (1992). Endemism in the vascular flora of the Juan Fernandez Islands. *Aliso,* **13,** 297–307.

Sun, B.Y., Stuessy, T. F., Humaña, A. M., Riveros G., M. & Crawford, D. J. (1996). Evolution of *Rhaphithamnus venustus* (Verbenaceae), a gynodioecious hummingbird-pollinated endemic of the Juan Fernandez Islands, Chile. *Pacific Science,* **50,** 55–65.

Sundberg, S. D. & Stuessy, T. F. (1991). Isolating mechanisms and implications for modes of speciation in Heliantheae (Compositae). In *Research Advances in the Compositae*, ed. T. J. Mabry & G. Wagenitz, pp. 77–98. Wien: Springer-Verlag.

Valdebenito, H. A., Stuessy, T. F., Crawford, D. J. & Silva O., M. (1992). Evolution of *Erigeron* (Compositae) in the Juan Fernandez Islands, Chile. *Systematic Botany*, **17**, 470–80.

Wagner, W. L. & Funk, V. A. (ed.) (1995). *Hawaiian Biogeography: Evolution on a Hot Spot Archipelago*. Washington DC: Smithsonian Institution Press.

Wagner, W. L., Herbst, D. R. & Sohmer, S. H. (1990). *Manual of the Flowering Plants of Hawaii*. Honolulu: University of Hawaii Press; Bishop Museum Press.

Wallace, A. R. (1880). *Island Life*. London: Macmillan & Co.

Author index

Note: Multi-authored works are listed by first authors.

Adjaye, A.E., Dobberstein, R.H., Venton, D.L. & Fong, H.H.S., 267
Adler, G.H. & Dudley, R., 122
Adsersen, A., Adsersen, H. & Brimer, L., 247, 282, 284
Adsersen, A., Brimer, L., Olsen, C.E. & Jaroszewski, J.W., 248
Albrecht, G.H., 188
Allan, H.H., 260
Anderson, E., 318
Anderson, G.J. & Symon, D.E., 159
Arano, H. & Saito, H., 196–7
Ardévol Gonzales, J.F., Borgen, L. & Pérez de Pax, P.L., 309, 314–15
Arnold, H.L., 281
Arslanian, R.L., Mondragon, B., Stermitz, F.R. & Marr, K.L., 235
Asami, S., 142, 310
Atchley, W.R. & Anderson, D., 188
Atchley, W.R., Gaskins, C.T. & Anderson, D., 188
Averett, J.E., Hahn, W.J., Berry, P.E. & Raven, P.H., 262

Baker, H.G., 62, 112, 155, 165
Baker, H.G. & Cox, P.A., 155
Baker, R.J., Chesser, R.K., Koop, B.F. & Hoyt, R.A., 39
Baldwin, B.G., 27, 31, 52, 58, 66, 68–9
Baldwin, B.G., Kyhos, D.W. & Dvořák, J., 32, 50, 57, 61, 68
Baldwin, B.G., Kyhos, D.W., Dvořák, J. & Carr, G.D., 2, 32, 50, 52, 57–8
Baldwin, B.G. & Robichaux, R.H., 27–8, 31, 50, 53, 57–8, 60–2, 65, 68–9
Baldwin, B.G., Sanderson, M.J., Porter, J.M., Wojciechowski, M.F., Campbell, C.S. & Donoghue, M.J., 61–2, 68–9
Baldwin, M.E., Bick, I.R.C., Komzak, A.A. & Price, J.R., 269
Barth, F.G., 159, 163

Bartlett, M.F., Korzun, B., Sklar, R., Smith, A.F. & Taylor, W.I., 235
Bate-Smith, E.C., 268–9, 271, 275, 280, 283
Baudouin, G., Tillequin, F., Koch, M., Pusset, J. & Sévenet, T., 271
Baumberger, H., 196
Bawa, K.S., 155–6, 165
Bawa, K.S. & Opler, P.A., 155
Bedi, Y.S., Bir, S.S. & Gill, B.S., 194
Bell, C.R. & Constance, L., 196
Bengtsson, B.O. & Bodmer, W.F., 39
Benko-Iseppon, A.M., 195
Berry, K.M., Perry, N.B. & Weavers, R.T., 264
Bévalot, F., Vaquette, J. & Cabalion, M.P., 271
Bittner, M., Silva O., M., Rozas, Z., Jakupovic, J. & Stuessy, T.F., 259
Bohm, B.A., 242–3
Bohm, B.A., Crins, W.J. & Wells, T.C., 244
Bohm, B.A. & Koupai-Abyazani, M.R., 246, 283
Bohm, B.A., Nicholls, K.W. & Ornduff, R., 268, 275, 282–3
Bohm, B.A., & Ornduff, R., 265
Boomsma, J.J., Mabelis, A.A., Verbeck, M.G.M. & Los, E.C., 122
Borgen, L., 313
Bramwell, D., 313
Bramwell, D. & Dakshani, K.M.M., 272
Bremer, K., 52–3
Bremer, K. & Humphries, C.J., 315
Brooker, S.G., Cambie, R.C. & Cooper, R.C., 265–6
Brophy, J.J., Goldsack, R.J. & Goldsack, G., 267
Brown, A.S., 316
Brown, A.S. & Nasmith, H., 316
Brown, J.H., 122
Brown, J.H. & Kodric-Brown, A., 122
Browne, J., 312
Buckley, R.C., 122
Buckley, R.C. & Knedlhans, S.B., 131
Budesinsky, M., Perez Soutoo, N. & Holub, M., 275

Bush, G., 39
Bushnell, O.A., Fakuda, M. & Makinodan, T., 235

Cambie, R.C., Cox, R.E. & Sidwell, D., 270
Cambie, R.C., Lal, A.R. & Pausler, M.G., 271
Cannon, J.R., Hughes, G.K., Ritchie, E. & Taylor, W.C., 235
Carlquist, S., 1–2, 6, 53, 57–8, 85, 98, 101, 107, 112, 128, 132, 155–6, 170–1, 231, 244, 259, 281, 326
Carr, G.D., 1, 6, 25, 27–8, 35, 53, 55, 60, 67, 98, 197, 244, 309, 319
Carr, G.D. & Baker, J.K., 29, 35
Carr, G.D., Baldwin, B.G. & Kyhos, D.W., 27, 31
Carr, G.D. & Kyhos, D.W., 1, 27–8, 50, 60, 68, 87, 244, 309–10, 317–18
Carr, G.D., Powell, E.A. & Kyhos, D.W., 40, 62–3
Carr, G.D., Robichaux, R.H., Witter, M.S & Kyhos, D.W., 28, 31, 87, 308
Carson, H.L., 93, 320
Carson, H.L. & Clague, D.A., 52–3, 61, 67, 115
Carson, H.L. & Kaneshiro, K.Y., 1
Carson, H.L. & Wisotskey, R.G., 53, 320
Case, T.J. & Cody, M.L., 122
Cauwet-Marc, A.M., 196
Chan, R.K.G., 58–9, 61, 69
Chan, R.K.G., Lowrey, T.K., Natvig, D. & Whitkus, R., 58–9
Chang, C.S., 186
Chang, C.S. & Giannasi, D.E., 186
Charlesworth, B. & Charlesworth, D., 155
Charnov, E.L., 155, 158
Charnov, E.L., Smith, J.M. & Bull, J.J., 158
Chesser, R.K. & Baker, R.J., 39
Cho, H.J., Kim, S., Suh, Y. & Park, C.W., 186–7
Choi, H.K., Kim, H.J., Shin, H. & Kim, Y., 196
Christie, D.M., Duncan, R.A., McBirney, A.R., Richards, M.A., White, W.M., Harpp, K.S. & Fox, C.G., 328
Chung, Y.H. & Sun, B.Y., 187, 194
Coello, J., Cantagrel, J.M., Hernan, F., Fúston, J.M., Ibarrola, E., Ancochea, E., Casquet, C., Jamond, C., Díaz de Téran, J.R. & Cendrero, A., 272, 313–14
Coffey, J.C., 25
Colinvaux, P.A. & Schofield, E.K., 328
Colwell, R.K., 87
Conn, B.J., 32
Conner, E.F. & Simberloff, D., 122

Constance, L. & Affolter, F., 25
Cooper-Driver, G.A., Swain, T. & Conn, E.E., 281
Corner, E.J.H., 217, 220
Corruccini, R.S., 188
Cowlishaw, M.G., Bickerstaffe, R. & Conner, H.E., 267
Cox, A., 312
Cox, P.A., 155
Craven, B.M., 265
Crawford, D.J., 60, 81, 98, 141, 147, 151–2
Crawford, D.J., Brauner, S., Cosner, M.B. & Stuessy, T.F., 88
Crawford, D.J. & Smith, E.B., 308
Crawford, D.J. & Stuessy, T.F., 278
Crawford, D.J., Stuessy, T.F., Cosner, M.B., Haines, D.W. & Silva O., M., 105, 115, 123
Crawford, D.J., Stuessy, T.F., Cosner, M.B., Haines, D.W., Silva O., M. & Baeza, M., 80, 91, 102, 114, 123, 308
Crawford, D.J., Stuessy, T.F., Cosner, M.B., Haines, D.W., Wiens, D., Peñailillo, P., 319
Crawford, D.J., Stuessy, T.F., Haines, D.W., Cosner, M.B., Silva O., M. & Lopez, P., 88, 109–11, 123, 142, 144, 146, 149
Crawford, D.J., Stuessy, T.F., Lammers, T.G. & Silva O., M., 329
Crawford, D.J., Stuessy, T.F. & Silva O., M., 88, 108, 110, 123, 142, 259, 308
Crawford, D.J., Whitkus, R. & Stuessy, T.F., 91, 144, 308
Crins, W.J. & Bohm, B.A., 244
Crins, W.J., Bohm, B. & Carr, G.D., 50, 245
Crowden, R.K., Wright, J. & Harborne, J.B., 262, 283
Crowell, K.L., 122
Cruden, R.W., 330
Cuddihy, L. & Stone, C.P., 67

Dalla-Favera, R., Bregni, M., Erikson, J., Patterson, D., Gallo, R.C. & Croce, C.M., 40
Darwin, C., 147, 155
Darwin, S.P. & Chaw, S.M., 32
Daushkevich, Y.V., Vasil'eva, M.G. & Pimenov, M.G., 196
Davis, J.I., 195
Defoe, D., 75
De Joode, D.R. & Wendel, J.F., 98, 109
de Pouques, M.L., 194
DeSalle, R., 1
DeVore, M.L. & Stuessy, T.F., 53
Diamond, J.M., 122
Ding Hou, 220

Ding Hou & van Steenis, C.G.G.J., 204–5
Dobzhansky, T., 39
Doyle, J.J., 68

Easterfield, T.H. & Aston, B.C., 265
Ehrendorfer, F., 155
Elavumootil, L., Garnier, J., Mahuteau, J.
& Plat, M., 270
Eliasson, U., 312
Erikson, J., Finnan, J., Nowell, P.C. &
Croce, C.M., 40
Espelie, K.E. & Hermann, H.R., 281
Estévez-Reyes, R., Estévez-Braun, A. &
Gonzalez, A.G., 274

Farris, J.S., 220
Fedorov, A., 25, 194–6
Felsenstein, J., 223
Ferakova, V., 194
Fernandez, C., Fraga, B.M. & Hernandez,
M.G., 273
Fish, F., Gray, A.I., Waigh, R.D. &
Waterman, P.G., 235
Fish, F., Gray, A.I. & Waterman, P.G., 235
Floyd, M.E., 241
Folkers, K. & Koniusy, F., 235
Foo, L.Y., 262
Foo, L.Y., Hrstich, L. & Vilain, C., 261–2
Fosberg, F.R., 34–5, 53, 59
Foster, R.C., 194
Fraga, B.M., Guillermo, R., Hernandez,
M.G., Mestres, T. & Artega, J.M., 274,
283
Fukuoka, N., 194
Funabiki, K., 195
Funk, V.A. & Wagner, W.L., 50, 52, 62, 65
Futuyama, D.J., 147

Gagne, W.C. & Cuddihy, L.W., 53
Gainsford, G.J., Russell, G.B. & Reay,
P.F., 267
Ganders, F.R., 52, 55, 58, 62, 281
Ganders, F.R., Bohm, B.A. & McCormick,
S.P., 243–4, 283
Ganders, F.R. & Nagata, K.M., 2, 29, 34,
50, 53, 55–6, 60, 66–7, 98, 242
Garber, E.D., 26
Gardner, R.C., 2, 25, 29, 35–6, 50, 55, 59,
245
Geh, S.Y. & Keng, H., 220
Giannasi, D.E., 280
Gilbert, F.S., 122, 133
Gillet, G.W., 25, 29, 37, 58
Gillet, G.W. & Lim, E.K.S., 1, 25, 29, 34,
50, 58, 60, 66
Gil, M.I., Ferreres, F., Marrero, A., Tomas-
Lorente, F. & Tomas-Barberan, F.A., 273

Gilpin, M.E. & Diamond, J.M., 122
Givnish, T.J., Sytsma, K.J. & Hahn, W.J.,
98, 115
Givnish, T.J., Sytsma, K.J., Smith, J.F. &
Hahn, W.J., 29, 36
Glennie, C.W., Harborne, J.B., Rowley,
G.D. & Marchant, C.J., 272, 279, 282
Goldblatt, P., 6, 25, 32
Goldblatt, P. & Johnson, D.E., 308
Gonzalez, A.G., Barrera, J.B., Mendez,
J.T., Martinez, J.E. & Sanchez, M.L., 273
Gonzalez, A.G., Fraga, B.M., Hernandez,
M.G., Larruga, F., Luis, J.G. & Ravelo,
A.G., 273–4
Gonzalez, A.G., Fraga, B.M., Hernandez,
M.G., Luis, J.G. & Larruga, F., 273–4
Gonzalez-Coloma, A., Hernandez, M.G.,
Perales, A. & Fraga, B.M., 274
Gorman, M., Neuss, N., Djerassi, C.,
Kutney, J.P. & Scheuer, P.J., 235
Gorovoy, P.G. & Volkova, S.A., 196
Gottlieb, L.D., 60, 144, 147, 308, 329
Govindachari, T.R. & Thyagarajan, B.S.,
235
Grant, V., 6, 82, 88, 90, 308, 317–19
Gupta, S. & Gillet, G.W., 25, 27, 33, 310
Gurzenkov, N.N., 194
Gurzenkov, N.N. & Gorovoy, P.G., 196

Hall, W.P., 39
Hamrick, J.L. & Godt, M.J.W., 98
Harborne, J.B., 251, 273, 281, 284
Harn, C.Y. & Lees, M.S., 195
Harris, S.A. & Ingram, R., 319
Hartley, T.G., Dunstone, E.A., Fitzgerald,
J.S., Jones, S.R. & Lambertson, J.A., 271
Hartley, T.G. & Stone, B.C., 240
Hasabe, M. & Iwatsuki, K., 222
Hashimoto, T. & Yamada, Y., 248
Haydon, D., Radtkey, R.R. & Pianka, E.R.,
122, 133
Hayman, A.R., Perry, N.B. & Weavers,
R.T., 265
Hedrick, P.W., 39
Hedrick, P.W. & Levin, D.A., 39
He, F. & Legendre, P., 122
Helenurm, K. & Ganders, F.R., 2, 50, 60,
62, 99, 142, 146, 149
Hemsley, W.B., 76, 80, 124
Hennion, F., Fiasson, J.L. & Gluchoff-
Fiasson, K., 279
Heslop-Harrison, J., 165
Heusser, C.J., 316
Higa, T. & Scheuer, P.J., 241, 283
Hillebrand, W., 327
Hills, M., 188
Hiroe, M. & Constance, L., 187, 196

Hoch, P.C. & Stephenson, A.G., 325
Hoeneison, M., Silva O., M., Wink, M.,
  Crawford, D.J. & Stuessy, T.F., 259
Holm, L., 241
Hooker, J.D., 278
Hounsell, R.W., 195
Hudgins, W.R. & Scheuer, P.J., 241
Humphries, C.J., 315

Isogai, A., Murakoshi, S., Suzuki, A. &
  Tamura, S., 274
Isogai, A., Suzuki, A., Tamura, S.,
  Murakoshi, S., Ohasi, Y. & Sasada, Y.,
  274
Ito, M., 176
Ito, M. & Ono, M., 146, 149, 151
Ito, M., Soejima, A. & Ono, M., 146, 149
Iwatsubo, Y. & Naruhashi, N., 196
Iwatsuki, K., Shimozono, F. & Hirai, K.,
  175

Jensen, S.R. & Nielsen, B.J., 264
John, B. & Lewis, K.R., 39
Johow, F., 76, 80, 124, 131
Jones, J.D.G., Dunsmuir, P. & Bedbrook,
  J., 40
Jordan, W. & Scheuer, P.J., 235
Juncosa, A.M., 205, 214, 227
Juncosa, A.M. & Tomlinson, P.B., 204–5,
  214, 220
Juvik, J.O. & Austring, A.P., 122

Kadmon, R. & Pulliam, H.R., 122
Kaneshiro, K.Y., Gillespie, R.G. & Carson,
  H.L., 2
Kato, M., 164, 166
Kawakubo, N., 156, 158, 310
Kiehn, M. & Lorence, D., 25, 27, 318–19
Kim, H.G., Keeley, S.C. & Jansen, R.K.,
  50, 52, 62
Kim, I., 308
Kim, I. & Carr, G.D., 25, 27, 33, 308, 317
Kim, J.W., 185
Kim, K.Y., 140
Kim, S.C., Crawford, D.J. & Jansen, R.K.,
  112
Kim, Y.K., 183
Kim, Y.S & Yoon, C.Y., 196
King, M., 82, 88
Kitagawa, M., 187, 196
Klein, G., 40
Kobayashi, S., 176
Kobayashi, S. & Ono, M., 170, 172
Komai, K., Iwamura, J. & Ueki, K., 241
Komai, K. & Tang, C.S., 242
Komai, K., Tang, C.S. & Nishimoto, R.K.,
  242, 283

Komai, K. & Ueki, K., 241
Koyama, H., 143
Kulanthaivel, P. & Benn, M.H., 264
Kunkel, G., 272
Kupchan, S.M., Meshulen, H. & Sneden,
  A.T., 265
Kurita, M., 196
Kyhos, D.W. & Carr, G.D. 39–41, 320
Kyhos, D.W., Carr, G.D. & Baldwin, B.G.,
  31, 41, 319

Lamboy, W.F., 329
Lammers, T.G., 25, 29, 36–7, 128
Lammers, T.G. & Lorence, D.H., 25, 29
Lammers, T.J., Stuessy, T.F. & Silva O.,
  M., 259
Lande, R., 39
Larson, A., Prager, E.M. & Wilson, A.C.,
  39
Lee, T.B., 186
Lee, T.B. & Yamazaki, T., 196
Lee, T.N., 186
Lee, W.T. & Yang, I.S., 140, 185, 197
Lee, Y.N., 186, 192–7
Leugmayr, E., 29, 35
Levin, D.A., 82–3, 281
Lewis, D., 155
Lewis, H., 39, 308
Lichius, J.J., Thoison, O., Montagnac, A.,
  Pais, M., Cosson, J.P. & Hadi, A.H.A.,
  270
Linney, G.K., 25
Littlejohn, M.J., 82
Liu, H.Y., 313–14
Li, X.Y., Song, W.Q. & Chen, R.Y., 196
Lloyd, D.G., 63, 155, 165
Louda, S.M. & Rodman, J.E., 281
Lowe, A. & Abbott, R.J., 319
Lowrey, T.K., 2, 25, 30, 36, 50, 52, 55,
  58–60, 62–3, 65, 67, 81, 98, 318
Lowrey, T.K., Chan, R., Daniels, D. &
  Whitkus, R., 58
Lowrey, T.K. & Crawford, D.J., 2, 50, 60,
  99, 141, 146, 149, 151, 171
Lucas, G., 171

Mabberley, D.J., 235, 257, 260–1, 265–7,
  275, 285
MacArthur, R.H. & Wilson, E.O., 121, 132
McAtee, W.L., 131
McCallion, R.F., Cole, A.L.J., Walker,
  J.R.L., Blunt, J.W. & Munro, M.H.G.,
  264
Macdonald, G.A., Abbot, A.T. & Peterson,
  F.L., 39, 127, 310, 328
Macedo, M. & Prance, G.T., 132
Macias, F.A. & Fischer, N.H., 248

Macias, F.A., Molinillo, J.M.G. & Fischer, N.H., 249
Majak, W., & Benn, M., 266
Mansour, R.M., Saleh, N.A. & Boulos, L., 273
Manuelidis, L., 41
Marchant, Y.Y., Ganders, F.R., Wat, C.K. & Towers, G.H.N., 243
Marchant, Y.Y. & Towers, G.H.N., 243, 281
Marchant, Y.Y., Turjman, M., Flynn, T., Balza, F., Mitchell, J.C. & Towers, G.H., 241
Markham, K.R. & Godley, E.J., 278
Markham, K.R., Vilain, C. & Molloy, B.P.J., 261–2
Markham, K.R., Webby, R.F., & Vilain, C., 261
Markham, K.R., & Whitehouse, L.A., 261
Marr, K.L. & Tang, C.S, 242
Marticorena, C., 130
Marticorena, C. & Quezada, M., 129
Massias, M., Carbonnier, J. & Molho, D., 260
Massiot, G., Vercauteren, J., Jacquier, N.J., Leux, J. & Le Men-Olivier, L., 271
Massiot, G., Vercauteren, J., Richard, B., Jacquier, N.J., & Le Men-Olivier, L., 280
Matthei, O., Marticorena, C. & Stuessy, T.F., 80, 124
Maxwell, C.A., Hartwig, U.A., Joseph, C.M. & Phillips, D., 284
Mayer, S., 27, 38, 310
Mayer, S. & Charlesworth, D., 156, 159
Medina, J.T., 75
Mehra, P.N., 194
Mehra, P.N. & Khosla, P.K., 194
Mehra, P.N., Khosla, P.K. & Sareen, T.S., 194
Melville, R., 171
Menadue, Y. & Crowden, R.K., 268
Mes, T.H.M. & Hart, H.T., 314
Miles, D.H.C, Chittawong, V., Hedin, P.A. & Kokpol, U., 246
Miyaji, Y., 197
Moore, D.M., 277
Moore, D.M., Harborne, J.B. & Williams, C.A., 277
Moore, R.J., 25, 27, 32
Morat, P., Veillon, J.M. & Mackee, H.S., 226
Mori, S.A. & Brown, J.L., 132
Moyer, B.G., Pfeffer, P.E., Valentine, K.M. & Gustine, D.L., 266
Muir, A.D., Cole, A.L.J. & Walker, J.R.L., 264, 266
Muller, C.H., 281

Muyard, F., Bevalot, F., Laude, B. & Vaquette, J., 271
Muyard, F., Bevalot, F., Regnier, A. & Vaquette, J., 271

Nagy, F., Morelli, G., Fraley, R.T., Rogers, S.G. & Chua, N.H., 40
Nakai, T., 140, 185–6, 194–5, 197
Napoli, C., Lemieux, C. & Jorgensen, R., 40
Nei, M., 108, 144, 149–50
Nei, M., Muruyama, T. & Roychoudhury, A.K., 146
Nicharat, S. & Gillet, G.W., 29, 37
Nishioka, T., 143
Nyananyo, B.L., 279, 283
Nyananyo, B.L. & Heywood, V.H., 278

Oginuma, K., Tanaka, R. & Suzuki, K., 195
Ohba, H., 144
Oh, B.U., 187
Ohwi, J., 196
Okada, H. & Tamura, M., 196
Okada, M., Whitkus, R. & Lowrey, T.K., 59
Ono, M., 132, 143, 170–1, 197, 323, 330
Ono, M. & Kobayashi, S., 310
Ono, M., Kobayashi, S. & Kawakubo, N., 143, 147, 172, 175–7
Ono, M. & Masuda, Y., 144, 310, 312
Ono, M. & Nagai, S., 143
Ornduff, R., 25
Ourecky, D.K., 195

Pacheco, P., Crawford, D.J., Stuessy, T.F. & Silva O., M., 80, 88, 123, 251–2, 254, 257, 282–3
Pacheco, P., Stuessy, T.F. & Crawford, D.J., 80, 318
Pacheco, P., Stuessy, T.F. & Marticorena, C., 80, 85
Pacheco, P., Stuessy, T.F., Marticorena, C. & Silva O., M., 80, 85
Paik, W.K. & Lee, W.T., 187
Pamilo, P. & Nei, M., 68
Panichanum, S., Bick, I.R.C. & Blunt, J.W., 267
Pan, Z., Chin, H., Wu, Z., Yuan, C. & Liou, S., 196
Parham, J.W., 226
Park, C.W., Oh, S.H. & Shin, H., 186, 194
Park, M.K., 195
Patama, T.T. & Widén, C.J., 246
Patterson, R., 29, 37, 246
Perry, N.B. & Weavers, R.T., 265, 283
Pojer, P.M., Ritchie, E. & Taylor, W.C., 272
Porter, D.M., 143

Preece, R.C., Bennett, K.D. & Carter, J.R., 276
Probatova, N.W. & Sokolovskaya, A.P., 194, 196
Proksch, P., Budzikiewicz, H., Tanowitz, B.D. & Smith, D.M., 244
Purdie, A.W., 261, 282
Putnam, A.R. & Tang, C.S., 281

Quader, A., Armstrong, J.A., Gray, A.I., Hartley, T.G. & Waterman, P.G., 269
Quinn, S.L., Wilson, J.B. & Mark, A.F., 122

Rabakonandrianina, E., 25, 29, 35, 59–60
Rabakonandrianina, E. & Carr, G.D., 2, 29, 35–6, 50, 59–60
Randriaminahy, M., Proksch, P., Witte, L. & Wray, V., 279
Real, L., 163
Reid, A.R. & Bohm, B.A., 263, 268
Rice, E., 281
Richards, A.J., 165
Rizvi, S.J.H. & Rizvi, V., 281
Roberts, A. & Stone, L., 122
Robichaux, R.H., Carr, G.D., Liebman, M. & Pearcy, R.W., 50, 52, 59
Rodríguez, R., R., 124
Rosenthal, G.A. & Janzen, D.H., 281
Ross, M.D., 155
Ross, M.D. & Weir, B.S., 155
Rothrock, P.E. & Reznicek, A.A,. 316
Rüdenberg, L. & Green, P.S., 194
Russell, G.B. & Reay, P.F., 267

St John, H., 327
Sakai, K., 196
Sanders, R.W., Stuessy, T.F. & Marticorena, C., 124
Sanders, R.W., Stuessy, T.F., Marticorena, C. & Silva O., M., 99, 101, 103, 107–8, 116, 123, 127, 130, 328
Sanders, R.W., Stuessy, T.F. & Rodríguez R., R., 6, 88, 90, 123, 197, 308, 310–11, 317
Sang, T., Crawford, D.J., Kim, S.C. & Stuessy, T.F., 80, 104–5, 114, 123
Sang, T., Crawford, D.J., Stuessy, T.F. & Silva O., M., 105–6, 115, 123
Santamour, F.S., 194
Sastrapradja, S., 25, 27, 32
Sax, K. & Kribs, D.A., 194
Scheuer, P.J., 241
Scheuer, P.J., Horigan, L.P. & Hudgins, W.R., 235
Scheuer, P.J. & Hudgins, W.R., 241
Scheuer, P.J. & Metzger, J.T.H., 235
Scheuer, P.J. & Pattabhiraman, T.R., 240

Schmid, R., 217
Schminke, H.U., 272, 313
Schultz, S.T. & Ganders, F.R., 63
Seberg, O., 312
Setoguchi, H. & Ohba, H., 222
Setoguchi, H., Ohba, H. & Tobe, H., 211
Setoguchi, H., Tobe, H. & Ohba, H., 216
Sévenet, T., Pusset, J., Bourret, D. & Potier, P., 284–5
Shimotomai, N., 195
Shimozono, F. & Iwatsuki, K., 175
Simberloff, D.S., 122
Skottsberg, C., 6, 25, 27, 32–4, 76, 80, 84, 86–7, 113, 117, 123–4, 126, 132, 277, 308, 317–18
Smith, A.C., 205, 226
Smith, E.B., 308
Smith, J.F., Burke, C.C. & Wagner, W.L., 35
Smith, R.M., Marty, R.A. & Peters, C.F., 270
Soejima, A., Nagamasu, H., Ito, M. & Ono, M., 144, 146, 149, 151
Soltis, D.E., 308
Soltis, D.E. & Soltis, P.S., 90, 308
Spooner, D., Stuessy, T.F., Crawford, D.J. & Silva O., M., 88, 90, 123, 197, 310–11
Stace, H.M. & Fripp, Y.J., 268
Steyermark, J.A. & Lisner, R., 212
Storey, W.B., 25
Strother, J.L., 36
Stuessy, T.F., 101, 329
Stuessy, T.F. & Crawford, D.J., 308
Stuessy, T.F., Crawford, D.J. & Marticorena, C., 80–1, 123–4, 128, 329
Stuessy, T.F., Foland, K.A., Sutter, J.F., Sanders, R.W. & Silva O., M., 76, 80, 99, 114–15, 123, 310, 317
Stuessy, T.F., Marticorena, C., Rodríguez, R., R., Crawford, D.J. & Silva O., M., 76, 80, 249, 329, 330
Stuessy, T.F., Sanders, R.W. & Silva O., M., 99, 113, 123
Stuessy, T.F., Swenson, U., Crawford, D.J., Anderson, G., Jensen, R. & Silva O., M., 127
Sugiura, M., Shinozaki, K., Zaita, N., Kasuda, M. & Kumano, M., 223
Sugiura, T., 195
Sun, B.Y., Park, J.H., Kwak, M.J., Kim, C.H. & Kim, K.S., 194
Sun, B.Y., Stuessy, T.F. & Crawford, D.J., 88, 90, 123, 197, 308, 310–11, 317, 320
Sun, B.Y., Stuessy, T.F., Humaña, A.M., Riveros, G.M. & Crawford, D.J., 87, 165, 191, 329
Sun, M. & Ganders, F.R., 50, 63

Sundberg, S.D. & Stuessy, T.F., 329
Suzuki, K., 81
Swanholm, C.E., St John, H. & Scheuer, P.J., 235
Swenson, U., Stuessy, T.F., Baeza, M. & Crawford, D.J., 123–4
Swofford, D.L., 220
Sykes, W.R. & Godley, E.J., 278
Sytsma, K.J. & Gottlieb, L.D., 308

Takhtajan, A., 269
Tanowitz, B., Leeder, G., Ross, P.N. & Proksch, P., 244
Taub, R., Kirsch, I., Morton, C., Lenoir, G., Swan, D., Tronick, S., Aaronson, S. & Leder, P., 40
Taylor, L.P., 284
Taylor, R.L. & Mulligan, G.A., 309, 314, 316–17
Templeton, A.R., 88–9
Thompson, M.M., 25
Tobe, H. & Raven, P.H., 220
Torres, A.M., 312
Torres, R., Delle Monache, F. & Marini-Bettolo, G.B., 267
Tuyama, T., 170

Uhl, C.H., 313, 319
Urzua, A. & Cassells, B.K., 267, 271

Valdebenito, H., Stuessy, T.F. & Crawford, D.J., 124, 276
Valdebenito, H., Stuessy, T.F., Crawford, D.J. & Silva O., M., 80, 86, 134, 249–50, 254, 256–7, 276, 283, 327
Vander Kloet, S.P., 30, 38
Van der Krol, A.R., Mur, L.A., Beld, M., Mol, J.N. & Stuitje, A.R., 40

van der Werff, H., 122
van Vliet, G.J.C.M., 204
Wagner, H., Geyer, B., Kiso, Y., Hikino, H. & Rao, G.S., 246
Wagner, W.L., 35, 53
Wagner, W.L. & Funk, V.A., 1, 327–8, 330
Wagner, W.L., Herbst, D.R. & Sohmer, S.H., 1, 6, 27, 29, 32–5, 37, 50, 52–3, 67, 143, 235, 240–1, 310, 327
Wagner, W.L., Weller, S.G. & Sakai, A., 98
Wallace, A.R., 330
Warwick, S.I. & Gottlieb, L.D., 60
Watanabe, K., Nishi, Y. & Tanaka, R., 195
Wendel, J.F. & Percival, A.E., 91, 142
Werny, F. & Scheuer, P.J., 240
White, M.J.D., 82, 88
Wiggins, I. & Porter, D., 170, 312
Williams, C.A. & Harborne, J.B., 272, 275
Williams, C.A., Harborne, J.B. & Tomas-Barberan, F.A., 264, 268
Williams, C.A. & Harvey, W.J,. 269
Williams, C.A. & Warnock-Jones, P.J., 282
Williamson, M., 122, 133
Wilson, H.D., 260
Wilson, M.F., 155, 165
Wilson, R.D., 260
Witter, M.S., 6, 60
Witter, M.S. & Carr, G.D., 50, 60, 62, 68, 87, 111, 142, 146, 149–51, 244
Woodward, R.L., 76, 85, 90, 133
Wu, J. & Vankat, J.L., 133

Yamazaki, T., 170
Yamazaki, T. & Tateoka, T., 196
Yoo, K.O., 194
Yoshikawa, K., 172

Taxon index

Note: References in bold indicate a photograph/illustration

Abelia biflora, 187
Abelia coreana, 187, 194
Abelia engleriana, 194
Abelia insularis, 184, 187, 191–2, 194
Abelia integrifolia, 187
Abelia schumannii, 194
Abelia trifolia, 194
Abelia uniflora, 194
Abutilon, 26, 34
Abutilon incanum, 16
Abutilon menziesii, 16
Acacia, 26
Acacia confusa, 171
Acacia koa, 13
Acacia simplicifolia, 303
Acacia spirobis, 302
Acaena, 19
Acaena argentea, 88
Acalypha, 312
Acer mono, 186
Acer okamotoanum, 184, 186
Acer pseudoplatanum, 194
Acer pseudosieboldianum, 186–7, 194
Acer takesimense, 184, 186–7, 191–2, 194
Achantia furica, 172–4, 177
Achyranthes, 26
Achyranthes splendens, 7
Achyrochaena, 244
Acradenia, 269
Acradenia euodiiformis, 269
Acradenia franklinieae, 269
Acronychia porteri, 270
Adenostemma lavenia, 8, 52
Adenothamnus, 244
Aeonium, 313–15, 319
Agathis atropurpurea, 271

Agathis lanceolata, 270–1, 296
Agathis macrophylla, 270
Agathis vitiensis, 270
Agrostis avenacea, 24
Aistopetalum viticoides, 271
Alangia bussyanum, 296
Albizia, 303
Alectryon macrococcus, 21
Allagopappus, 314
Allagopappus viscosissimus, 273
Allium victorialis var. *platyphyllum*, 192
Alphitonia, 19
Alpinia, 310–12
Alpinia bilamellata, 312
Alpinia boninensis, 312
Alsinidendron, 98
Alsinidendron trinerve, 12
Alstonia boulindaensis, 296
Alstonia coriacea, 296
Alstonia deplanchei, 297
Alstonia lanceolata, 297
Alstonia lanceolifera, 297
Alstonia legouixieae, 297
Alstonia lenormandii, 297
Alstonia odontofera, 297
Alstonia plumosa, 297
Alstonia plumosa var. *communis*, 271
Alstonia quaternata, 297
Alstonia sphaerocapitata, 297
Alstonia undulata, 297
Alstonia vieillardii, 297
Alstonia vitiensis, 297
Alyxia, 26
Alyxia oliviformis, 7
Amaranthus brownii, 7
Ampelocera cubensis, 280
Anodopetalum biglandulosum, 268
Anoectochilus, 24
Anthriscus sulvestris, 193

Antidesma, 13
Aphananthe philippensis, 280
Apis melifera, 161, 166
Arabis takesiana, 184
Argemone glauca, 17
Argyranthemum, 315
Argyroxiphium, 26–7, 51
  biogeography, 61–2
  phenotype, 53
  phylogenetic relationships, 64
  secondary compounds, 244
Argyroxiphium caliginis, 8, 64
Argyroxiphium grayanum, 8, 64
Argyroxiphium kauense, 8, 64
Argyroxiphium sandwicense, 8, 54–5, 64, 67, 244
Arisaema takesimense, 184
Arnica, 317
Artemisia, 50–1
Artemisia australis, 8
Artemisia mauiensis, 8
Asperula odorata, 193
Asplenium nidus, 175, 178
Asplenium tenerum, 177
Astelia, 26
Astelia argyrocoma, 24
Astelia menziesiana, 24
Aster spathifolius, 192
Astragalus, 266
Austrotaxus spicata, 296

Babcockia, 273
Bacopa monnieri, 21
Beilschmiedia oreophila, 302
Belliolum, 269–70
Berberis, 83
Berberis amurensis, 192
Berberis corymbosa, 84–5
Berberis masafuerana, 84–5
Bidens, 1, 50–1
  breeding system diversity, 63
  chromosomal stasis, 29, 34
  genetic diversity/similarity, 60–2, 146, 149
  hybridization, 66–7
  phenotypic diversity, 53, 55
  radiation, 58, 98
  secondary compounds, 242–4, 283
  sect. *Greenmania*, 58
Bidens conjuncta, 8
Bidens cosmoides, 8, **54**, 55, 61
Bidens forbesii, 8
Bidens hawaiiensis, 8
Bidens hillebrandiana, 8
Bidens macrocarpa, 8, 243
Bidens mauiensis, 8
Bidens menziesii, 8, **54**
Bidens micrantha, 8

Bidens molokaiensis, 8
Bidens sandvicensis, 8, 63, 243
Bidens torta, 9
Bidens wiebkei, 9
Bishofia javanica, 171
Blepharipappus, 244
Bobea, 26–7, 32
Bobea elatior, 19, 32
Bobea timonioides, 19, 32
Boehmeria grandis, 22
Boerhavia repens, 17
Bolboschoenus maritimus, 23
Bonamia menziesii, 12
Boninia, 143, 170, 311
Boronella pancheri, 271
Boronella verticillata, 271, 305
Brachyglottis, 262–3
Brachyglottis cassinioides, 263
Brachyglottis compacta, 263
Brachyglottis greyi, 263
Brachyglottis huntii, 263
Brachyglottis monroii, 263
Brachyglottis repanda, 263
Brighamia insignis, 11
Broussaisia arguta, 15
Bubbia, 269–70
Bupleurum latissimum, 185, 193, 196
Bupleurum longiradiatum, 196
Bupleurum salicifolium, 274

Cabucala, 280
Cabucala caudata, 280
Caesalpina bonduc, 13
Caesalpina kavaiensis, 13
Calamagrostis, 24
Calanthe hattorii, **173**, 175, 177
Calanthe hoshii, 172, **173**, 175, 177
Callicarpa, 143, 172
  chromosome levels, 310–12
  evolution of cryptic dioecy, 155–68
Callicarpa dichotoma, 165
Callicarpa formosana, 165
Callicarpa glabra, 156, 158, 177, 310
Callicarpa nishimurae, 156, 158, 174, 177, 310
Callicarpa shikokiana, 165
Callicarpa subpubescens, 177
  chromosome levels, 310
  dioecious features, 156, **157**
  pollen/pseudo-pollen, 158–9
  pollinator behaviour, 160–2
Calophyllum inophyllum, 300
Calycadenia, 244
Calyptocarya glomerulata, 275
Calystegia soldanella, 192
Campanula punctata, 194
Campanula takesimana, 184, 192, 194

Canavalia, 13
Canthium, 20
Capparis sandwichiana, 11
Carallia, 220, 222
Carex, 316
Carex blepharicarpa var. *insularis*, 184
Carex echinata, 23
Carex macloviana, 23
Carex montis-eeka, 23
Carex secta, 264
Carex takesimensis, 184
Carex wahuensis, 23
Carmichaelia, 261
Carmichaelia exsul, 261, 267, 282
Cassinia, 263
Cassinia amoena, 263
Cassinia fulvida, 263
Cassinia leptophylla, 263
Cassinia vauvilliersii, 263, 268
Cassytha filiformis, 16
Casuarina equisetifolia, 171
Celtis trinerva, 280
Cenchrus, 24
Centaurea, 315
Centaurium sebaeoides, 14
Centaurodendron dracaenoides, 86
Centaurodendron palmiforme, 86
Chamaesyce arnottiana, 13
Chamaesyce celastroides, 13
Chamaesyce clusiifolia, 13
Chamaesyce multiformis, 13
Charpentiera obovata, 240
Charpentiera ovata, 7
Charpentiera tomentosa, 7
Cheirodendron trigynum, 8
Chelidonium majus, 193
Chenopodium crusoeanum, 86
Chenopodium oahuense, 12
Chenopodium sanctae-clarae, 86
Chinochloa, 266–7
Chinochloa flavescens, 266
Chinochloa pallens, 266
Chinochloa rigida, 266
Chinochloa rubra, 266
Chrysanthemum zawadskii var. *lucidum*, 184, 192, 195
Chrysopogon aciculatus, 24
Cinnamonum zelanicum, 274
Cirrhopetalum boninense, 175
Cirsium, 172, 317
Claoxylon centenarium, 173
Claoxylon sandwicense, 13
Clarkia, 308
Cleistocalyx fullageri, 267
Cleome spinosa, 11
Clermontia, 26, 36
Clermontia calophylla, 10

Clermontia clermontioides, 11
Clermontia drepanomorpha, 11
Clermontia grandiflora, 11
Clermontia kakeana, 10
Clermontia montis-loa, 11
Clermontia oblongifolia, 11
Clermontia parviflora, 11
Cocculus trilobus, 17
Coelospermum billardieri, 304
Colubrina oppositifolia, 19
Compositae, 1
    evolution on Hawaiian Islands, 49–73
        breeding system evolution, 62–3
        endemic/indigenous, 51–2
        genetic similarity, 60–2
        hybridization, 66–8
        morphological/ecological shifts, 64–6
        origin and monophyly, 57–60
        phenotypic diversity, 53–7
Comptonella, 305
Convolvulus, 315
Coprosma, 26, 34, 260–1
Coprosma baueri, 261
Coprosma ernodeoides, 20
Coprosma foliosa, 20
Coprosma longifolia, 20
Coprosma montana, 20
Coprosma ochracea, 20
Coprosma petiolata, 261
Coprosma rhynchocarpa, 20
Coreopsis, 308
Coriaria, 265–6, 282
Coriaria angustissima, 265
Coriaria arborea, 265
Coriaria intermedia, 265
Coriaria japonica, 265
Coriaria lurida, 265
Coriaria myrtifolia, 265
Coriaria ruscifolia, 265
Coriaria sarmentosa, 265
Cornus macrophylla, 192
Coronilla, 266
Corydalis filistipes, 184, 187
Corydalis speciosa, 192
Corymborchis subdensa, 175
Corynocarpus, 266, 282
Corynocarpus laevigatus, 266
Cotoneaster wilsonii, 184
Crepidiastrum, 143, 172
    chromosome number, 143, 311
    genetic differentiation, 151
    genetic diversity, 144–5, 147–50
Crepidiastrum ameristophyllum, 143, 147, 148, 150, 177
Crepidiastrum grandicollum, 143–4, 177
    genetic identity, 145, 147–50
Crepidiastrum keiskeanum, 145

Crepidiastrum lanceolatum, 145
Crepidiastrum linguaefolium, 143, **173**, 177
  genetic identity, 145, 147–8, 150
Crepidiastrum platyphyllum, 145
Cressa truxillensis, 12
Crossostylis, 203–29
  distribution, 206–8
  morphological diversity
    flowers/floral traits, 208–16
    seed coat structure, 216–20
  phylogenetic analysis, 220–6
  secondary compounds, 304
  speciation, 226–7
Crossostylis biflora
  distribution, 204, 206–7
  floral traits, 208–9, 211, **212**, **213**, 214, 216
  phylogeny, 222–5, 226
  seed coat structure, 217, **219**, 220
Crossostylis cominsii, **205**
  distribution, 204, 206, 208
  floral traits, 208, 209, **210**, 211, **213**, 214, **215**
  phylogeny, 222–5
  seed coat structure, **217**
Crossostylis dimera
  distribution, 204, 206, 208
  floral traits, 208–9, 211, 214
  phylogeny, 222
  seed coat structure, **217**
Crossostylis grandiflora, 226–7
  distribution, 204, 206–7
  floral traits, **207**, 208, 209, 211, **212**, **213**, 214, **215**, 216
  phylogeny, 222–5
  seed coat structure, 216–17, **219**, 220
Crossostylis multiflora, 227
  distribution, 204, 206–7
  floral traits, **207**, 208–9, **210**, 211–12, 214, **215**, 216
  phylogeny, 222–5
  seed coat structure, 216–17, **219**
Crossostylis pachyantha
  distribution, 204, 206–7
  floral traits, 208–9, 211, **212**, **213**, 214
  phylogeny, 222–5
  seed coat structure, 217, **218**
Crossostylis parksii
  distribution, 204, 206–7
  floral traits, 208–9, **210**, 211, 214
  phylogeny, 222–5
  seed coat structure, **216**, 217, **218**
Crossostylis richii
  distribution, 204, 206–7
  floral traits, 208–9, 211, 214, **215**, 216
  phylogeny, 222–5
  seed coat structure, 217, **218**
Crossostylis sebertii, 227

  distribution, 204, 206–7
  floral traits, **207**, 208–9, 211, **212**, **213**, 214, **215**, 216
  phylogeny, 222–5
  seed coat structure, 217, **219**, 220
Crossostylis seemannii
  distribution, 204, 206–7
  floral traits, 208–9, 211, **212**, 214
  phylogeny, 222–5
  seed coat structure, 217, **218**
Cryptocarya, 16
Cryptocarya longifolia, 302
Cryptocarya odorata, 302
Cryptocarya oubatchensis, 302
Cryptocarya phyllostemon, 302
Cryptocarya velutinosa, 302
Cryptostylis, 296
Cuminia eriantha, 87
Cuminia fernandezia, 87
Cunonia pulchella, 269
Cunonia purpurea, 269
Cuscuta sandwichiana, 12
Cyanea, 26, 36, 98
Cyanea angustifolia, 11
Cyanea degeneriana, 11
Cyanea leptostegia, 11
Cyanea remyi, 11
Cyperaceae, 264
Cyperus laevigatus, 23
Cyperus rotundus, 241–2, 283
Cyrtandra, 29, 35, 327
Cyrtandra calpidicarpa, 14
Cyrtandra confertiflora, 14
Cyrtandra cordifolia, 14
Cyrtandra garnotiana, 14
Cyrtandra grandiflora, 14
Cyrtandra grayana, 14
Cyrtandra hashimotoi, 14
Cyrtandra kauiensis, 14
Cyrtandra kaulantha, 14
Cyrtandra laxiflora, 14
Cyrtandra longifolia, 14
Cyrtandra lysiosepala, 15
Cyrtandra oenobarba, 15
Cyrtandra paludosa, 15
Cyrtandra propinqua, 15
Cyrtandra sandwicensis, 15

Dacrycarpus dacrydioides, 261
Dacrycarpus viellardii, 270
Dacrydium, 264–5
Dacrydium biforme, 265
Dacrydium cupressinum, 264–5
Dacrydium intermedium, 265, 283
Datura stramonium, 248
Decussocarpus comptonii, 270
Degeneria vitiensis, 269, 271

Delissea, 36
Delissia rhytidosperma, 11
Dendrocacalia, 143
Dendroseris, 82, 84
  comparative biology and phylogeny,
    97–119
    adaptive radiation/phylogenetic
      hypotheses, 101–8
    chromosome number, 311
    dispersal and radiation, 115–17
    genetic diversity and divergence, 88,
      108–14
    molecular evolution, 114–15
    representative species, **100**
  mode of speciation in, 89, 91–2
  secondary compounds, 251–2, 282
  subg. *Dendroseris*, 84
    genetic identities in, 109–10
    phylogenetic relationships, 92, 102–3,
      105
    secondary compounds, 252
  subg. *Phoenicoseris*, 102–3, 105, 109–10,
    252
  subg. *Rea*
    genetic identity in, 109–10
    phylogenetic relationships, 92, 102–3,
      105
    secondary compounds, 252
Dendroseris berteroana, 84, 116
  genetic identity and divergence, 88, 90–1,
    108–10
  phylogenetic relationships, 103–5
  secondary compounds, 252
Dendroseris gigantea, 84, 102–3
Dendroseris litoralis, 84, 87, **100**
  genetic identity and divergence, 92, 108,
    110, 113–14
  phylogenetic relationships, 103–4, 107
Dendroseris macrantha, 84, 92, 103–4, 110
Dendroseris macrophylla, 84, 103, 107
Dendroseris marginata, 84, 92, 103–4, 110
Dendroseris micrantha, 84, 116
  genetic identity and divergence, 92,
    108–10
  phylogenetic relationships, 102–5
Dendroseris neriifolia, 84
  genetic identity and divergence, 92,
    108–10, 112, 114
  phylogenetic relationships, 102–5
Dendroseris pinnata, 84, **100**
  genetic identity and divergence, 88, 90–2,
    108–10
  phylogenetic relationships, 103–5
Dendroseris pruinata, 84, **100**
  genetic identity and divergence, 92,
    107–8, 110, 113–14
  phylogenetic relationships, 102–5, 107

Dendroseris regia, 84, 103–5
Deplanchea speciosa, 300
Derris eliptica, 171
Deschampsia nubigena, 24
Desmos tiebaghiensis, 296
Dianella, 26
Dianella sandwicensis, 24
Dicarpellum pronyensis, 300
Dichanthelium cynodon, 24
Dichanthelium hillebrandianum, 24
Dichanthelium isachmoides, 25
Digitaria setigera, 25
Diospyros sandwichensis, 12
Diplacrum longifolium, 275
Diplazium longicarpum, 175, 178
Discaria toumatou, 267
Dissochondrus biflorus, 25
Dodonaea viscosa, 21
Drimys, 269–70
Drimys winteri, 264
Drosera anglica, 12
Drosophila, 1
Dryadodaphne novoguineensis, 271
Dryopteris, 246–7
Dryopteris fusco-atra, 247
Dryopteris hawaiiensis, 247
Dubautia, 51
  chromosome levels, 6, 26–7, 149, 309–10
  dysploidy, 31
  genetic similarity, 60, 149–50
  hybridization, 31, 50, 68
  phenotypic diversity, 55
  phylogenetic relationships, 64–5
  secondary compounds, 244–5
  sect. *Dubautia*, 64
  sect. *Raillardia*, 60, 64
Dubautia arborea, 9, 64–5
Dubautia ciliolata, 9, 64, 245
Dubautia dolosa, 9
Dubautia herbstobatae, 9, 64
Dubautia imbricata, 9, 64
Dubautia knudsenii, 9, 64–5, 68
Dubautia laevigata, 9, 28, 64
Dubautia latifolia, 9, **54**, 55, 64
Dubautia laxa, 9, 64
Dubautia linearis, 9, 64
Dubautia menziesii, 9, 64, 67
Dubautia microcephala, 9, 64–5
Dubautia paleata, 9, 64
Dubautia pauciflorula, 9, 55, 64, 68
Dubautia plantaginea, 9, 64–5
Dubautia platyphylla, 9, 64
Dubautia raillardioides, 9, 64
Dubautia reticulata, 9, 64–5
Dubautia scabra, 9, 28, 64, 245
Dubautia sherffiana, 9, 64
Dubautia waialealae, 9

Dutaillyea, 271
Dysoxylum roseum, 302
Dystaenia ibukiensis, 187, 196
Dystaenia takesimana, 185, 187, 193, 196

Echium, 315
Elaeocarpus, 13
Elaeocarpus persicaefolius, 300
Eleocharis calva, 23
Eleocharis obtusa, 23
Embelia, 17
Embergeria, 252, 273
Emmenospermum pancherianum, 304
Empetrum rubrum, 277
Entada phaseoloides, 13
Epacris impressa, 268
Eragrostis, 26
Eragrostis grandis, 25
Eragrostis variabilis, 25
Erigeron, 82, 84–6, 89, 134, 327
  chromosome number, 311, 317
  secondary compounds, 249–51
Erigeron annuus, 192
Erigeron campanensis, 250
Erigeron fernandezianus, 85–6, 249–50
Erigeron ingae, 86, 249–50
Erigeron luteoviridis, 86, 249–50
Erigeron rupicola, 86, 249–50
Erigeron stuessyi, 249–50
Erigeron turricola, 86, 249–50
Eriophorum vaginatum, 272
Ervatamia lifuana, 297
Eryngium bupleuroides, 318
Eryngium fernandezianum, 318
Eryngium inaccessum, 318
Erythrina sandwicensis, 13
Eugenia, 17
Euphorbia, 13, 315
Eurya, 21
Exocarpus gaudichaudii, 21
Exospermum, 269–70

Fagara, 83
Fagara externa, 85
Fagara mayu, 85
Fagara semiarticulata, 235
Fagus multinervis, 184
Falcatifolium taxoides, 270
Fauria, 275
ferns, 173, 175, 178
Festuca, 25
Ficus, 143, 149
Fimbristylis cymosa, 23
Fimbristylis dichotoma, 23
Flindersia fournieri, 301
Flueggea, 13
Fragaria chiloensis, 19

Freycinetia, 24
Fuschia, 262
Fuschia colensoi, 262
Fuschia cyrtandroides, 262, 282
Fuschia excorticata, 262, 283
Fuschia perscandens, 262
Fuschia procumbens, 262

Gahnia beecheyi, 23
Gardenia brighamii, 20
Gardenia remyi, 20
Garnieria spathulaefolia, 304
Geijera balansae, 305
Geissois primosa, 269
Geniostoma rupestre, 32
Gentiana corymbifera, 260
Geranium arboreum, 14
Gillbeea papuana, 271
Gironniera celtidifolia, 280
Gironniera hirta, 280
Gleichenia lepidota, 124
Gnaphalium, 9, 51
Gonospermum, 314
Gossypium tomentosum, 16
Gouania, 26
Gouania hillebrandii, 19
Gouania meyenii, 19
Greenovia, 314
greenswords, 244
Griselinia, 264
Griselinia alata, 264
Griselinia jodinifolia, 264
Griselinia littoralis, 264
Griselinia lucida, 264
Griselinia ruscifolia, 264
Guettarda eximia, 304
Guettarda heterosepala, 304
Gunnera, 15, 84, 257–9, 283
Gunnera bolivari, 257–8
Gunnera boliviana, 258
Gunnera bracteata, 88, 257–8, 318
Gunnera magellanica, 259
Gunnera margaretae, 258
Gunnera masafuerae, 257–9
Gunnera mexicana, 258
Gunnera peltata, 88, 257–9, 318
Gunnera peruviana, 258
Gunnera tinctoria, 257–9
Gymnadenia camtschatica, 191, 193
Gynotroches, 220, 222–3

Halophylla, 23
Haloragis, 83
Haloragis masafuerana, 85
Haloragis masatierrana, 85
Haplostachys, 15
Hectorella, 278

Hectorella caespitosa, 278
Hedycarya baudouinii, 303
Hedycarya parvifolia, 303
Hedyotis, 26, 311
Hedyotis centranthoides, 20
Hedyotis hillebrandii, 20
Hedyotis terminalis, 20
Helichrysum, 279–80
Heliotropium, 26, 34
Heliotropium anomalum, 11
Heliotropium curassavicum, 11
Hemizonia, 244
Hepatica, 187–91
Hepatica asiatica, 187–8, 190, 195–6
Hepatica insularis, 188, 190, 195
Hepatica maxima, 184, 187–8, 190, 193, 195–6
Hepatica nobilis var. *japonica*, 188, 190
Hepatica nobilis var. *pubescens*, 188
Hernandia cordigera, 301
Hernandia peltata, 301
Hesperocnide, 22
Hesperomannia, 50–1, 62, 282
Hesperomannia arbuscula, 9
Heteropogon contortus, 25
Heywoodiella oligocephala, 273
Hibiscadelphus, 29, 35
Hibiscadelphus brackenridgei, 16
Hibiscadelphus clayi, 16
Hibiscadelphus distans, 16
Hibiscadelphus furcellatus, 16
Hibiscadelphus giffardianus, 16, 35
Hibiscadelphus hualalaiensis, 16, 35
Hibiscadelphus kokio, 16
Hibiscadelphus tiliaceus, 16
Hibiscadelphus waimeae, 17
Hibiscus, 26, 34, 170
Hieracium, 102–3, 252, 317
Hieracium umbellatum, 192
Hillebrandia sandwicensis, 10
Holocarpha, 244
Homalium, 301
Hugonia oreogena, 302
Hugonia penicillanthemum, 302
Hunteria, 235
Hylaeus, 166
Hypochaeris, 252

Ilex, 143, 310–11
Ilex anomala, 7
Illicium munipurense, 269
Indigofera, 266
Ipomoea imperati, 12
Ipomoea indica, 12
Ipomoea pes-caprae, 12
Isachne, 25
Ischaemum, 25

Isodendron laurifolium, 22
Isodendron longifolium, 22
Isophysis, 268
Isophysis tasmanica, 268

Jacquemontia ovalifolia, 12
Jaegeria, 312
Joinvillea, 24
Juncus, 316

Kleinia nerifolia, 272, 279
Knightia deplanchei, 304
Knightia strobilina, 304
Kokia cookei, 17
Kokia drynarioides, 17
Korthalsella, 22

Labordia, 26, 32
Labordia degeneri, 16
Labordia helleri, 16, 32
Labordia hirtella, 16
Labordia waiolani, 16
Lactoris fernandeziana, 259, 319
Lagenifera helenae, 9
Lagenifera maviensis, 9
Lagenophora, (= *Lagenifera*), 50–1
Lagophylla, 244
Lamium takeshimense, 184
Lasianthera austrocaledonica, 302
Laurelia, 267
Laurelia novae-zelandiae, 267
Layia, 60, 244
Lecocarpus, 248
Lecocarpus darwinii, 248
Lecocarpus lecocarpoides, 248–9
Lecocarpus pinnatifidus, 248
Lepechinia hastata, 15
Lepidium, 26–7, 32
Lepidium bidentatum, 11, 32
Lepidium graminifolium, 32
Lepidium latifolium, 32
Lepidium serra, 11
Lepidosperma tortuosum, 264
Leptospermum polygalifolium subsp. *howense*, 267
Lepturus repens, 25
Leucaena leucocephala, 171
Libocedrus yateensis, 296
Ligusticum, 187, 196–7
Ligustrum foliosum, 184
Lilium hansonii, 184, 192
Limonium, 315
Lindenia austrocaledonica, 304
Liparis, 24, 296
Liparophyllum, 275
Liparophyllum gunnii, 268, 275, 282
Lipochaeta, 2, 51

chromosomal stasis, 26, 29, 35–6
hybridization, 66–7
phenotypic diversity, 55
phylogenetic relationships, 59–61, 69
polyploidy, 34
representative species, **56**
secondary compounds, 245
Lipochaeta connata, 9, **56**
Lipochaeta heterophylla, 9
Lipochaeta integrifolia, 9, **56**, 67, 245
Lipochaeta kamolensis, 10
Lipochaeta lavarum, 10
Lipochaeta lobata, 10
Lipochaeta micrantha, 10
Lipochaeta remyi, 10, 55, **56**
Lipochaeta rockii, 10
Lipochaeta subcordata, 10
Lipochaeta succulenta, 10, 245
Lipochaeta tenuifolia, 10, **56**
Lipochaeta tenuis, 10
Lipochaeta venosa, 10
Lipochaeta waimeaensis, 10
Litsea lecardii, 302
Litsea triflora, 302
Lobelia, 172
Lobelia grayana, 11
Lobelia hypoleuca, 11
Lobelia yuccoides, 11
lobelioids, 29, 36–7
Lonicera, 194
Lonicera insularis, 184, 191–2, 194
Lotus, 315
Luzula, 26
Luzula hawaiiensis, 24, 31
Lyallia, 278–9, 283
Lyallia caespitosa, 278–9
Lyallia kerguelensis, 278
Lycium sandwicense, 21
Lysimachia, 26, 34
Lysimachia glutinosa, 19
Lysimachia hillebrandii, 19, 34
Lysimachia mauritiana, 19
Lythrum, 16

Macaranga vedeliana, 301
Machaerina angustifolia, 23
Machilus, 143, 149, 170
Madia, 31, 50, 58, 66, 244
Madia sativa, 52
Magyracaena skottsbergii, 88
Maianthemum dilatatum, 192
Malaxis, 310–12
Malaxis boninensis, **173**, 175, 312
Malaxis hakajimensis, 312
Margyricarpus digynus, 88
Mariscus, 6, 23
Megalachne, 84

Megalachne masafueranus, 87
Melaleuca howeana, 267
Melastoma tetramerum, 172, **173**, 176
Melicope, 6, 241
Melicope elliptica, 20
Melicope lasioneura, 305
Melicope leptococca, 305
Melicope leratii, 305
Melicope mauii, 241
Melodinus aemeus, 297
Melodinus balansae, 298
Melodinus celastroides, 298
Melodinus guillauminii, 298
Melodinus insulaepinorum, 298
Melodinus phylliraeoides, 298
Melodinus polyadenus, 298
Melodinus reticulatus, 298
Melodinus scandens, 298
Melodinus tiebaghiensis, 298
Menyanthes, 275
Metrosideros, 139
Metrosideros boninensis, **173**, 177
Metrosideros nervulosa, 267
Metrosideros polymorpha, 17
Metrosideros sclerocarpa, 267
Metrosideros tremuloides, 17
Metrosideros waialealae, 17
Monanthes, 315
Morinda trimera, 20
Morus, 170
Morus bombysis, 193
Morus boninensis, 176
Mucuna, 13
Munroidendron racemosum, 8
Myoporum, 17
Myrceugenia, 83
Myrceugenia fernandeziana, 85, 320
Myrceugenia schulzei, 85
Myrsine, 6, 17
Myrtopsis, 305

Nama sandwicensis, 15
Neisosperma brevituba, 298
Neisosperma kilneri, 299
Neisosperma lifuana, 299
Neisosperma miana, 299
Neisosperma oppositifolia, 299
Neisosperma sevenetii, 299
Neisosperma thiollierei, 299
Nemuaron vieillardii, 300
Neraudia ovata, 22
Nertera granadensis, 20
Nesoluma, 21
Nestegis sandwicensis, 17
Nothocestrum longifolium, 21
Nototrichium, 7
Nymphoides, 275

Nymphoides fallax, 275
Nymphoides geminata, 275
Nymphoides grayana, 275
Nymphoides indica, 275, 283

Ochrosia, 7, 170, 235
Ochrosia balansae var. *excelsior*, 299
Ochrosia elliptica, 299
Ochrosia moorei, 299
Ochrosia mulsanti, 299
Ochrosia sandwicensis, 235
Ochrosia silvatica, 299
Olearia, 51
Ophioderma pendulum, 175, 178
Opocunonia nymanii, 271
orchids, 24, 172–3, 175, 177
Oreobolus furcatus, 23
Orobanche coerulescens, 193
Osmanthus austrocaledonica, 303
Osteomeles, 311
Osteomeles anthyllidifolia, 19
Oxera morieri, 306

Pachygone vieillardii, 302
Pagiantha cerifera, 300
Pandanus, 26
Pandanus tectorius, 24
Panisum torridum, 25
Paraixeris denticulata, 148
Parasponia andersonii, 280–1
Parasponia rugosa, 281
Parolinia, 314
Paspalum scrobiculatum, 25
Passiflora biflora, 248
Passiflora colinvauxii, 248
Passiflora punctata, 248
Paucheria sebertii, 269
Pelea, 240–1, 283
Pelea anisata, 241
Pelea barbigera, 241
Pelea parviflora, 241
Peperomia, 26–7, 31–3, 84
    chromosome number, 311
    secondary compounds, 254–7, 276–7
Peperomia alata, 255
Peperomia alternifolia, 18
Peperomia berteroana, 254–6, 276–7
Peperomia cookiana, 18
Peperomia coquimbensis, 255
Peperomia eekana, 18
Peperomia elipticibacca, 18
Peperomia expallescens, 18
Peperomia fernandeziana, 254–7
Peperomia galiodes, 255
Peperomia globulanthera, 18
Peperomia hesperomannii, 18
Peperomia hypoleuca, 18

Peperomia kokeana, 18
Peperomia latifolia, 18
Peperomia leptostachya, 18
Peperomia macraeana, 18
Peperomia margaritifera, 254–6
Peperomia membranacea, 18
Peperomia olens, 255–6
Peperomia oahuensis, 18
Peperomia ovatilimba, 18
Peperomia remyi, 18
Peperomia sandwicensis, 18
Peperomia skottsbergii, 254–6
Peperomia tetraphylla, 18
Peperomia tristanensis, 276
Peripterygia marginata, 300
Perrottetia, 12
Persea indica, 274
Peucedanum sandwicense, 7
Phaius, 296
Phelline billardieri, 303
Phelline brachyphylla, 303
Phelline comosa, 303
Phelline lucida, 303
Phellodendron insulare, 184
Phormium tenax, 265
Phyllanthus, 13
Phyllocladus, 261–2
Phyllocladus alpinus, 262
Phyllocladus aspleniifolius, 262
Phyllocladus glaucus, 262
Phyllocladus hypophyllus, 262
Phyllocladus trichomanoides, 262
Phyllostegia, 26
Phyllostegia glabra, 15
Phyllostegia grandiflora, 15
Phyllostegia mollis, 15
Phyllostegia velutina, 15
Phyllostylon brasiliensis, 280
Physocarpus insularis, 184
Phytolacca, 17
Phytolacca esculenta, 195
Phytolacca insularis, 184, 193, 195
Phytolacca japonica, 195
Phytolacca sessilifolia, 195
Pilea, 22
Pinus lutchuensis, 171
Piper kazura, 152
Piper postelsianum, 174–5
Pipturus, 29, 37
Pipturus albidus, 22
Pipturus kauaiensis, 22
Pisonia brunoniana, 17
Pisonia umbellifera, 17
Pittosporum, 143–4
    chromosome levels, 144, 310–11
    genetic identity/diversity, 141, 145,
    147–51

Pittosporum beecheyi, 144–5, 176
Pittosporum boninense, 144–5, 148, 176
Pittosporum chichijimense, 144–5, 148, 176
Pittosporum glabrum, 18
Pittosporum parvifolium, 144–8, 172, **173**, 174, 176
Pittosporum tobira, 145, 148, 151
Plantago, 26–7, 32–3
Plantago asiatica, 193
Plantago pachyphylla, 18
Plantago princeps, 18, 33
Plantago princeps var. *anomala*, 18
Platanthera, 24
Platydesma, 240
Platydesma campanulatum, 240
Platydesma cornuta, 20
Platydesma remyi, 240
Platydesma rostrata, 20
Plectranthus parviflorus, 15
Plectronia odorata, 305
Pleomele, 23
Pleurostylia opposita, 300
Plumbago zeylanica, 19
Poa, 25
Poa takeshimana, 184
Podocarpus, 261
Podocarpus affinis, 270
Podocarpus falcatus, 270
Podocarpus koordersii, 270
Podocarpus nivalis, 261
Podocarpus ustus, 270
Polyathia nitidissima, 296
Portulaca, 26–7, 31, 33
Portulaca lutea, 19
Portulaca molokiniensis, 19
Portulaca oleracea, 33
Portulaca sclerocarpa, 19
Portulaca villosa, 19
Potamogeton, 316
Potamogeton foliosus, 25
Potamogeton nodosus, 25
Potentilla dickinsii var. *glabrata*, 184
Pouteria, 170
Pouteria sandwicensis, 21
Pritchardia, 6, 23
Prumnopitys ferruginoides, 270
Prunus takesimensis, 184
Pseudolysimachion, 196
Pseudowintera, 269
Pseudowintera axillaris, 264
Pseudowintera colorata, 264
Pseudowintera traversii, 264
Psidium cattleyanum, 171
Psychotria, 26–7, 32–3, 305
Psychotria greenwelliae, 20, 33
Psychotria hobdyi, 20, 33
Psychotria kaduana, 20, 33

Psychotria mariniana, 20, 33
Pteralyxia, 7
Pteridium esculentum, 266
Pteridophyta, 178
Pterobium arboreum, 278
Pycreus polystachyos, 23

Raillardella, 244
Raillardiopsis, 31, 50, 58, 66
Ranunculaceae, 184
Ranunculus biternatus, 279
Ranunculus hawaiensis, 19
Ranunculus mauiensis, 19
Ranunculus moseleyi, 279
Ranunculus pseudotrullifolius, 279
Raoulia, 263–4
Rauvolfia balansae, 300
Rauvolfia sandwicensis, 7
Rauvolfia semperflorens, 300
Remya, 50–1, 62, 282
Remya mauiensis, 10
Reynoldsia, 8
Rhaphithamnus, 87
Rhaphithamnus spinosus, 191
Rhaphithamnus venustus, 165, 191
Rhetinodendron, 252–3
Rhododendron boninense, 172, 176–7
Rhus, 7
Rhynchospora chinensis, 23
Richea, 268
Richea angustifolia, 268
Robinsonia, 82, 84
   comparative biology and phylogeny, 97–119
      adaptive radiation/phylogenetic hypotheses, 101–8
      chromosome number, 149, 311
      dispersal and radiation, 115–17
      genetic diversity/divergence, 88, 108–14, 145
      molecular evolution, 114–15
      representative species, **102**
      secondary compounds, 252–4
   subg. *Rhetinodendron*, 103, 105–6, 115–16
   subg. *Robinsonia* and sections
      genetic identity, 110–11
      phylogentic relationships, 103, 105–6, 115–17
Robinsonia berteroi, **102**, 103, 105–6, 113, 116
   secondary compounds, 252–3
Robinsonia evenia
   genetic identity and divergence, 92, 108–11, 145
   phylogenetic relationships, 103, 105–6
   secondary compounds, 253
Robinsonia gayana, 86, **102**

genetic identity and divergence, 92,
  108–11, 113, 145
phylogenetic relationships, 103, 106
secondary compounds, 253
Robinsonia gracilis, **102**
genetic identity and divergence, 92,
  108–11, 113, 145
phylogenetic relationships, 103, 105–6
secondary compounds, 253
Robinsonia macrocephala, 86, 103, 105, 113,
  117
secondary compounds, 253
Robinsonia masafuerae, 103, 105–6, 113
secondary compounds, 253
Robinsonia thurifera, 86
genetic identity and divergence, 92,
  108–11, 113, 145
phylogenetic relationships, 106
secondary compounds, 253
Rollandia, 36
Rollandia lanceolata, 11
Rubus crataegifolius, 187, 196
Rubus hawaiensis, 19
Rubus takesimensis, 184, 187, 191, 193, 196
Rumex, 26
Rumex albescens, 19
Rumex giganteus, 19
Rumex skottsbergii, 19
Ruppia maritima, 25
Ryania speciosa, 274

Salix ishidoyana, 185
Sambucus sieboldiana var. *pendula*, 185,
  191–2, 195
Sanicula buergeriana, 195
Sanicula racemosa, 195
Sanicula sandwicensis, 7
Santalum, 139
Santalum ellipticum, 21
Santalum freycinetianum, 21
Santalum paniculatum, 21
Sapindus, 21
Sarcomelicope argyrophylla, 305
Sarcomelicope dogniensis, 305
Sarcomelicope glauca, 305
Sarcomelicope leiocarpa, 305
Sarcomelicope pembaiensis, 305
Sarcomelicope simplifolia subsp. *neo-scotia*,
  306
Scaevola, 26, 29, 34, 37
secondary compounds, 246
Scaevola chamissoniana, 15
Scaevola coriacea, 246
Scaevola gaudichaudii, 15
Scaevola gaudichaudiana, 15
Scaevola glabra, 15, 37
Scaevola kilaueae, 15

Scaevola mollis, 15
Scaevola montana, 301
Scaevola procera, 15
Scaevola racemigera, 301
Scaevola sericea, 15
Scalesia, 312
Schefflera digitata, 266
Schiaphila, 174
Schiedea, 98
Schiedea verticillata, 12
Schizogyne, 314
Schoenoplectus juncoides, 23
Schoenoplectus lacustris, 23
Scirpus frondosa, 264
Scrophularia takesimensis, 185
Sedum takesimense, 184
Senecio, 102–3, 106, 112, 254
  chromosome number, 315, 317
Senecio angulatus, 272
Senecio glaber, 106
Senecio kleinia, 272, 279
Senecio longiflorus var. *madagascariensis*,
  279
Senecio longiflorus var. *violacea*, 279
Senecio poeppigii, 106
Senecio radicans, 272, 279, 282
Senecio teneriffae, 319
Senna gaudichaudii, 13
Sesbania tomentosa, 13
Sesuvium portulacastrum, 7
Sicyos pachycarpus, 12
Sicyos waimanoloensis, 12
Sida fallax, 17
Sideritis, 273–4
Sideritis bolleana, 273
Sideritis canariensis, 274, 283
Sideritis cystosiphon, 273
Sideritis dasygnaphala, 273
Sideritis gomerae, 273
Sideritis infernalis, 273
Sideritis lotsyi, 273
Sideritis sventenii, 273
Silene, 26
Silene hawaiiensis, 12
Silene struthioloides, 12
Silene takesimensis, 184, 192, 195
silversword alliance, 1, 51
  breeding systems, 62–3
  chromosome evolution, 28, 31
  genetic identities, 60–1, 149–50, 318
  hybridization, 66–8
  phenotypic diversity, 53, **54**, 55, 98, 308
  phylogenetic relationships, 57–8, 64–6
  secondary compounds, 244
  see also Argyroxiphium; *Dubautia*;
    Wilkesia
Sisyrinchium acre, 23

Smilax, 25
Smilax riparia var. *ussuriensis*, 191–2
Solanum, 159
Solanum americanum, 21
Solanum nelsonii, 21
Solanum sandwicense, 21
Solidago virga-aurea var. *gigantea*, 192
Sonchus, 252, 272–3, 315
Sonchus brachylobus, 273
Sonchus brachyotus, 192
Sonchus canariensis, 273
Sonchus fruticosus, 104
Sophora, 83, 259–60
Sophora chrysophylla, 13
Sophora fernandeziana, 85, 259
Sophora linearifolia, 259
Sophora macrocarpa, 259
Sophora masafuerana, 85, 259
Sophora microphylla, 259–60, 278
Sorbus asurensis, 192
Soulamea fraxinifolia, 306
Soulamea muelleri, 306
Soulamea panceri, 306
Spartocytisus, 314
Spermolepis hawaiiensis, 7
Spiraea blumei, 191, 193
Spiraeanthemum samoense, 271
Spiraeopsis brassei, 271
Sporobolus virginicus, 25
Stenogyne calaminthoides, 15
Stenogyne kaalae, 15
Stenogyne microphylla, 16
Stenogyne purpurea, 16
Stephanomeria, 308
Sterigmapetalum, 212
Streblus pendulinus, 17
Strongylodon, 13
Styphelia tameiameiae, 13
Sventenia bupleuroides, 104
Symplocos, 143–4
  genetic identity and divergence, 141, 145,
  148–51
Symplocos boninensis, 144–5, 148–9
Symplocos kawakamii, 144–5, 148–9, 174
Symplocos kuroki, 145, 148
Symplocos makaharae, 145
Symplocos naharae, 145
Symplocos pergracilis, 145, 148
Syringa velutina var. *venosa*, 184
Syzygium sandwicensis, 17

Taeckholmia, 315
Taeckholmia canariensis, 273
Tanacetum, 317
tarweeds, 2, 31, 57–8, 244
Tasmannia, 269–70
Tephrosa noctiflora, 303

Tetramolopium, 2, 51, 171
  breeding system, 63
  chromosomal stasis, 26, 30, 37–8, 318
  genetic similarity/diversity, 60–1, 146, 149
  hybridization, 66–7
  phenotypic diversity, 55, 98
  phylogenetic relationships, 49–50, 58–9,
  65
  representative species, **57**
  sect. *Alpinum*, 59
  sect. *Sandwicense*, 63
Tetramolopium alinae, 59
Tetramolopium consanguineum, 10, **57**
Tetramolopium filiforme, 10
Tetramolopium humile, 10, 37, **57**, 59
Tetramolopium lepidotum, 10
Tetramolopium remyi, 10
Tetramolopium rockii, 10
Tetramolopium sylvae, 10
Tetrapathaea tetrandra, 267
Tetraplasandra oahuensis, 8
Thespesia populnea, 17
Tiarella polyphylla, 193
Tilia insularis, 185
Toddalia, 235
Tolpis, 315
Torulinium, 23
Touchardia, 22
Trachelospermum asiaticum, 152
Traversia baccharoides, 264
Trema, 281
Trematolobelia kauiensis, 11
Trematolobelia macrostachys, 11
Trematolobelia singularis, 11
Tribulus cistoides, 22
Trisetum, 25

Uncinia, 264
Uncinia brevicaulis, 86
Uncinia douglasii, 86
Uncinia uncinata, 23
Urera kaalae, 22

Vaccinium, 30, 38, 246, 283
Vaccinium calycinum, 13, 246
Vaccinium dentatum, 246
Vaccinium reticulatum, 13, 246
Valeriana fauriei, 193
Valeriana officinalis var. *integra*, 185
Vernonia, 51, 62, 275–6
Vernonia acunnae, 275
Vernonia angusticeps, 275
Vernonia menthaefolia, 275–6
Vernonia moaensis, 275
Veronica insulare, 185, 193, 196–7
Veronica schmidtiana, 196
Vicia menziesii, 14

Vigna adenantha, 14
Vigna marina, 14
Vigna o-wahuensis, 14
Villarsia, 275
Villarsia exaltata, 268, 275
Viola, 26
Viola chamissoniana, 22
Viola grypoceras, 197
Viola lanaiensis, 22
Viola maviensis, 22
Viola takesimana, 185, 193, 197
Viola verecunda, 193
Vitex rotundifolia, 22

Wahlenbergia, 84, 90, 310–11
Wahlenbergia berteroi, 90, 92, 128
Wahlenbergia fernandeziana, 87, 92
Wahlenbergia grahamae, 87
Wahlenbergia masafuerae, 90, 92
Waltheria indica, 21
Wasabia japonica, 192
Wedelia, 245–6
Wedelia biflora, 51, 59, 245
Wedelia calendulacea, 246
Weinmannia, 271, 280
Weinmannia hirta, 275
Weinmannia laevis, 280
Weinmannia macrostachys, 280
Weinmannia serrata, 269
Weinmannia tinctoria, 280
Wikstroemia, 26–7, 31, 33–4, 38

chromosome levels, 309–10, 318–19
Wikstroemia forbesii, 21
Wikstroemia furcata, 21
Wikstroemia monticola, 21
Wikstroemia oahuensis, 21
Wikstroemia phillyreifolia, 21
Wikstroemia pulcherrima, 22, 33
Wikstroemia sandwicensis, 22
Wikstroemia uva-ursi, 22, 33
Wilkesia, 26–7, 51, 55, 64, 149
  secondary compounds, 244
Wilkesia gymnoxiphium, 10, **54**, 64
Wilkesia hobdyi, 10, 64
Wollastonia biflora, 36, 51, 59

Xylocopa ogasawaraensis, 161, 166
Xylopia pancheri, 296
Xylopia vieillardii, 296
Xylosma hawaiiense, 14

Zanthoxylum, 26, 235, 242
Zanthoxylum dipetalum, 235, 242
Zanthoxylum hawaiiense, 20, 242
Zanthoxylum kauaense, 235, 242
Zanthoxylum oreophilum, 306
Zanthoxylum sarasinii, 306
Zelkova serrata, 193
Zieridium ignambiensis, 270
Zieridium pseudoobtusifolium, 270
Zygogynum, 269–70
Zygogynum pauciflorum, 306

Subject index

abietic acid, 270
acacetin, 254–5, 277
acetophenones, 269
acetylenes, 243, 266
acylhistamines, 235
African agate snail, 172–4, 177
agathic acid, 270–1
aliphatic ketones, 242
alkaloids
   comparisons with mainland species, 283
   function of, 281
   of Galapagos species, 248
   of Hawaiian species, 235, 240–1
   of Juan Fernandez species, 259–60
   of Madagascan species, 280
   of New Caledonian species, 270–1,
      296–306
   of New Zealand species, 267
   see also under named species; named
      compounds
allopolyploidy, 89–91, 308
allozymes, 97–8, 101, 108–14
amentoflavone, 268
androdioecy, 165
anethole, 241–2
aneuploidy, 89–91, 197, 307–8
   variation betweeen species, 312, 315–16,
      319
Anijima, 174, 176–8
anthochlors, 243, 278
anthocyanins, 283
apigenin compounds
   of Hawaiian species, 243, 245–6
   of Juan Fernandez species, 249, 251,
      253–4, 277
   of Madagascan species, 279
   of New Caledonian species, 269
aporphine, 271
aromadendrene, 265, 267

p-aspidin, 247
aurones, 243, 268
autogamy, 83, 86–8
autopolyploidy, 90, 307–8

Banks Group Islands, 206
benchequiol, 274
benzylisoquinolones, 267
ent-beyerene, 265
biflavonoids, 268
biogeography, 329–30
   of Juan Fernandez species, 121–38
biscodarine, 271
Bonin Islands
   chromosome evolution, 310–12
   conservation of endemic species, 169–80
   endangered species, 170–2, 175–8
   factors leading to endangerment, 172–5
   present status, 175–8
   evolution of cryptic dioecy in Callicarpa,
      155–68
   genetic diversity, 141–54
      among congeners, 147–51
      differentiation from ancestors, 151
      of endemics, 144–7
      of widespread species, 151–2
   geography, 139, 142, 169–70
   levels of endemism, 143
   representative flora, 173
breeding system evolution, 62–3
butein glycosides, 243
butyrolactone, 304

cabufiline, 280
Canary Islands
   chemistry of island species, 272–4
   chromosome levels in flora, 313–15, 318
canthin-6-one, 235
Caribbean Islands, 275–6

caryophyllene, 242, 265
catechin, 262
chalcones, 243, 263, 279
chasnarolide, 274
Chatham Islands, 260, 263, 278
chelerythrine, 235
Chichijima, 171
    flora on, 175–8, 310, 312
    geography, 142, 170
chloroplast DNA (cpDNA), 61, 68, 102,
    104–5, 114, 222–5
chromosome evolution
    in Bonin Island species, 310–12
    in Galapagos Island species, 312
    in Hawaiian species, 5–47
        chromosome numbers (listed), 7–25
        chromosome stasis, 29–30, 34–8,
            309–10, 318
        chromosome variation, 26–34
            dysploid evolution, 29, 32
            polyploid evolution, 5–6, 31–4
            natural selection for, 39–41
    in Juan Fernandez species, 310–11
    in Ullung Island endemics, 191–7
        chromosome numbers (listed), 192–3
chromosome stasis, 29–30, 34–8
    during speciation, 307–24
        chromosome levels in Canary Island
            species, 313–15, 318
        chromosome levels in Queen Charlotte
            Island species, 314–17
        chromosome survey of oceanic
            archipelagos, 309–13
        hypothesis for absence of chromosome
            variation, 317–20
chrysoeriol, 274
cibarin, 266
cichoriin, 272
cieole, 267
cinnceylanol, 274
cirsimaritin, 273
communic acids, 270
Comoro Islands, 280
condensed tannins, 246
conservation, 330
    of Bonin Island endemic species,
        169–80
Cook Islands, 58–9
corollin, 266
coronarian, 266
corydine, 267
corynocarpin, 266
coumarins, 235, 241, 272–4, 305
cryptic dioecy, *see* dioecy
Cuba, 275–6
curculoside, 272
cyanidin, 262, 264, 268–9, 271, 275, 280

cyanogenic glucosides, 247–8, 281, 283
    reduction in, 281–2

daucosterol, 259
dehydroabietic acid, 270
delphinidin, 262
diaryl epoxide, 302
dihydroamentoflavone, 268
dihydrochalcones, 263
dihydrochelerythrine, 235
dihydrofuroquinolones, 271
dihydrokaempferol compounds, 261
dihydroquercetin compounds, 261, 270
5,3'-dihydroxy-3,6,7,8,4'-pentamethoxy-
    flavone, 270
dihydroxyursene, 259
2,6-dimethoxybenzoic acid, 272
dioecy
    evolution in *Callicarpa*, 155–68
    dioecious features, 156–9
    hypothesis of, 163–6
    pseudo-pollen grains, 158–63
diosmetin, 243, 254–6, 277
diploid (recombinational) hybrid speciation,
    89, 91
diterpenes, 265, 274, 283, 296
drought, 174
dysploid evolution, 29, 32
dysploidy, 26, 29, 32, 315

ecological isolation, 83–6, 88–9
edulinine, 241
elemenes, 265
ellagic acid, 268–9, 271, 275, 280
environmental isolation, 83–6, 88–9
    and speciation, 89, 91
epicatechin, 246, 262
eriodictyol compounds, 252–3, 262–4, 273
erythrodiol, 259
esculin, 272
essential oils, 242, 267, 302
estragole, 241–2
eudesmols, 267
euploidy, 90–1, 197, 307, 316, 319
evolitrine, 270
evolutionary patterns and processes, 231–2
    see also chromosome stasis; secondary
        compounds

Faeroe Islands, 272
falcarindiol derivatives, 266
Falkland Islands, 276–7
Fiji Islands, 204
    chemistry of island species, 271–2
    Crossostylis on, 206–7, 226–7
filixic acid, 247
flavaspidic acid, 247

flavonoids, 280–1
  of Canary Island species, 272–4
  of Caribbean species, 275
  comparisons with mainland species, 283
  of Faeroe Island species, 272
  function, 284
  of Hawaiian species, 243–7
  of Juan Fernandez species, 249–59
  glycosylation, 258–9
  of Kerguelen Island species, 279
  of Mauritian species, 280
  of New Caledonian species, 269–70
  of New Zealand species, 260–4, 265–6
  reduction in, 282–3
  of South Atlantic Island species, 276–8
  of Tasmanian species, 268
  see also under named species
flavonoid sulfate derivatives, 255–7
fossil data, 328
furoquinolones, 240–1, 269–71

galangin, 263
Galapagos Islands
  chemistry of island species, 247–9
  chromosome evolution in island species, 312
genkwanin, 275
gentiobiosides, 251
gentiopicrin, 260
geographical isolation, 83–4, 86, 88
  and speciation, 89–91
geological data, 328
germacranolides, 249, 273
glucosinolates, 281
glucosylacecetin glycosides, 249–50
gluobulol, 267
C-glycosylflavones
  in Canary Island species, 272
  in Caribbean species, 275
  in Juan Fernandez species, 254, 257
  in Kerguelen Island species, 279
  in New Zealand species, 260–1, 264
C-glycosylxanthones, 260
goats, 172, 177
Gomera, 313
Gough Island, 276–8
Gran Canaria, 313–14
Grande Terre Island, 206
griselinoside, 264
gynodioecy, 63, 164–6
gynomonoecy, 63

habitat isolation, *see* ecological isolation
Hahajima, 171
  flora on, 174, 176, 178, 310, 312
  geography, 142, 170
Hawaii, 60–2, 65, 235

Hawaiian Islands, 1–2
  chemistry of island species, 234–47
  chromosome evolution in island species, *see* chromosome evolution
  evolution in endemic Compositae, *see* Compositae in Taxon index
  geography, 38–9, 53, 234
hesperidin, 235
n-hexadecanol, 267
Hierro, 313
hinokiflavone, 268
histamine derivatives, 235
holeinine, 235
humulene, 265
hunterburnine methochloride, 235
hybridization
  in Hawaiian flora, 28, 31–3, 60, 318
  importance in, 66–9
  in Juan fernandez flora, 88, 318
  and speciation, 89, 91
hydrogen cyanide, 247, 284
2-hydroxy-6-methoxybenzoic acid, 272
hypolaetin, 273

Inaccessible Island, 276
inbreeding depression, 63, 165–6
incompatibility barriers, 87–8
Indian Ocean islands, 278–80
intergenic spacer (IGS) region data, 102, 105–6
internal transcribed spacer (ITS) region data, 61, 65, 104–7, 114–16
ionone, 259
iridoids, 260, 264
  in New Caledonian species, 300–6
Isla Alejandro Selkirk, *see* Masafuera
Isla Española, 248
island biogeography, equilibrium theory, 121–2
Isla Robinson Crusoe, *see* Masatierra
Isla Santa Cruz, 248
Isla Santa Maria, 248
isocucurbitacin-D, 265
isoetin glycosides, 273
isoflavones, 246, 261
isolating mechanisms
  in flowering plants, 82–3
  on Juan Fernandez Islands, 83–8
isoörientin, 2″-rhamnosides, 261
isopimpinellin, 241
isoplatydesmine, 241
isorhamnetin compounds, 246
  of Juan Fernandez species, 252–3, 257–9
  of New Zealand species, 261, 263
isoscutellarein, 273
isovitexin, 261

Juan Fernandez Islands, 75–6
  chemistry of island species, 249–60
  chromosome evolution, 310–11, 317–18
  geography, 80, 99, 122–3
  island biogeography, 121–38
    species diversity, 124–6
      factors influencing, 126–8
      Masafuera, 125, 127–8, 130–3
      Masatierra, 124–5, 127–30
    species number and distribution, 124–6
  isolating mechanisms, 82–8
  levels of endemism, 81–2, 125–6
  speciation, modes of, 88–92
  see also Dendroseris; Robinsonia in Taxon
    index

kaempferol compounds
  of Atlantic island species, 272–3, 275
  of Hawaiian species, 243
  of Juan Fernandez species, 252–4, 256–9
  of Madagascan species, 279
  of Mauritian species, 280
  of New Caledonian species, 269
  of New Guinean species, 271
  of New Zealand species, 261–5, 268
karakin, 266
Kauai, 56
  flora
    genetic similarity, 60–2
    habitat shifts, 65
    secondary compounds, 235, 241–2
ent-kaurene, 265
Kerguelen Islands, 278–9
Kermadec Islands, 260–1
kokusaginine, 241

Lanzarote, 313
La Palma, 313
lauterine, 267
lignans, 274, 296, 301
limonene, 241
limonoids, 269
lineage sorting, 68
liriodenine, 267
loliolide, 259
longibornyl acetate, 265
longifolene, 265
α-longipinene, 265
Lord Howe Island, 260–1, 265, 267
lupeol, 259
luteolin compounds
  of Canary Island species, 272–4
  of Cuban species, 275–6
  of Hawaiian species, 243, 245
  of Juan Fernandez species, 249–54, 259,
    277
  of New Caledonian species, 269

of New Zealand species, 261–2, 264

Madagascar, 279–80
Madeira, 272–4
malvidin, 262
margaspidin, 247
maritimetin glycosides, 243
marmesin, 241
Marquesas Islands, 204, 206–7, 226
Masafuera
  age, 128
  flora, 113
    chemistry of island species, *see* Juan
      Fernandez Islands
    modes of speciation, 89–90
    species diversity, 125, 127–8, 130–2
  geography, 80, 99, 122–3, 127
  habitat, 85, 128
Masatierra
  age, 128
  flora, 113
    chemistry of island species, *see* Juan
      Fernandez Islands
    modes of speciation, 89–90
    species diversity, 124–5, 127–8, 129–30
  geography, 80, 99, 122–3, 127
  habitat, 85, 128
Maui, 61–2, 65
Maui Nui, 60, 62, 65
Mauritius, 280
mechanical isolation, 83, 86–7
melampolides, 248–9
4-methoxy-6-canthinone, 240
p-methoxypropenylbenzene, 241
8-methoxypsoralen, 241
methylcoclaurine, 267
methyleugenol, 241
methylisoeugenol, 241
monoecy, 63
myricetin compounds, 261, 263, 268–9
myristicin, 272

naringenin compounds, 252–3, 263, 265,
  273
neoabietic acid, 270
neochlorogenic acid, 146
New Caledonia, 204
  chemistry of island species, 269–71,
    296–307
  Crossostylis on, 206–7, 226–7
New Guinea, 58–9, 271–2
New Zealand, chemistry of island species,
  260–7, 278
Nightingale Island, 276
nitidine, 235
3-nitropropionic acid compounds, 266
Norfolk Island, 260–1, 265

nuclear ribosomal DNA (rDNA), *see* intergenic spacer (IGS) region data; internal transcribed spacer (ITS) region data

Oahu, 34, 62, 67, 235
ochrasandwine, 235
Ogasawara Islands, *see* Bonin Islands
okanin glycosides, 243
oleanolic acid, 259
orientin, 261
Otohtojima, 176
Ovalau Island, 206
oxoaporphine, 271

passibiflorin, 248
pelargonidin, 262
penduletin, 249
5,7,3',4',5'-pentahydroxyflavone, 261
peonidin, 262
peptides, 303
petchoulenyl acetate, 242
phenolic compounds, 246, 270, 273, 279–81, 283
C-phenylated acetophenones, 269
phenylcoumarines, 300
phenylpropanoids, 242, 300
phloraspidinol, 247
phloroglucinol derivatives, 246–7
phyllocladene, 265
phylloflavan, 262
phylogenetic analyses, 328–9
pimpinellin, 241
pinene, 267
pinoresinol, 259
plant collectors, illicit, 175
plant inventories, 327
plumocraline, 271
podocarpusflavone A, 268
pollen grains, female flowers, 158–9
polyacetylenes, 243, 281
Polynesia, *see* Marquesas Islands; Samoa; Society Islands
polyploidy, 5–6, 31–4, 308, 318–19
procyanidin, 246, 270
pseudo-pollen grains, 158–9
as reward for pollinators, 159–63
psoralen, 241
pukateine, 267
pyranoquinolones, 271

quantum speciation, 89
quassinoids, 306
Queen Charlotte Islands, 314–17
quercetagetin glycosides, 249
quercetin compounds
of Atlantic island species, 272–3, 275

of Hawaiian species, 243, 245–6
of Juan Fernandez species, 249, 251–4, 256–8
of Madagascan species, 279
of Mauritian species, 280
of New Guinean species, 271
of New Zealand species, 261, 263–5
of Tasmanian species, 268
quinolones, 240–1, 271

rats, 174
reproductive isolation, 83, 86–8
and speciation, 89, 91
research protocols, 325–32
rhamnazin, 264
rhamnetin, 264
rhamnosylapigenin, 279
ribenol, 274
ribosomal DNA (rDNA), *see* intergenic spacer (IGS) region data; internal transcribed (ITS) region data
rimuene, 265
road construction, 174
Robinson Crusoe Islands, *see* Juan Fernandez Islands
romneine, 267
ent-rosadiene, 265
ryanodol, 274

St Helena, 276, 278
Samoa, 204
chemistry of island species, 271–2
Crossostylis on, 206–7, 226
sandarocopimaric acid, 270
Santa Clara, 80, 99, 249
Santa Cruz Group Islands, 206, 248
sclarene, 265
scopoletin, 241, 272
secondary compounds, 233–306
chemistry of island species, 234–81
function of, 281–4
selinenes, 265
sesquiterpenes, 241–2, 248, 264–5, 267, 275, 283
siderin, 274
sitosterol, 259, 270–1
snails, 172–4, 177
Society Islands, 204, 206–7, 226
Solomon Islands, 204, 206, 208, 226
sophorosides, 251
South Atlantic Islands, 276–8
South Pacific Islands, evolution on, *see* *Crossostylis* in Taxon index
spatial isolation, 83–4, 86, 88
speciation, 88–9, 328–9
and chromosomal stasis, 307–24
on Juan Fernandez Islands, 89–92

staminodia, 214
Stewart Island, 260
stilbene, 301
sugeonyl acetate, 242
sulfuretin glycosides, 243

Tahiti, 262
tannins, 246
Tasmania, 262, 268–9
taxane, 296
taxifolin glycosides, 252–3
tembetarine, 235
temporal isolation, 83, 86
Tenerife, 313
teraphyllin-B, 267
terpenoids, 281, 305
    of Canary Island species, 274
    of Caribbean species, 275–6
    of Hawaiian species, 241–2
    of Juan Fernandez species, 259
    of New Caledonian species, 270
    of New Zealand species, 264–5
    of Tasmanian species, 269
tetrahydropyran, 243
2′,4′,6′,4-tetrahydroxychalcone, 263
1,2,8-tetramethoxy-4,5-dihydroxyxanthone,
    260
thalictricine, 235
thiophenes, 243
Tierra del Fuego, 277
totarol, 270
tricetin, 261
tricin compounds, 264

2-tridecanone, 242
Tristan da Cunha Island group, 276–7
triterpenoids, 235, 265, 269, 302, 304
tropinone, 248
truxillic acid, 264
tutin, 265

Ullung Island, 140
    evolution of endemics on, 181–202
        chromosomal survey, 191–7
        morphological variation, 186–91
        flora (listed), 184–5
        geography, 181–3
    2-undecanone, 242
    urban development, 174
ursolic acid, 259
uvaol, 259

Vanikoro Island, 208
Vanua Lava Island, 206, 208
Vanuatu, 204, 206, 208, 226
velutin, 275
vitexin, 261
Viti Levu Island, 206
vomifoliol, 259

waxes, 267, 281
Western Pacific Islands, *see* Bonin Islands;
    Ullung Island

xanthomicrol, 273
xanthones, 260